Lecture Notes of the Institute for Computer Sciences, Social Informatics and Telecommunications Engineering   348

More information about this series at http://www.springer.com/series/8197

Shuai Liu · Liyun Xia (Eds.)

Advanced Hybrid Information Processing

4th EAI International Conference, ADHIP 2020
Binzhou, China, September 26–27, 2020
Proceedings, Part II

Springer

Editors
Shuai Liu 🆔
Hunan Normal University
Changsha, China

Liyun Xia
Hunan Normal University
Changsha, China

ISSN 1867-8211   ISSN 1867-822X (electronic)
Lecture Notes of the Institute for Computer Sciences, Social Informatics
and Telecommunications Engineering
ISBN 978-3-030-67873-9   ISBN 978-3-030-67874-6 (eBook)
https://doi.org/10.1007/978-3-030-67874-6

This Springer imprint is published by the registered company Springer Nature Switzerland AG
The registered company address is: Gewerbestrasse 11, 6330 Cham, Switzerland

Preface

We are delighted to introduce the proceedings of the fourth edition of the European Alliance for Innovation (EAI) International Conference on Advanced Hybrid Information Processing (ADHIP 2020). This conference brought together researchers, developers and practitioners around the world who are leveraging and developing hybrid information processing technology for smarter and more effective research and applications. The theme of ADHIP 2020 was "Industrial applications of aspects with big data".

The technical program of ADHIP 2020 consisted of 190 full papers, with acceptance ratio about 46.8%. The conference tracks were: Track 1 –Industrial application of multi-modal information processing; Track 2 –Industrialized big data processing; Track 3 –Industrial automation and intelligent control; and Track 4 –Visual information processing. Aside from the high-quality technical paper presentations, the technical program also featured two keynote speeches. The two keynote speakers were Dr. Khan Muhammad from Sejong University, Republic of Korea, who is currently working as an Assistant Professor at the Department of Software and Lead Researcher of the Intelligent Media Laboratory, Sejong University, Seoul, Republic of Korea, and is an editorial board member of the Journal of Artificial Intelligence and Systems and Review Editor for the Section "Mathematics of Computation and Data Science" in the journal Frontiers in Applied Mathematics and Statistics; as well as Dr. Gautam Srivastava from Brandon University in the Canada, who has published a total of 143 papers in high-impact conferences in many countries and in high-status journals (SCI, SCIE) and has also delivered invited guest lectures on Big Data, Cloud Computing, Internet of Things and Cryptography at many Taiwanese and Czech universities. He is an Editor of several international scientific research journals.

Coordination with the steering chairs, Imrich Chlamtac, Guanglu Sun and Yun Lin, was essential for the success of the conference. We sincerely appreciate their constant support and guidance. It was also a great pleasure to work with such an excellent organizing committee team for their hard work in organizing and supporting the conference. In particular, the Technical Program Committee, led by our TPC Chair, Dr. Shuai Liu, completed the process of peer-review of technical papers and made a high-quality technical program. We are also grateful to the Conference Manager, Natasha Onofrei, for her support and to all the authors who submitted their papers to the ADHIP 2020 conference and workshops.

We strongly believe that the ADHIP conference provides a good forum for all researchers, developers and practitioners to discuss all scientific and technical aspects that are relevant to hybrid information processing. We also expect that future ADHIP conferences will be as successful and stimulating, as indicated by the contributions presented in this volume.

Shuai Liu

Conference Organization

Steering Committee

Imrich Chlamtac	University of Trento
Yun Lin	Harbin Engineering University
Guanglu Sun	Harbin University of Science and Technology

Organizing Committee

General Chairs

Shuai Liu	Hunan Normal University
Yun Lin	Harbin Engineering University

General Co-chair

Gautam Srivistava	Brandon University

TPC Chair and Co-chair

Xunli Zhang	Binzhou University

Sponsorship and Exhibit Chair

Zhaoyue Zhang	Civil Aviation University of China

Local Chairs

Ligang Chen	Binzhou University
Aixue Qi	Binzhou University

Workshops Chair

Gautam Srivastava	Brandon University

Publicity and Social Media Chair

Weina Fu	Hunan Normal University

Publications Chair

Khan Muhammad	Sejong University

Web Chair

Wei Wei	Xi'an University of Technology

Posters and PhD Track Chair

Liyun Xia   Hunan Normal University

Panels Chair

Xiaojun Deng   Hunan University of Technology

Demos Chair

Jie Gao   Hunan Normal University

Tutorials Chair

Qingxiang Wu   Yiyang Vocational and Technical College

Technical Program Committee

Hari M. Srivastava	University of Victoria
Guangjie Han	Hohai University
Amjad Mehmood	University of Valencia
Guanglu Sun	Harbin University of Science and Technology
Gautam Srivastava	Brandon University
Guan Gui	Nanjing University of Posts and Telecommunications
Yun Lin	Harbin Engineering University
Arun K Sangaiah	Vellore Institute of Technology
Carlo Cattani	University of Tuscia
Bing Jia	Inner Mongolia University
Houbing Song	Embry-Riddle Aeronautical University
Qingxiang Wu	Yiyang Vocational and Technical College
Xiaojun Deng	Hunan University of Technology
Zhaojun Li	Western New England University
Weina Fu	Hunan Normal University
Han Zou	University of California
Xiaochun Cheng	Middlesex University
Wuyungerile Li	Inner Mongolia University
Huiyu Zhou	University of Leicester
Weidong Liu	Inner Mongolia University
Juan Augusto	Middlesex University
Jianfeng Cui	Xiamen University of Technology
Xuanyue Tong	Nanyang Institute of Technology
Qiu Jing	Harbin University of Science and Technology
Mengye Lu	Inner Mongolia University
Heng Li	Henan Finance University
Lei Ma	Beijing Polytechnic
Mingcheng Peng	Jiangmen Vocational and Polytechnic College
Wenbo Fu	Datong university
Yafei Wang	Ping dingshan University

Yanning Zhang	Beijing Polytechnic
Guangzhou Yu	Guangdong Ocean University
Dan Zhang	Xinyang Vocational and Technical College
Fuguang Guo	Henan Vocational College of Industry and Information Technology
Weibo Yu	Changchun University of Technology
Dan Sui	Califoraia State Polytecnic University-Pomona
Juan Wang	Zhengzhou Institute of Technology
Xinchun Zhou	Baoji University of Arts and Sciences
Qingmei Lu	University of Louisville
Hong Tian	Baotou lron steel vocational technical college
Yuling Jin	Chizhou Vocational And Technical College
Yongjun Qin	Guilin Normal College
Wen da Xie	Jiangmen Polytechnic
Shuai Yang	Changchun University of Technology

Contents – Part II

Industrial Automation and Intelligent Control

Visual Information Processing

Contents – Part I

Industrialized Big Data Processing

Industrial Automation and Intelligent Control

Industrial Automation and Intelligent Control

A Communication Channel Selection Algorithm Considering Equilibrium

Yu-jie Zhao[1] and Han-yang Li[2,3(✉)]

[1] Lanzhou City University, Lanzhou 730070, China
zhaoyujie315@163.com
[2] School of Humanities, Tongji University, Shanghai 200092, China
lihanyang1685@163.com
[3] School of Journalism and Communication, Minnan Normal University,
Zhangzhou 363000, China

Abstract. In order to overcome the problems of accuracy and low efficiency of the communication channel selection algorithm, a communication channel selection algorithm considering the balance is proposed. The communication channel selection algorithm considering the balance first needs to collect the data of the communication network, and extract the channel impulse response and power delay spectrum from the original data, and then select the noise floor of the channel impulse response and power delay spectrum And multipath search and analysis of channel fading characteristics, and finally through the communication channel selection algorithm for optimal channel selection to realize the communication channel selection algorithm considering the balance. The comparison of experiments verifies that the channel selection accuracy and efficiency of the communication channel selection algorithm considering the balance is always higher than the ALOHA algorithm.

Keywords: Equilibrium · Communication channel · Selection algorithm

1 Introduction

Mobile communication is premised on the use of radio waves. In this field, the available spectrum is limited, so it must be used effectively. This requires starting from the three basic factors of frequency, time and space [1]. For this purpose, it is necessary to adopt a dense method represented by narrowband transmission in the frequency domain, a multiplex scheme in the time domain, and a frequency reuse scheme in the space domain. The multiplex mode of the time domain is called channel selection, and it is widely used in various mobile communication systems [2]. With the increasing demand for communication services such as wireless Internet access and mobile TV, it is becoming the direction of wireless communication development to increase the channel transmission rate as much as possible under limited spectrum resources [3]. However, the overhead of the OFDM guard interval makes the improvement in spectral efficiency not particularly noticeable. Researchers have found another technique for improving spectral efficiency from the airspace: MIMO. The research of information theory shows that when working in a rich scattering wireless environment, if multiple

S. Liu and L. Xia (Eds.): ADHIP 2020, LNICST 348, pp. 3–14, 2021.
https://doi.org/10.1007/978-3-030-67874-6_1

antennas are used at the receiving end and the transmitting end of the communication system, the obtained multi-input and multi-output system can make full use of airspace resources. Under the premise of not increasing the system bandwidth and transmit power, the channel capacity and spectrum efficiency of the system are greatly improved, so as to effectively counter the influence of wireless fading [4]. MIMO and OFDM seek solutions to improve spectrum utilization from the airspace and frequency domain, respectively, so they can be combined to further increase the information transmission rate with limited resources. However, in actual communication, the channel capacity of the feedback channel is limited, and sometimes the channel change is fast, so that the channel information cannot be transmitted back to the transmitting end in time, so the transmitting end cannot completely obtain the CSI. At the same time, it is hoped that the smaller the amount of backhaul required by the channel selection algorithm, the better. The existing channel selection algorithm has a large amount of backhaul information, and some channel selection algorithms have low balance. Therefore, a communication channel selection algorithm considering equalization is designed.

2 Data Collection

The communication network consists of N nodes, each with multiple RF communication interfaces [5]. There are n channels in the network, one of which is dedicated to control information, and the other n − 1 channels are used for data packet transmission. The control channel is used to resolve contention for the data channel, and the data channel is used to transmit data packets and acknowledgement frames. A radio communication interface in a node is assigned to a control channel, which is called a control interface. In addition to receiving and sending broadcast routing messages, the control interface also sends and receives information such as available channels needs to be switched, the broadcast of the local update message is triggered [6]. At the same time, the control channel also sends and receives RTS and CTS messages. The remaining RF communication interfaces are assigned to data channels, called data interfaces. The data interface performs data collection through data transmission and data reception. The node consists of four RF communication interfaces, one control interface, one data receiving interface, and two data sending interfaces.

3 Data Processing

After completing the data collection, the first step in data processing is to extract the channel impulse response from the original data [7]. The channel impulse response reflects the small-scale propagation characteristics of the wireless fading channel, and is also the basic premise for extracting other channel parameters, and plays an extremely important role in data analysis. By collecting the raw data, we can get the impulse response CIR of the channel, which is the first step in our analysis of the data. The second step is to analyze the fading characteristics of the channel through CIR.

The measurement data collected from the receiving end in the actual channel measurement is called original data, and the baseband signal is obtained after down-conversion, analog-to-digital conversion and quantization. After preliminary correlation processing, it is converted into a large number of independent CIRs, and accurate extraction of multipath component information from CIR is the basis of channel selection [8]. Since the system response is included, in order to more accurately obtain the channel characteristics of the current measurement scenario, our first step is to remove the system response. Extracting the multipath component information mainly includes two steps. First, the noise floor is accurately selected, the noise sampling points are removed, and then the remaining sampling points are multipath searched to determine the number and position of the multipath. Since the electromagnetic propagation environment experiences reflection, diffraction and scattering during electromagnetic wave propagation, the channel impulse response is a set of multipath components dispersed in the time domain and the delay domain, defined as $h(t, r_1)$. It reflects the propagation characteristics of the wireless channel, and r1 represents the corresponding multipath component. The average processing in the time domain for $h(t, r_1)$ results in a channel impulse response that is only a function of delay, as shown in Eq. 1.

$$h_{av}(r_1) = E_1[h(t, r_1)] \tag{1}$$

The power delay spectrum, also called the multipath intensity spectrum, represents the power at the multipath delay. The power delay spectrum can be obtained by squaring $h_{av}(r_1)$, as shown in Eq. 2.

$$p(r_1) = |h_{av}(r_1)|^2 \tag{2}$$

4   Noise Floor Selection and Multipath Search

The obtained channel impulse response and power delay spectrum contain a large amount of noise in addition to the multipath component. Therefore, the noise floor selection and multipath search of the data are continued. In order to calculate the appropriate noise floor, the setting of the noise threshold is very important [9, 10]. Because the noise threshold is too low or too high, it will cause inaccuracy in the number of multipaths extracted later. For example, when the selected noise threshold is high, due to more random factors in the actual measurement process, a certain portion of the lower power multipath component may be misidentified as noise and filtered out. When the selected noise threshold is low, the higher decibel noise may be misjudged as multipath component information, resulting in an estimated number of multipaths. Generally, the methods for selecting the noise threshold in the post-measurement data analysis mainly include the static threshold setting method and the dynamic threshold setting method [11, 12]. The static threshold setting method is to select a specific fixed constant as the noise threshold between the average noise and the power of the strongest path, which is an estimate of the difference between the two. However, this method has a large error when there is an interference signal or a shock hum. Another method of dynamically setting the threshold is to dynamically select the noise threshold as the channel noise power changes. Combining these two methods to improve, and based on practical experience to set the engineering parameters, it is more suitable for the correct selection of noise floor in the actual channel measurement data analysis.

After the noise threshold is determined by the above method, the number of multipaths detected by different multipath search methods will be different, that is, the correct selection of the multipath search method determines the accuracy of the multipath extraction analysis in the later stage. Generally, according to the signal bandwidth of the channel, different methods of detecting multipath are classified into local maximization, threshold limiting method and cluster identification method. The first method of local maximization is to perform a peak search for a signal whose power is higher than the threshold after setting the noise threshold by a specific method, and define each signal peak of the search cable as a multipath component. This method is more suitable for channel measurement when the signal bandwidth is narrower, and its multipath resolution is low. In the 5G communication system standard, in order to meet the requirements of high speed and large capacity, the measurement bandwidth is selected to be large, and the resolution of multipath is very high. Therefore, after the noise threshold is accurately set, the noise sampling point is filtered out, and the signal higher than the threshold can be used as the multipath component. This method is the threshold limiting method. Finally, in the cluster identification multipath component method, the path components having characteristic parameters such as the same delay and leaving angle are considered to be in the same cluster, and are generally used for channel measurement of ultra-wideband signals. Through the above introduction, according to the conditions such as the bandwidth required for measurement, it is reasonable to use the threshold limit method for multipath chords.

5   Analysis of Channel Fading Characteristics

The distance between the transceiver antenna, antenna loss and height, carrier frequency and other factors are the influencing factors of large-scale fading. Large-scale fading can usually be divided into two parts: path loss and shadow fading. When electromagnetic waves propagate in space, path loss occurs due to the propagation of the apex. Path loss also includes penetration loss, line-of-sight loss, non-line-of-sight loss, and the like. The free-space propagation mode is suitable for the completely unobstructed line-of-sight between the transceivers. Usually, the free-space propagation loss is defined as the ratio of the transmitted power to the received power, as shown in Eq. (3):

$$p_r = \frac{p_t G_t G_r \lambda^2}{(4\pi d)^2} \tag{3}$$

Where G_t, G_r respectively represent the transmit antenna gain and the receive antenna gain, P_t, P_r respectively refer to the transmit power and receive power, d denotes the distance between the transmit and receive antennas, and λ denotes the electromagnetic wave wavelength. It can be seen from Eq. (3) that the free-space propagation path loss is proportional to the square of the distance d between the transmitting and receiving antennas, and inversely proportional to the square of the wavelength λ. For a narrowband system, the wavelength range of the electromagnetic wave is small, so the free space path loss can be considered to be only related to the distance; However, for broadband systems, the influence of carrier frequency on path loss cannot be ignored due to the large range of wavelength variation. In general, frequency and distance are irrelevant.

6 Algorithm Description

$G(V, E)$ represents the undirected graph corresponding to the communication network, where V is the set of nodes in the network, E is the set of links in the network; $d(u, v)$ represents the Euclidean distance between node u and node v. R_T represents the communication radius of the node; R_I represents the interference radius of the node; ch_i represents the i-th channel; $CAS(v)$ represents the set of available channels of node v.

Where node u and node v satisfy:

$$d(u, v) \leq R_T \tag{4}$$

When any data interface in node u works on channel ch, the data interface of node v also works on channel ch; then node u and node v are said to be neighbors.

A set consisting of all accessible neighbor nodes is called a set of neighboring neighbor nodes. The neighbor node formula can be expressed as:

$$v|d(u, v) \leq R_T(x(u_i^{ch})) = 1 \tag{5}$$

Where i and j are data interfaces and 1 indicates that node u is operating on channel ch.

When two links use the same channel, and one endpoint of at least one link is within the interference range of one or both endpoints of the other link, then the two links are said to interfere with each other.

When the channel is not occupied by the data transmitting interface and the data receiving interface and is allowed to be used by the node, such a channel is referred to as an available channel.

If each element in the set is an available channel of a node, then such a set is said to be a set of available channels for that node.

To simplify the analysis, make the following assumptions.

1) The nodes in the network are randomly distributed in the two-dimensional plane area;
2) All data interfaces can both send and receive data;
3) All data channel transmission bandwidth and other transmission performance are the same;
4) The channels in the network are all orthogonal channels.

In the case of a large network topology, when each node in the network needs to know the topology information of the entire network, the entire network needs to pay a large price to broadcast and update the information of the node. If the topology of the network changes from time to time, the cost of maintaining the node information of the entire network is unacceptable due to the limited storage and computing power of the node. Therefore, the paper adopts the theory of relative balance, that is, only considers the local range of nodes and their neighbor nodes, and the cost is relatively small. At the same time, it can be seen that the higher the number of available channels, the greater the probability of being used, and the more conflicts that occur. Based on this probability theory, all nodes select the optimal channel each time they send data, which can avoid channel selection conflict between adjacent links and improve data transmission efficiency. The overall steps are as follows:

1) Establishing a usable neighbor node available channel list according to the set of available neighbor nodes and the set of available channels;
2) Establishing a priority queue and selecting an optimal channel by using a channel list available to the neighbor node;
3) Channel negotiation and data transmission and reception are performed using the improved RTS/CTS protocol.

The available neighbor node available channel statistics table is obtained based on the number of available neighbor node sets, counting the number of times the available channel appears in the available channel set of the neighboring node. When any node joins the network, the source node sends a broadcast message through the control channel to notify the surrounding available channel set of the neighbor node. After the neighboring node receives the broadcast message of the source node, if the source node is not in the set of existing neighbor nodes of the destination node, the source node is added to the set of accessible neighbor nodes of the destination node, and the set of available channels of the source node is added to the set of available channels of the source node of the accessible neighbor node.

During the data transmission, when the destination node receives the CTS message. If the destination node receives a type of CTS packet, the destination node deletes the channel contained in the CTS packet from the available channel set of the destination node storage source node. And subsequently updating the available channel list of the neighboring node of the destination node; When the destination node receives the second-class CTS packet, the destination node adds the selected channel included in the second type CTS message to the available channel set of the source node stored in the destination node, and then uses the available channel list of the neighboring node of the new destination node. The node counts the number of available channels in the available neighbor nodes through the set of available neighbor nodes and the available channel sets of the neighboring nodes, and establishes and updates its own available neighbor node available channel list. The available channel table for the neighboring nodes is a line structure, and the elements in the table are represented by structures, each with two fields: the channel and the number of occurrences.

In the local topology diagram of the example network, the available channel set of node a is $\{ch_1, ch_2, ch_3\}$, the available channel set of node b is $\{ch_1, ch_2, ch_3\}$, and the available channel set of node c is $\{ch_1, ch_2\}$, the set of available channels for node d is

$\{ch_1\}$. Node a and nodes b, c, and d occupy channels ch_4, ch_5, and ch_6, respectively, for communication. For the establishment of the available channel list of the neighboring node, the result of the local topology of the example network that can pass the available channel list of the neighbor node is shown in Table 1.

Table 1. Available channels of the neighboring nodes of the example network

Channel	Number of occurrences
ch_1	3
ch_2	2
ch_3	1

Suppose the total number of channels in the network is n and the total number of network nodes is N. Since the available channels of the neighboring nodes of each node do not exceed the actual number of channels of the network at most, and the worst case number of neighboring nodes of a single node in the network is the number of remaining nodes in the network. Therefore, when the statistics of the available channels of the neighboring nodes are known, the worst time complexity of the available channel statistics is O(nN); the channel priority queue is used by the heap sorting algorithm, and the time complexity is O(logN). When the network is initialized, the time complexity of the network is O(nN). After each node receives the broadcast message of the neighbor node, it modifies and reorders the existing available channel statistics, and the time complexity is O(logN).

7   Optimal Channel Selection

The available channel list of the neighboring nodes is only the number of times the source node appears the available channels, but the available channels with the fewest occurrences in the table are not necessarily in the set of available channels of the source and destination nodes. Therefore, the priority queue of the set of available channels of the destination node is established according to the available channel statistics table, and the source node selects the optimal channel from the priority queue.

The establishment process of the available channel priority queue: According to the information in the available channel table of the neighboring node, the source node establishes the available channel priority queue according to the principle that the available channel in the destination node has fewer occurrences and higher priority. If the queue is empty, the source node has to wait for the appropriate available channel to arrive. The source node dequeues the priority queue squad head element each time to see if it is in the source node's available neighbors available channel set. If not, the queue head element of the queue is dequeued, and the queue head element of the queue is taken until the acquired channel is in the available channel set of the source node. Figure 1 is a flow chart of the optimal channel selection process.

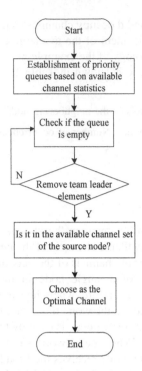

Fig. 1. Optimal channel selection process flow

When the optimal channel is selected for data transmission, the improved RTS/CTS protocol is used for channel negotiation and data transmission and reception control. The improved RTS/CTS protocol uses an improved RTS message and two types of improved CTS messages. The node negotiates the channel using the improved RTS and CTS messages before transmitting the data. After the transmission is complete, the channel is released using the improved CTS.

The above process mainly includes 4 steps:

1) The source node sends an RTS message through the control channel, where the RTS message includes the selected channel to stipulate the channel to be transmitted.
2) After receiving the RTS message, the destination node sends a CTS packet to indicate that the destination node will select the channel for data transmission if it agrees to use the channel for transmission.
3) After receiving the CTS packet, the source node also sends a CTS packet to notify the neighbor node that the channel will be used for data transmission.
4) When the data transmission is completed, the source node and the destination node sequentially send the second type CTS packet to release the occupied channel.

During the above data transmission process, all nodes that receive CTS messages modify the set of available neighbor nodes and the set of available channels. When a type of CTS message is received, the selected channel included in the CTS message is deleted from the set of available channels, and the available channel list of the neighboring node of the destination node is updated. When receiving the second-class CTS packet, the destination node adds the selected channel included in the second-class CTS packet to the available channel set of the source node stored in the destination node, and updates the available channel list of the neighboring node of the destination node.

8 Experimental Results and Analysis

This simulation experiment uses NS3 as the simulation tool, and uses C++ language to write the simulation code. The orthogonal channel (OFDM) is used as the multichannel, that is, the data is transmitted simultaneously for every two channels without interference; Randomly distributed nodes are used, that is, neighbor nodes of nodes are randomly generated. The number of packets per node is set to 2,500. Through the ALOHA algorithm as a comparison, each experiment was repeated 10 times, and the average value was taken as the final result. When the data buffer queues of all nodes in the network are empty, the experiment stops running and is considered as one experiment completion. The simulation parameters are shown in Table 2.

Table 2. Simulation parameter Table

Parameter	Numerical value	Parameter	Numerical value
SIFS	10 μs	DIFS	20 μs
Symbol time	32 μs	Slot time	60 μs
Packet length	1460 bytes	Basic data rate	6 Mbps

According to the set value of the simulation parameters, the simulation results are shown in Fig. 2 and Fig. 3.

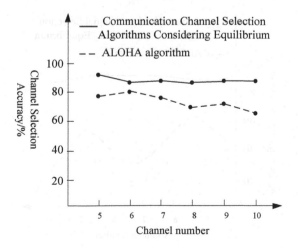

Fig. 2. Channel selection accuracy when the number of network nodes is 10.

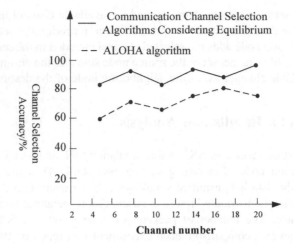

Fig. 3. Channel selection accuracy when the number of network nodes is 20.

A comparative analysis of Fig. 2 and Fig. 3 shows that under the condition that the number of network nodes is 10 or 20, the selection accuracy of the proposed communication channel selection algorithm is higher than that of the comparative ALOHA algorithm, which fully explains the proposed consideration of balance The communication channel selection algorithm has been greatly improved in performance, which can meet the needs of network communication channel selection.

In order to further verify the performance of the proposed channel selection algorithm, in the above experimental environment, the channel selection efficiency is used as the experimental comparison index, and the proposed channel selection algorithm is compared with the ALOHA algorithm. The comparison results of channel selection efficiency when the number of network nodes are 10 and 20 are shown in Fig. 4 and Fig. 5.

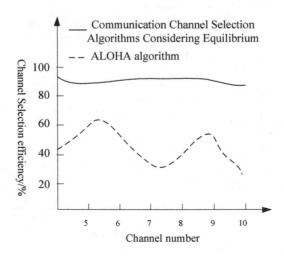

Fig. 4. Channel selection efficiency when the number of network nodes is 10

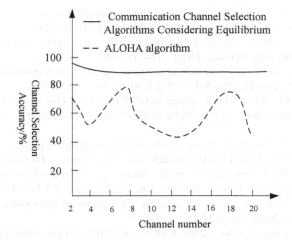

Fig. 5. Channel selection efficiency when the number of network nodes is 20

It can be seen from Fig. 4 and Fig. 5 that the performance of the proposed channel selection algorithm is relatively stable, and the proposed channel selection algorithm has a higher level under the two network node numbers. The ALOHA algorithm has strong volatility and is difficult to maintain a stable communication state. Therefore, it is fully proved that the proposed channel selection algorithm has high channel selection performance.

9 Conclusion

Aiming at the problems of traditional communication channel selection algorithms, the paper proposes a new communication channel selection algorithm considering balance. The following conclusions are proved from both theoretical and experimental aspects. The algorithm has a higher selection when selecting communication channels. Accuracy and selection efficiency. Therefore, it is fully proved that the proposed communication channel selection algorithm has high application performance.

References

1. Dun, C., Yulong, T., Zhengbao, L.: A channel access backoff algorithm considering residual time in DSRC. Comput. Eng. Appl. **52**(13), 126–130 (2016)
2. Yiwen, T., Yang, L., Chunyan, A., et al.: Idle channel detection scheme under impulsive noise environments. J. Beijing Univ. Posts Telecommun. **42**(1), 41–46 (2019)
3. Zhenghua, M., Menglin, Z., Yang, C., et al.: Channel equalization algorithm based on sparse source reconstruction. Comput. Eng. Des. **40**(9), 2489–2493 (2019)
4. Jifeng, L.: A design of speech communication system based on crosstalk decision feedback equalizer. Sci. Technol. Eng. **19**(25), 231–237 (2019)

5. Ling, Y., Bin, Z., Liang, C., et al.: Online blind equalization algorithm for satellite channel based on echo state network. J. Electron. Inf. Technol. **41**(10), 2334–2341 (2019)
6. Xi, J., Yan, S., Xu, L., et al.: Bidirectional turbo equalization for underwater acoustic communications. Acta Acustica **43**(5), 771–778 (2018)
7. Quansheng, L., Shan, L.: Research on channel equalization technology of mobile ship communication network. Ship Sci. Technol. **41**(6), 160–162 (2019)
8. Liu, S., Liu, G., Zhou, H.: A robust parallel object tracking method for illumination variations. Mobile Netw. Appl. **24**(1), 5–17 (2018). https://doi.org/10.1007/s11036-018-1134-8
9. Miao, C., Li, T., Lyu, J., et al.: Research on dynamic frequency hopping channel access algorithm based on hidden Markov model. Syst. Eng. Electron. **41**(8), 1873–1880 (2019)
10. Wenqing, Z., Hu, J., Jianpeng, G., et al.: Backoff algorithm for MAC protocol based on channel occupancy and priority. Comput. Eng. Appl. **55**(11), 80–84+116 (2019)
11. Liu, S., Li, Z., Zhang, Y., et al.: Introduction of key problems in long-distance learning and training. Mobile Netw. Appl. **24**(1), 1–4 (2019)
12. Gui, G., Yun, L. (eds.): ADHIP 2019. LNICST, vol. 301. Springer, Cham (2019). https://doi.org/10.1007/978-3-030-36402-1

Research on Intelligent Investment Prediction Model of Building Based on Support Vector Machine

Yuan-ling Ma, Run-lin Li, and Xiao Ma$^{(\boxtimes)}$

CCTEG Chongqing Engineering Co., Ltd., Chongqing 400042, China
mayuanling2367@163.com, mxiao2546@163.com

Abstract. In view of the imperfection of intelligent construction cost specification, the complexity of cost influencing factors and the lack of historical cost data, the expert system and support vector machine theory are combined to achieve knowledge acquisition and data integration. By using the expert system module, the regression calculation, the establishment of project cost prediction model and the model test of parameter setting and optimization are realized. In addition, the investment prediction speed of the model is faster. Finally, through the empirical data analysis, the accuracy and effectiveness of the model are verified, which provides the economic indicators and reference materials for the design stage of intelligent building projects.

Keywords: Building intelligence · Expert system · Support vector machine · Project cost prediction

1 Introduction

The forecast of building intelligent project investment is the basis to control the project cost of the whole process, the important index used in the project fund reserve management, the project system construction and the project economic benefit analysis. Predicting project investment with high efficiency and accuracy is of great significance for improving project work efficiency, strengthening project cost management and promoting project economic benefits. However, there are very few researches on investment prediction of intelligent building engineering in China, and most project information management is still in a traditional way, which makes it difficult to realize project information exchange [1]. Traditional engineering investment forecasting generally uses simple function regression model, least square method, quota calculation, exponential smoothing method, fuzzy mathematics, gray forecasting and other methods. The above traditional investment prediction methods have simple mathematical principle and fast speed on prediction. But for the investment estimation of large and complex intelligent building projects, their prediction quality is uneven, and the accuracy, effectiveness and practicability are difficult to guarantee [2].

Recently, machine learning technology has been gradually applied in the field of engineering cost predictions. For data mining and model prediction, this technology has higher reliability and accuracy than traditional ones. With the deepening of the

S. Liu and L. Xia (Eds.): ADHIP 2020, LNICST 348, pp. 15–24, 2021.
https://doi.org/10.1007/978-3-030-67874-6_2

research on machine learning technology, support vector machine, a method developed from hinge loss function based on the principle of minimizing structural risk, has been widely used in pattern recognition, text classification, data mining and regression analysis [3].

Due to the late starting time of intelligent building engineering, historical data of project design, construction and cost is relatively less. Also, the construction cost of intelligent building project varies with the difference of construction content, region and type, which brings out the characters like, small number of data sample, multiple impact factors and nonlinear regression of data in intelligent building system. Nevertheless, support vector machine algorithm works well in solving this type of problems.

2   Materials and Methods

2.1   Support Vector Regression Algorithm

The input sample x is firstly mapped to the high-dimensional feature space through nonlinear function $\phi(x)$, and a linear model is established in this feature space to calculate the regression function, as shown in the following formula, where w is weight vector and b is threshold [4].

$$f(x, w) = w \times \phi(x) + b \tag{1}$$

For a given training data set $(y_1, x_1), (y_2, x_2), \cdots (y_i, x_i)$, insensitivity loss function ε is adopted, and the corresponding support vector machine is called ε- support vector machine, then its constrained optimization problem can be expressed by the following formula [5].

$$\min_{w} \frac{1}{2} \times w^2 + C \sum_{l}^{i} (\xi_l + \xi_l^*), l = 1, 2, \cdots n$$
$$\text{s.t.}$$
$$y_l - w \cdot \phi(x) - b \leq \varepsilon + \xi_l^* \tag{2}$$
$$w \cdot \phi(x) + b - y_l \leq \varepsilon + \xi_l$$
$$\xi_l^*, \xi_l \geq 0$$

Where, C is the penalty coefficient, and ξ_l, ξ_l^* is the relaxation variable, the optimization problem in formula (2) is converted to dual problem to solve formula (1) after Lagrange function imported:

$$f(x) = \sum_{i=1}^{n_{sv}} (a_i - a_i^*) K(x_i, x) + b \tag{3}$$

Lagrange multipliers are $a_i, a_i^* (i = 1, 2, \cdots l)$, a small part of them are not 0, which correspond to support vectors in sample, n_{sv} is the number of support vector, $K(x_l, x)$ is Radial Gaussian kernel function, λ is kernel parameter.

$$K(x_l, x) = \exp(-\lambda \times x_l - x^2) \tag{4}$$

Kernel parameter λ is used to control the degree of sample division; Penalty coefficient C is used to get command of empirical risk and confidence range of SVR model; Insensitive loss function ε controls the width of insensitive area which is used in regression function acting on sample data [6, 7].

2.2 Application Steps of Support Vector Regression (SVR)

The prediction model of building intelligent engineering investment can be regarded as a nonlinear regression function problem: in the i^{st} year, the index value of project cost influencing factor is the independent variable $X_i = \{x_{i1}, x_{i2}, \cdots x_{in}\}$; in the j^{st} year, the prediction value of project cost is assumed to be y_j. The input is X_i, the output is Y_j; the relation between X_i and Y_j is assumed to be a nonlinear function mapping $F(x)$, which makes $y_i = F(x_{i1}, x_{i2} \cdots x_{in})$, where input is $X_i = \{x_{i1}, x_{i2}, \cdots x_{in}\}$ and output is y_j.

Then, SVR is used to solve the regression equation $f(x) = \sum_{l=1}^{n_{sv}} (a_l - a_l^*) K(x_l, x) + b$, to obtain the predicted value of project investment 错误!未找到引用源。. The application ideas are as follows:

(1) Determine the engineering cost impact factor, obtain the original cost data, extract the index value of the engineering cost impact factor, and preprocess the data with Maximum and minimum normalization method [8].

(2) Choosing a proper type of SVR function, reasonable kernel function and kernel parameter, the optimal parameter (C, λ) is obtained.

(3) The optimal parameter (C, λ) is substituted into the prediction model, and the training set is extracted from the sample data. The kernel function $K(x_l, x) = \exp(-\lambda \times x_l - x^2)$ and kernel parameter λ are substituted into formula (3) to train the sample data training set [9].

(4) Use the trained prediction model to verify its accuracy through the test set.

3 Construction of Expert System in Investment Prediction Model

The investment prediction expert system integrating expert knowledge, engineering data and prediction algorithm (as shown in Fig. 1) can be roughly composed of human-computer interaction interface, knowledge acquisition module, database, knowledge base, inference machine and explanation organization [10].

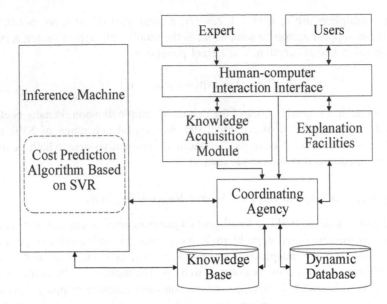

Fig. 1. Expert system architecture

In the process of system operation, users realize knowledge acquisition, data logging and prediction requests through human-computer interaction interface. The interactive system enables the inference machine to answer the request. This system USES the prediction algorithm based on support vector machine regression (*SVR*) to replace the traditional inference machine, which has higher prediction efficiency, better generalization ability and accuracy [11, 12]. The knowledge base and database are called in the prediction process [13–15]; Prediction results are output to the user through the interpretation mechanism and interactive interface.

4   Application of Investment Prediction Model

4.1   The Determination of Project Cost Prediction Index

In order to select the prediction index of building intelligent engineering cost accurately, this paper consulted a lot of literature and summarized dozens of prediction indexes. By means of questionnaire survey, combined with the opinions of 30 experts in intelligent building industry and the experience in design and construction, 5 factors that have the greatest influence on the cost are finally selected as prediction indicators: x_1 Construction Area, x_2 Building Type, x_3 Area of Structure, x_4 Construction Cost and x_5 Construction Demand. Both the quantization of qualitative index and magnitude difference between quantitative index needs data preprocessing.

The construction areas and different types of buildings are respectively quantified as natural numbers. The quotation index is determined by the construction time. According to the integration result of experts' opinions in the intelligent building industry and details of *Construction Project Pricing Quotation of Chongqing in 2018*, the labor cost and equipment cost index are calculated according to the weights of 0.15 and 0.85 respectively. The quantification of the prediction index of "construction demand" will affect the accuracy of the prediction result due to the variety of construction systems of building intelligence specialty. According to the *Intelligent Building Design Standard (GB50314-2015)*, for different types of buildings, there are three system construction schemes: "must be built", "should be built" and "can be built". The construction demand is simplified into: basic type (including items that must be built), expanded type (including items that must be built and should be built), and high-end type (including items that must be built, should be built and can be built) to facilitate data preprocessing.

4.2   Acquisition and Preprocessing of Sample Data

This paper selects 90 sets of cost data from the completed intelligent building projects in a certain area in recent years, covering different years, regions, building types, construction requirements and building area. According to the quantitative standard of prediction indicators (shown as Table 1), it is stored in the dynamic database as the original data of the investment prediction model in this paper (shown as Table 2). Data will be invoked while the prediction model is operating.

Table 1. Quantitative standard of engineering characteristics

Serial number	Predictive indicators	Value of feature vector	Quantized value
1	Construction area	Downtown	1
		Suburb	2
		Country	3
2	Building type	Residence	1
		Office	2
		School	3
		Hospital	4
		Business	5
		Hotel	6
		Factory	7
		Tourism	8
		Conference&Exhibition	9
3	Area of structure	The actual number	
4	Construction cost index	2016	0.9463
		2017	0.9781
		2018	1.0000
		2019	1.0204
5	Construction demand	Basic	1
		Expanded	2
		High-end	3

Table 2. Original sample data

Serial number	Construction area	Building type	Area of Structure (m^2)	Construction time (Year)	Construction demand (Type)	Cost of unilateral (RMB/m^2)
1	Downtown	Business	57000	2018	Basic	172
2	Downtown	Business	60000	2018	Expanded	224
3	Suburb	Business	89400	2017	High-end	292
4	Downtown	School	45000	2019	High-end	330
5	Downtown	School	80000	2018	Expanded	225
7	Suburb	Hotel	60109	2017	Expanded	353
8	Country	Hospital	64000	2016	Basic	273
⋮	⋮	⋮	⋮	⋮	⋮	⋮
86	Downtown	Office	100338	2019	Basic	221
87	Downtown	Office	97600	2018	Basic	215
88	Suburb	Factory	280000	2017	Basic	60
89	Suburb	School	69034	2018	High-end	287
90	Country	School	27800	2016	Basic	175

Pre-processing is required after the input data is quantized, in order to avoid the input data appearing in the saturated region of the program function, improve the influence of the data convergence speed and data magnitude difference in the program on the prediction result, and reduce the prediction error. Data normalization (normalization) is a common method for data preprocessing. The maximum and minimum value method can flexibly specify the value interval after normalization, eliminate the weight difference between different attributes, and normalize the sample data to [0, 1]. The calculation method is maximum and minimum normalization $X' = \frac{X_i - X_{min}}{X_{max} - X_{min}}$, where $i = 1, 2, \cdots n$. X_{max}, X_{min} are the maximum and minimum values of the data in a prediction indicator.

4.3   Interface and Function Construction of Expert System

This paper uses software development environment to build an investment prediction platform based on the expert system.

(1) Achieve data acquisition through interface input or file import; Through interface input or automatic knowledge acquisition and update; Realize the visualization of database and knowledge base.
(2) Based on the good prediction results of the *SVR* prediction model, it is embedded into the expert system as a kind of inference machine and combined with the expert knowledge to make the project cost prediction.
(3) After clicking the extension button on the right, the interface displays important parameters of the model, comparison of test results and other information.

(4) New project cost information prediction is input from the interface, and finally the project cost estimation book is generated automatically with the import of expert knowledge.

4.4 Selection of SVR Parameters

(1) SVR function type selection: In support vector machine regression model, nu-SVR and ε-SVR are commonly used functions. In nu-SVR method, the number of support vectors have to be determined in advance, and the optimal model should be obtained by slowly adjusting the values of (optimization parameters). In ε-SVR method, it's necessary to determine the value of ε in advance, that is the value of loss function (limit error), which can determine the bandwidth of the fitting function, calculation error effectively, and the prediction model function has a good performance. As a result, this paper use ε-SVR.

(2) Selection of Kernel Functions: In SVR model, linear kernel function is not mapped to the high-dimensional space for linear regression problem; the complexity of model is increased by polynomial kernel function due to its parameter being large in number; The radial Gaussian kernel function performs well in solving nonlinear regression problems, owing to its rather small characteristic dimension, which makes it highly desirable when the number of samples is relatively small; Sigmoid kernel function generates neural network, its generalization ability is relatively weak, and some parameters are invalid. Eventually, this paper use radial Gaussian kernel function (RBF).

(3) Parameter Finding and Optimization: According to the content of chapter 2.1, parameter (C, λ) plays a job of vital importance in performance of prediction model. In this model, we use usual k fold cross validation method to calculate each set of (C, λ) within a specified range to obtain the best solution. There are 3 kinds of optimization methods of (C, λ) which are frequently-used. Firstly, Genetic neural network algorithm (GA) is complex and weak in generalization. Secondly, Particle Swarm Optimization algorithm (PSO) is a heuristic algorithm. It can find the global optimal solution without traversing all points in the grid, but it is easy to fall into the local optimal solution. Last but not the least, Grid-search algorithm is to traverse all possibilities (C, λ) through the range of attempts and add a two-layer loop before the cross-validation program; the complexity of it is not high, but the accuracy is improved obviously. Therefore, grid search algorithm is used to optimize the (C, λ) parameters.

4.5 Model Training and Testing

For the determination of limit error parameters, 80 groups of data are used as training samples to establish a prediction model, then the value of ε with the best prediction performance will be selected. In this process, coefficient of determination is $R^2 = 1 - \frac{SSE}{SST}$, SSE is the sum of squares of the difference between the fitting value and the mean value of the original data, the sum of squares of the original data and its mean. R^2 is to determine the degree of fitting optimization of the model, the closer to 1, the more the model fits the training data; When the value of R^2 is far from 1, the model fitting degree is poor (Table 3).

Table 3. The model information corresponding to the change value of ε

Limit error ε	Number of training samples	Number of support vectors	Mean square error MSE	Coefficient of determination R^2
0.01	30	17	0.01402	0.96514
0.05	30	20	0.00817	0.97833
0.1	30	25	0.00685	0.93322
0.2	30	26	0.01887	0.94745
0.5	30	25	0.00463	0.98680

As shown in table, the model has high fitting degree and the best performance when the value of $\varepsilon = 0.5$. After 80 sets of data imported into the expert system as training samples, the optimal value of (C, λ) are obtained through Grid-search algorithm and cross validation, where $C - 16$, $\lambda = 0.3125$ and $n_{SV} = 25$ (n_{SV} is the number of support vectors); the remaining 10 groups of data are tested in the model, and the comparison and deviation between the predicted results of intelligent building engineering investment and the actual values are shown in Fig. 2.

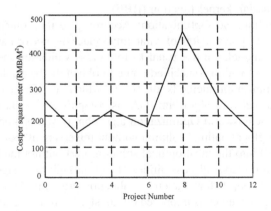

Fig. 2. *SVR* prediction model result comparison chart

4.6   Analysis of Prediction Results

The prediction model of intelligent building investment established in this paper is mainly aimed at the cost prediction in the design stage of intelligent engineering projects. Therefore, when the relative error between the predicted value and the actual value of the prediction model is less than or equal to 10%, the calculation of the model can be viewed as accurate and reliable one.

According to the test results, the regression model based on the expert system has a prediction error of less than 10% for the cost of intelligent building engineering, high accuracy and relatively strong generalization performance. For individual samples, such as item No. 1, the reason for the large error lies in the insufficient similarity between the training sample and original sample. With the further acquisition of data and knowledge by the expert system, the prediction error will be further reduced. This model has a good applicability for intelligent building engineering where the prediction index is in the middle dimension and the influence relationship between each index is rather complex.

5 Conclusion

The effective combination of expert system and support vector machine regression model can make up for the shortcomings of traditional estimation methods, such as weak data, weak knowledge acquisition ability, insufficient generalization ability and low prediction accuracy. The parameter setting of the model plays a key role in the final prediction results. Through reasonable selection, selection and optimization of kernel function, limit error and penalty factor, the fitting degree and prediction accuracy of the model are improved.

In view of the current situation that the cost standard of intelligent building specialty is not perfect, the cost influencing factors are complex, and the cost historical data are few, the regression investment prediction model based on expert system can reasonably predict the project cost in the design stage, and obtain the construction cost of new construction and reconstruction project. It can provide assistance and reference for the owners, investors and design units, and the prediction results can be used as the basis for project decision-making. At the same time, the cost prediction of this paper enriches the project management method of intelligent building engineering, saves manpower and material resources, and brings certain economic benefits for the project.

References

1. Zhou, Q., Ding, S., Feng, Y., et al.: The elevator brake intelligent monitoring and fault early warning system based on SVM. China Special Equip. Safety **34**(5), 22–27 (2018)
2. Li, Q.: Discussion on the integration method of weak building system in intelligent buildings. Modern Sci. Instrum. **12**(5), 80–84 (2018)
3. Yang, T., Yue, C.: Research on hydrometeor classification of convective weather based on SVM by dual linear polarization radar. Torrential Rain Disas. **38**(4), 297–302 (2019)
4. Kong, J., Cao, X., Xiao, F.: Research on cost prediction model of power transmission and transformation engineering based on support vector machine. Modern Electron. Technol. **41**(04), 27–130 (2018)
5. Goren, L.: Ten strategies for building emotional intelligence and preventing burnout. Family Pract. Manage. **25**(1), 11–14 (2018)
6. Lee, J.-Y.J., Miller, J.A., Basu, S., et al.: Building predictive in vitro pulmonary toxicity assays using high-throughput imaging and artificial intelligence. Arch. Toxicol. **92**(6), 2055–2075 (2018)

7. Wang, N., Wang, F., Yin, Y., et al.: Research on cost prediction of substation engineering based on support vector machine. Constr. Econ. **37**(05), 48–52 (2016)

8. Jin, Z., Hu, J., Jin, H., et al.: Analysis of traditional Chinese medicine prescriptions based on support vector machine and analytic hierarchy process. China J. Chin. Materia Med. **43**(13), 25–32 (2018)

9. Rios, D., Haake, J.: Getting off the ground floor of the smart building revolution. Appl. Des. **67**(3), 28–29 (2019)

10. Goldschmidt, I.: Intelligent building integration. Eng. Syst. **35**(2), 23–26 (2018)

11. Zheng, P., Shuai, L., Arun, S., Khan, M.: Visual attention feature (VAF): a novel strategy for visual tracking based on cloud platform in intelligent surveillance systems. J. Parallel Distrib. Comput. **120**, 182–194 (2018)

12. Car, J., Sheikh, A., Wicks, P., et al.: Beyond the hype of big data and artificial intelligence: building foundations for knowledge and wisdom. BMC Med. **17**(1), 36–42 (2019)

13. Bhandari, M., Reddiboina, M., et al.: Building artificial intelligence-based personalized predictive models. BJU Int. **12**(12), 25–31 (2019)

14. Fu, W., Liu, S., Srivastava, G.: Optimization of big data scheduling in social networks. Entropy **21**(9), 902 (2019)

15. Liu, S., Bai, W., Zeng, N., et al.: A fast fractal based compression for MRI images. IEEE Access **7**, 62412–62420 (2019)

Research on Electricity Characteristic Recognition Method of Clean Heating Based on Big Data Model

Xin-lei Wang[1], Jia-song Luo[1], Tong Xu[1], and Guo-bin Zeng[2(✉)]

[1] State Grid Economic and Technological Research Institute Co. LTD.,
Beijing 102200, China
wangxinlei5525@163.com
[2] Haikou University of Economics, Haikou 571127, China
zengguobin9965@163.com

Abstract. Because the traditional coal-fired heating mode consumes a lot of energy and is harmful to the environment, it produces a clean heating mode using electric energy, which is realized by energy storage heating equipment. The operation of energy storage heating equipment needs to be planned according to the electricity characteristics of clean heating. Therefore, a method based on large data model is proposed to integrate the electricity characteristics of clean heating using Hadoop platform. Then, according to the integrated data, the electricity load characteristics, electricity consumption characteristics, electricity consumption cycle characteristics and regional characteristics are identified to complete the electricity characteristics of clean heating. Farewell. Through experimental demonstration, it is proved that this method can effectively identify the electrical characteristics of clean heating and accurately predict the future heating data.

Keywords: Clean heating · Electricity forecasting · Seasonal cycle · Electricity characteristics

1 Introduction

Energy is the lifeblood of the national economy in modern society. Any resident and industrial activities can not do without the support of energy supply. Nowadays, China's energy production and consumption are increasing rapidly with the development of economy and society. The problems of backward energy structure, unreasonable energy utilization mode, over-reliance on fossil fuels and low energy utilization rate caused by rapid growth are becoming increasingly prominent [1]. Traditional coal-fired heating mode has a huge demand for coal mining, while coal-fired produces waste gas, which greatly aggravates environmental pollution. Therefore, clean heating has been respected. Electric energy storage heating equipment uses grain power or wind power to heat solid regenerator. The temperature reaches 750 °C. Heat is stored through insulation materials. When the fan is heated, the heat in the regenerator is blown out, and the water is transferred through the heat exchanger, and then through

S. Liu and L. Xia (Eds.): ADHIP 2020, LNICST 348, pp. 25–35, 2021.
https://doi.org/10.1007/978-3-030-67874-6_3

the heating pipeline to the resident's home [2]. This method has no emissions, no pollution, and achieves clean heating.

Under the background of environmental protection concept and realistic energy problems, this clean heating mode is worth being fully implemented. The operation method of energy storage heating equipment needs to be designed in detail according to the heating and electricity characteristics. Since the concept of big data was put forward, domestic research has been in full swing. In theory, the concept, technology and application of big data have gradually formed a scientific, systematic and industrialized research system. In terms of theoretical system, the analysis of large data is mainly divided into five aspects: visual analysis, data mining algorithm, predictive analysis ability, semantic engine, data quality and data management [3]. Therefore, a large data model platform was born. This paper uses Hadoop platform based on big data model to integrate the characteristic data of clean heating data, and then analyses the characteristics of clean heating data, so as to provide a reasonable scheme for the allocation of energy storage heating equipment, and ensure that the clean heating mode achieves the optimal technology-economy ratio, the best sustainable development, the greatest economic benefit, the most reliable power supply and the best environmental protection [4].

2 Electricity Characteristic Recognition of Clean Heating Based on Large Data Model

2.1 Feature Recognition Data Integration

Before identifying the characteristics of clean heating electricity data, it is necessary to collect the electricity data, which often comes from different data sources, including databases, data warehouses and documents. At this time, it is necessary to integrate the data, that is, data integration. Because an attribute of an object may have different names in different databases, data integration may lead to inconsistencies and a large number of redundant data. So data integration can improve data quality better, reduce the occurrence of redundancy of result data sets and inconsistency of entity names, and improve the accuracy of data in the mining process [5]. At present, data integration still has the following problems.

Entity recognition. The same attribute in the clean heating data may have different attribute names in different databases, which may involve entity recognition in data integration. Generally, metadata can be used to avoid entity recognition problems that may arise in data integration.

Redundant data. Redundancy is a common problem in data integration. The main situation of data redundancy includes repeated occurrences of the same attribute value and duplication caused by inconsistent naming of the same attribute. Relevance analysis can be used to detect whether the attribute is redundant.

Numeric conflicts. The actual power consumption data may be stored in different measurement units because of different storage scenarios. For example, the time in a power data set may be 12 h in one system and 24 h in another system. If these two sets of data are integrated together, there will be a numerical conflict. For such numerical conflicts, normalization and other operations can be used to remove the impact of unit conflicts.

In view of the problems that may arise in the above data, Hadoop platform is selected for data processing. Hadoop is a distributed system infrastructure developed by Apache Foundation. It has high throughput and fault tolerance, and solves the problem of huge amounts of clean heating and power data. The Hadoop software framework is implemented in Java language, which supports cluster operations of thousands of data computing nodes and data processing at PB level [6]. Hadoop is composed of several independent systems, which are interdependent and constitute the whole of Hadoop. The operation of Hadoop platform system is shown in Fig. 1.

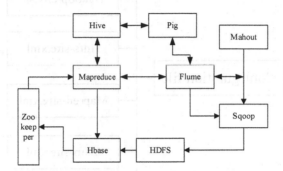

Fig. 1. Hadoop platform system operation diagram

MapReduce is a distributed computing framework with many data analysis components. Hive is a data warehouse that can provide a query language similar to SQL. It converts SQL to MapReduce and executes on Hadoop. It can be used for offline analysis. Pig is a data stream processing framework. Mahout is a data mining algorithm library. It mainly implements some classical algorithms in machine learning field, which can help designers build effective models better. HBase is a column-oriented database, which uses Google's BigTable data model and enhanced coefficient ranking mapping table. In order to facilitate the unified management of clusters, a distributed collaboration service Zookeeper is used. Sqoop is a data synchronization tool. It is mainly used to transfer data between traditional database and Hadoop. Its essence is MapReduce program. Flume is an open source log collection system [7, 8].

The data analysis platform in this paper uses four Aliyun servers to build Hadoop cluster. The Hadoop version is Apache version of Hadoop 2.6.4. The server system is Centos 6.5, 64-bit operating system, single-core CPU. The memory of each server is 1G, and the size of hard disk is 40G. The high availability architecture of Hadoop is used to ensure the stability of the data processing cluster. When the primary node has problems, the standby nodes can be activated to ensure the normal operation of the cluster.

First, the host name is changed to the host name that receives heating and power data, and then Java files are installed on each server to define the configuration environment variables. Then configure password-free login, install SSH for each server, so that password-free communication between each host, and synchronize the time to ensure that the cluster time is consistent. Download hadoop.2.6.4. tar.gz, copy the compressed package to the same location on each server and decompress it. Then we begin to configure Hadoop. The data files in the etc./hadoop directory of the Hadoop folder are XML edited, and the editing process is shown in Fig. 2.

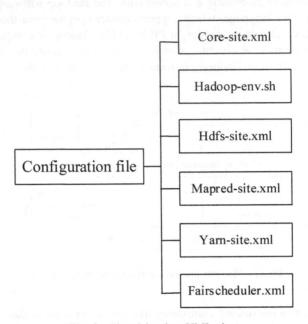

Fig. 2. Electricity data XML chart

Because of the huge amount of data of clean heating power, the time of collecting and processing data by other methods is long and the workload is too large, so the Hadoop method can also save the time needed for data loading and improve the performance of data warehouse construction. In the process of data integration, based on the detailed understanding of power consumption data, index is used to improve the performance of analysis and query, and index the column attributes often involved in query in dimension tables and fact tables, which further improves the performance of data analysis. Considering the data warehouse based on Hive, data analysis is carried out in the form of MapReduce, so the number of Map and Reduce can be set to ensure high performance in the process of data integration.

After integrating the data of clean heating power consumption, the identification of the characteristics of clean heating power consumption is gradually completed by acquiring the corresponding identification data. The identification of electric charac-teristics of clean heating includes: identification of electric load, identification of electric consumption, periodic identification of electric consumption, identification of regional characteristics and so on. Through the interpretation of data, the identification of electric characteristics of clean heating is carried out.

2.2 Identifying Electric Load

In the process of clean heating and power consumption, there is a certain rule that the power consumption of heating and power changes with time, which can be described by the change of power load with time [9, 10]. According to the different time periods, the load curve can be divided into daily load curve, weekly load curve, monthly load curve and annual load curve. The identification of power load characteristics is an important index for power supply side planning and management. By acquiring load data and identifying load characteristics, the index data of load characteristics of demand are shown in Table 1.

Table 1. Identification indicators of power load characteristics

Description class	Comparison class	Curve class
Daily maximum (minimum) load	Daily load rate	Daily load curve
Daily average load	Daily minimum load rate	Weekly load curve
Peak valley difference	Daily peak-valley difference rate	Annual load curve
Monthly maximum (minimum) load	Monthly average daily load rate	Annual continuous load curve

Among them, the calculation methods of various characteristic indexes of electric load are as follows:

Load rate (ratio of average load to maximum load) = heating power average load (KW)/heating power maximum load (KW)] * 100%; average daily load rate (%) = heating power period daily load rate/heating power period calendar day; peak-valley difference (%) = heating power period peak-valley difference maximum value (KW)/maximum load on the day (KW) * 100%; monthly load balance rate (%) = reporting monthly average daily power consumption (KWH)/Report Maximum daily power output (KWII) * 100%; annual load balance rate (%) = (maximum monthly load/maximum monthly load of 12 * maximum) × 100%; maximum load utilization hour (H) = maximum power output (supply and use) during reporting period (KWH)/maximum power output (supply and use) during reporting period (KW); simultaneous rate (%) = [maximum power system load (KW)/absolute maximum load of each component unit of power system (KW) * 100%.

Through calculation, the load characteristic curve of clean heating electricity is drawn. Usually, the load characteristic can be divided into high load rate type, continuous type, peak-up type, peak-avoidance type, temperature sensitive type and so on. High load rate type means that the load characteristic curve of heating power is stable, the difference between peak and valley is small, and the load rate is over 90%. The load curve of continuous type is between 80% and 90%. The load curve is relatively stable. Because of the limited adjustment ability of the existing heat storage heating equipment, the load curve will be continuous, but there will be a certain peak-valley difference. The peak time point of heating load in a day; the peak avoidance type shows the trough time of heating power.

2.3   Identifying Electricity Consumption

Accurate identification of electricity consumption is one of the keys to ensure the operation of clean heating mode. Therefore, multi-dimensional features of heating data are identified based on large data model.

Based on the massive data of clean heating power consumption, a multi-dimensional evaluation index system of electricity consumption characteristics is established. The time and space of clean heating power consumption are clustered and analyzed, and the characteristics of clean heating power consumption are mined and identified. Develop the process of power consumption identification, and use the evaluation method to avoid the adverse impact of other factors on the accuracy of the identification algorithm. The data identification process of clean heating power consumption based on large data model is shown in Fig. 3.

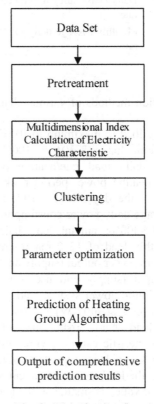

Fig. 3. Identification flow

Due to the difference of electricity consumption characteristics among different heating users, the heating objects are classified according to different electricity consumption modes, and the changing rules of electricity consumption of different users are studied. The key to realize the identification is to use large data analysis. The identification methods and results of electricity consumption characteristics are closely related to the selection of electricity consumption characteristics indicators. Therefore, it is necessary to define reasonable evaluation indicators of electricity consumption characteristics to assist the acquisition and identification of electricity consumption data.

The steps of power consumption identification are as follows: firstly, effective data preprocessing is carried out for m large data sets of heating power consumption, and target analysis data set Vd is generated, in which the first user's power consumption sequence is Vi (xt1, xt2, ... , xtn), I = 1, 2, ... , m; by calculating the multi-dimensional evaluation index of power consumption of mass users, the data set VD of user's multi-dimensional characteristics is established, in which the first user's characterization sequence is Vi (xt1, xt2, ... , xtn), I = 1, 2, ... , M Vi (xt1, xt2, ... , xtn, 1, 2, W), I = 1, 2, ... , m; Then clustering the data sets of heating multi-dimensional electricity characteristics, mining and identifying electricity consumption patterns, and getting the K user groups set Gk of M users. Finally, according to the characteristics of user groups, the parameters are trained separately. On the basis of parameter selection, an algorithm recognition model is established to obtain the identification values of electricity consumption of each heating group and all heating groups, and to identify and evaluate them. The formulas for evaluating the recognition results are as follows:

$$\beta_i = \sum_{j=1}^{w} a_{ij} x_j = a_i^T x \tag{1}$$

The parameters in the above formulas are derived from the data of power consumption groups. When the evaluation result is beta > 8, the evaluation result is effective. This method is suitable for the analysis and processing of large data platforms, and has high recognition accuracy and stability. The sources of big data mainly include internet, physical information system, scientific experiments and so on. Usually, physical information system data and scientific data are obtained through a sensor network composed of sensors or observation devices. The rapid popularity of the Internet and mobile Internet provides the most convenient and lasting carrier for data recording. Because heating is a periodic process, the identification of clean heating power consumption can get the overall data of annual electricity consumption. In order to better understand the time rule of heating power consumption, it is necessary to identify the periodicity of clean heating power consumption.

2.4 Identifying Periodicity of Electricity Consumption

The fluctuation of electricity consumption for clean heating has obvious periodic regularity with time, which is called seasonal effect [11, 12]. According to a certain mathematical method, time series with seasonal effects are decomposed into components that show periodic changes with time series and components that are basically independent of time changes, so that seasonal electricity consumption law can be more clearly presented. The seasonal adjustment algorithm is used to obtain monthly or quarterly time series data, which contains four components, namely, long-term trend component, fluctuation cycle component, seasonal component and irregular component [13, 14]. The long-term trend component represents the long-term trend characteristics of time series. The fluctuation cyclic component is a cyclic change in units of several years. They describe the basic change law of time series. The variation of time series is an average value, and the formula is as follows:

$$\overline{Y}_{T+1} = \frac{1}{N}(Y_T + \ldots + Y_{T-N+1}) \tag{2}$$

Among them, T is the number of seasonal cycles, N is the statistic, t is the initial quantity of seasonal cycles, and its standard error S formula is:

$$S = \sqrt{\frac{\sum_t^T N + 1(\overline{Y} - Y_t)^2}{T - N}} \tag{3}$$

According to the above calculation method, seasonal periodic variation characteristics of clean heating power consumption are obtained with the support of large data.

2.5 Recognition of Regional Characteristics

The purpose of dividing the data of clean heating power consumption by heating area is to qualitatively identify the characteristics of clean heating power consumption in different areas, grasp the trend of heating power consumption in different areas, and provide guidance for accurate heating power supply.

In order to identify the regional characteristics of heating power consumption, it is necessary to subdivide the type of electricity consumption data in the identified area, and to refine the type of electricity consumption in the area as far as possible, so as to ensure the accuracy and validity of the identification results [15]. In the conventional classification of regional electricity consumption data, the industrial characteristics of the region are positioned according to the step quantity of electricity consumption. For example, some industries with high heating demand will inevitably lead to the increase of heating power consumption in the region. In addition, different regions have different climate temperatures and different demands for heating, so the data generated are necessarily different [16]. In view of the above two aspects, the data characteristics of regional clean heating power consumption are identified, as shown in Fig. 4.

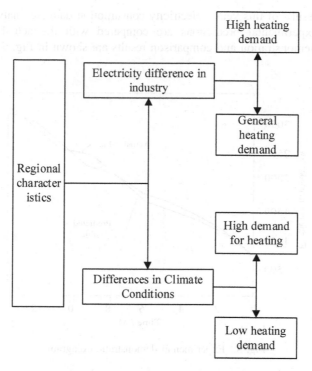

Fig. 4. Regional characteristic recognition classification chart

So far, the identification of electricity characteristics of clean heating based on large data model is basically completed.

3 Analysis of Experimental Demonstration

In order to verify the validity of the method designed in this paper for identifying the electrical characteristics of clean heating, an experimental demonstration is carried out. Taking S city of North China, which is heated by clean heating mode, as the experimental object, the characteristics of its previous heating data are identified, and the electricity characteristics and seasonal periodic characteristics of clean heating power are obtained. The results are shown in Table 2.

Table 2. S identification result of electricity characteristic of clean heating in city

Features	Warm winter condition	Cold winter	Normal behavior
Electricity consumption	2800	3100	3000
Seasonal cycle	3 months	4 months	3.5 months

Then the results of this year's electricity consumption data are analyzed and estimated. The experimental predictions are compared with the actual values. The experimental demonstration and comparison results are shown in Fig. 5.

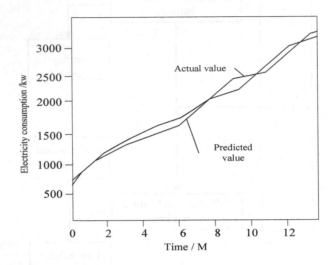

Fig. 5. Experimental demonstration diagram

It can be seen from Fig. 5 that the identification results of electrical characteristics of clean heating by using the identification method proposed in this paper are reliable, and the power consumption of clean heating is predicted according to the identification results. The predicted value is very close to the actual value, which proves the effectiveness of the proposed method.

4   Concluding

Based on the large data model, this paper uses Hadoop platform to integrate the characteristics of the clean heating data, and then identifies the electricity consumption characteristics, periodic characteristics and regional characteristics of the clean heating data, and proves the accuracy and effectiveness of the design through experiments. At the stage of full implementation of clean heating mode, it is hoped that the design can identify the electricity characteristics of clean heating more accurately, and make a more scientific storage and discharge plan for power supply and storage equipment.

References

1. Tang, J.: Precise prediction of monthly electricity consumption based on multi-dimensional feature analysis. Power Syst. Protection Control **45**(16), 145–150 (2017)

2. Peng, S.: Mining method model and application of online learning behavior research in big data era. Audiovis. Educ. Res. **40**(1), 70–79 (2017)
3. Anonymous: Simulation of intelligent data acquisition, operation and maintenance Mining Model for power consumption information under large data. Comput. Simul. **35**(10), 412–415 (2018)
4. Anonymous: Prediction and analysis of electric heating power consumption based on big data platform. Appl. Electron. Technol. **44**(11), 67–69 (2018)
5. Chai, Q., Zheng, W., Pan, J., et al.: Research on condition monitoring and fault handling methods of intelligent distribution network based on large data analysis. Modern Electron. Technol. **41**(4), 105–108 (2018)
6. Anonymous: Performance evaluation of wind turbines based on mutual information association of large wind data. Sci. Technol. Eng. **18**(29), 216–220 (2018)
7. Guo, J., Zheng, X.: A summary of learning analysis based on big data. Audiovis. Educ. China **1**, 121–130 (2017)
8. Vincenti, G., Bucciero, A., Helfert, M., Glowatz, M. (eds.): E-Learning, E-Education, and Online Training. LNICST, vol. 180. Springer, Cham (2017). https://doi.org/10.1007/978-3-319-49625-2
9. Li, T., Tan, W., Liu, Z.: Key technologies of smart grid based on big data. Power Supply Technol. **41**(8), 1195–1197 (2017)
10. Yu, Y., Zhu, Z., Huang, C., et al.: Distribution network fault trend judgement based on big data analysis. Power Supply Technol. **42**(1), 132–134 (2018)
11. Cong, H., Zhao, L., Meng, H., et al.: Applicability evaluation of biomass pyrolytic poly-generation technology on clean heating in northern rural of China. Trans. Chinese Soc. Agric. Eng. **34**(1), 8–14 (2018)
12. Ke, G., Tao, D., Qiao, J.-F., et al.: Learning a no-reference quality assessment model of enhanced images with big data. IEEE Trans. Neural Netw. Learn. Syst. **29**(4), 1301–1313 (2018)
13. Fu, W., Liu, S., Srivastava, G.: Optimization of big data scheduling in social networks. Entropy **21**(9), 902–903 (2019)
14. Liu, S., Li, Z., Zhang, Y., et al.: Introduction of key problems in long-distance learning and training. Mobile Netw. Appl. **24**(1), 1–4 (2019)
15. Fer, I., Kelly, R., Moorcroft, P.R., et al.: Linking big models to big data: efficient ecosystem model calibration through Bayesian model emulation. Biogeosci. Discus. **12**(1), 1–30 (2018)
16. ur Rehman, M.H., Ahmed, E., Yaqoob, I., et al.: Big data analytics in industrial IoT using a concentric computing model. IEEE Commun. Mag. **56**(2), 15–21 (2018)

Study on the Dynamics of Virus Propagation in Combination with Big Data and Kinetic Models

Guo-bin Zeng[✉] and Yan-ni Chen

Haikou University of Economics, Haikou 571127, China
zengguobin9965@163.com

Abstract. With the continuous development of science and technology, in the context of current big data, the research on the law of traditional virus propagation dynamics had been developed to the bottleneck. The traditional law of virus propagation dynamics was less sensitive and the mathematical model was not easy to operate. Therefore, it was proposed to study the dynamics of viral propagation based on the combination of big data and kinetic models. The model was established by using differential equations and so on, and the accurate prediction law of virus propagation dynamics was completed by experimental tracking control. A graph of the number of patients over time was obtained by bringing the problem into the model, and the changes in the model results were derived from this graph. In this way, corresponding countermeasures was drawn based on the changes in the results. Finally, through simulation experiments, it was proved that the combination of big data and kinetic model of viral propagation kinetics scientifically and accurately studied the laws of viral propagation dynamics. The established mathematical model was easy to operate and had a good guiding significance for practice.

Keywords: Viral transmission · Big data · Kinetic model · Differential equation

1 Introduction

The research on the law of virus propagation dynamics has been developed to the bottleneck. The traditional law of virus propagation dynamics is less sensitive, and the mathematical model is not easy to operate. Therefore, we need to expand the field of view to see the virus propagation dynamics. With the continuous development of science and technology, in the context of current big data, a study is proposed to combine the big data and dynamics models to study the dynamics of virus propagation. The model is established by using differential equations and so on, and the accurate prediction law of virus propagation dynamics is completed by experimental tracking control. By establishing a differential equation model, the process of virus spread and propagation is described. Finally, the dynamics of virus propagation is studied through analysis. We bring the required questions into the model to get a graph of the number of patients over time, and based on this graph, we get the changes in the model results. In this way, according to the change of the result, the corresponding countermeasures

S. Liu and L. Xia (Eds.): ADHIP 2020, LNICST 348, pp. 36–43, 2021.
https://doi.org/10.1007/978-3-030-67874-6_4

can be obtained [1]. Finally, through the sensitivity analysis of the incubation period and the healing period of the infectious disease, it is found that the proposed virus propagation dynamics law combined with big data and kinetic model can scientifically and accurately study the law of virus propagation dynamics, and the established mathematical model is easy and has a good guiding significance for practical operation.

2 Theoretical Analysis of Big Data and Dynamics Models

Considering the total number of people in the region, we turn the problem into how to find the correct relational expression to express the total number of patients per day [2], to find out the normal number of people per unit time, the incubation period per unit time, the change of the number of people, the change of the number of patients diagnosed per unit time, the number of people who withdrew within the unit time, and the number of suspected patients per unit time, and the big data to establish the differential equation model and dynamics models were obtained.

①Treat all routes of transmission of the virus as contact with the pathogen;② The total number N of the areas examined during the spread of the disease is considered constant, that is, the number of people flowing into the area is equal to the number of people flowing out, and the time is measured in days;③ The virus in which the virus is in an incubation period is not contagious [3];④ the probability of a second infection of the healer is 0, they withdraw from the infection system, so they are classified as "exitors";⑤ Regardless of the birth rate and natural mortality rate during this period, the number of deaths caused by the virus is also classified as "exit";⑥ The isolated population completely severes contact with the outside world and is no longer contagious [4].

Changes in the normal number of people per unit time:

$$\frac{dS_1}{dt} = -\alpha_1 I(t)(1-\varpi)S_1 - \alpha_2 \left[A(t)(1-\varpi) + A(t)\varpi\frac{1}{d_3} \right] S_1 \tag{1}$$

According to the above algorithm, the number of patients in the incubation period per unit time can be obtained:

$$\frac{dE}{dt} = -\alpha_1 I(t)(1-\varpi) \left[S_1 + A(t)(1-\varpi) + A(t)\varpi\frac{1}{d_3} \right] - \frac{2}{d_1+d_2}E \tag{2}$$

Changes in the number of patients diagnosed per unit time:

$$\frac{dI}{dt} = \frac{2}{d_1+d_2}E\frac{1}{d_3}I \tag{3}$$

Changes in the number of people who quit during the unit time:

$$\frac{dR}{dt} = \frac{1}{d_3}I \tag{4}$$

Changes in the number of suspected patients per unit time:

$$\frac{dA}{dt} = -\alpha_2 \left[A(t)(1 - \varpi) + A(t)\varpi \frac{1}{d_3} \right] S_1 - \alpha_1 I(t)(1 - \varpi) \left[A(t)(1 - \varpi) + A(t)\varpi \frac{1}{d_3} \right] \quad (5)$$

Where I_0, S_{10}, R_0, A_0, E_0 are initial values. The average incubation period of the infectious virus is $\frac{d_1 + d_2}{2}$, that is, the patient in the incubation period per unit time is converted to the infected person by the proportional constant $\frac{2}{d_1 + d_2} > 0$; The average course of death or recovery of the confirmed patient is d_3, that is, the recovery rate of the infected person per unit time is $\frac{2}{d_3} > 0$; The average course of treatment of suspected patients is d_3, that is, the recovery rate of suspected patients per unit time is $\frac{1}{d_3} > 0$; The contact rate parameter of each susceptible person and patient per unit time is $r > 0$.

The contact rate parameter $\alpha_2 > 0$ of the susceptible person and the suspected patient; Considering that the suspected patient's infection is converted into a latent patient, but the latent patient does not turn into a suspected patient; p is the strength of the isolation measure; $A(t)p\frac{1}{d_3}$ is a recovered suspected patient who has been quarantined; initial value setting: (These data are an estimate based on the total population and medical knowledge). $I_0 = 1$, $s_{10} = 1.1$ ten million, without considering the floating population;

R0 = 0; E0 = 0; A0 = 100; parameters setting: $\frac{2}{d_1 + d_2} > \frac{1}{5}$, The average incubation period for infectious diseases is 5; $\frac{1}{d_3} = \frac{1}{20}$, The average course of death or recovery for a confirmed patient is 20.

$\frac{1}{d_3} = \frac{1}{20}$, Set the average course of suspected patients to 20; $\alpha_2 = 1.0 \times 10^{-11}$, The contact rate parameters of suspected patients and susceptible persons are also assumed to be fixed. Establish the calculus equation as follows:

$$\begin{cases} \dfrac{dS_1}{dt} = -\alpha_1 I(t)(1 - \varpi)S_1 - \alpha_2 \left[A(t)(1 - \varpi) + A(t)\varpi \dfrac{1}{d_3} \right] S_1 \\[2mm] \dfrac{dE}{dt} = -\alpha_1 I(t)(1 - \varpi) \left[S_1 + A(t)(1 - \varpi) + A(t)\varpi \dfrac{1}{d_3} \right] - \dfrac{2}{d_1 + d_2} E \\[2mm] \dfrac{dI}{dt} = \dfrac{2}{d_1 + d_2} E \dfrac{1}{d_3} I \\[2mm] \dfrac{dR}{dt} = \dfrac{1}{d_3} I \\[2mm] \dfrac{dA}{dt} = -\alpha_2 \left[A(t)(1 - \varpi) + A(t)\varpi \dfrac{1}{d_3} \right] S_1 - \alpha_1 I(t)(1 - \varpi) \left[A(t)(1 - \varpi) + A(t)\varpi \dfrac{1}{d_3} \right] \end{cases} \quad (6)$$

$I_0 = 1$, $s_0 = 1.1$ ten million, without considering the floating population, $R_0 = 0$, $E_0 = 0, A_0 = 100$.

3 Virus Propagation Model

Since the spread of the virus is influenced by social, economic, cultural, customary habits and other factors [5], the most direct factors affecting the development trend of the epidemic are: the number of infected persons, the form of transmission, and the ability of the virus itself to spread, isolation strength, admission time, etc. When we build a model, we can't and don't have to consider all the factors. We can only grasp the key factors and make reasonable assumptions and modeling. Assume that the incubation period of an infective virus that is not completely known is d1–d2 days, and the healing time of the patient is d3 days. The virus can spread and spread through direct contact, oral droplets [6], and the number of people per day in this population is r. In order to control the spread and spread of the virus, the population is divided into four categories: normal population, patient population, cured human and death population, represented by H(t), X(t)R(t) and D(t), respectively. The controllable parameter is the isolation measure strength P (percentage of patients isolated during the incubation period). Therefore, the law of virus transmission can be divided into two stages: "pre-control" and "post-control".

The virus pre-control model is a virus model similar to that of natural propagation [7], and the post-control model is a differential equation model after interventional isolation intensity [8]. The transformation relationship of various types of people in the two models is shown in the following Fig. 1.

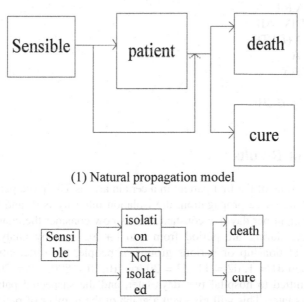

(1) Natural propagation model

(2) Differential equation model after isolation strength

Fig. 1. Virus propagation model

Based on the above model, use the model you created for the following data: d1 = 1, d2 = 11, d3 = 30, r = 10; numerical simulation [9, 10], assuming the initial number of cases is 890, suspected patients are 2000 isolation measures intensity p = 60%. The patients were hospitalized 1.5 days later, and the simulated results of the suspected patients were isolated after 1.5 days and the simulation results of isolation measures p = 40%. The sensitivity of the above parameters to the calculation results was analyzed and used for subsequent improvement of the model [11, 12]. The model program is as follows.

```
ORG0000H
LJMPSTART
ORG0003H
LJMPPINT0
ORG001BH
LJMPPTI1
START:MOVTMOD,#10H
MOVSCON,#00H
MOVTH1,#03CH
MOVTL1,#0B0H
MOVDPTR,#TAB
SETBEA
SETBEX0
SETBIT0
DISP:MOVA,R4
MOVB,#10DIV  AB
MOVCA,@A+DPTR
MOVSBUF,A
LCALLDELAY
JIXU:
MOV   TH1,#03CH
```

4 Analysis of Results

Suppose that the time of the first patient in a certain area is T0, in the period of (T0, T), it is a period of near-free propagation, the isolation intensity is 0, and the number of infections per patient per day is a constant. Let us now consider the changes in several groups of people during the period from t to t + Δt. And by analyzing the state transformation relationship of various groups of people, the differential equation is established. When d1 = 1, d2 = 11, d3 = 30, r = 10, I0 = 890, A0 = 20, p = 60%,the patient was admitted to hospital two days later, and the suspected patient was quarantined two days later. This will give you a graph of the number of patients over time: (see below) (Fig. 2).

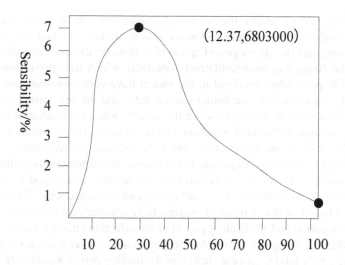

Fig. 2. Graph of the number of patients over time

From the figure we can see that the number of patients changes sharply with time. This shows that this is the development trend of the initial stage of virus transmission. Then you can see the highest point (Day 12.37) when the number of patients reached a maximum of 6803,000 people. By taking patient admission treatment and suspected patient isolation measures, we can clearly see from the figure that the number of patients is decreasing, and the number of patients has dropped to 540,800 after 100 days.

In order to further verify the sensitivity of the study of viral propagation kinetics, sensitivity simulation experiments were carried out on the dynamics of viral propagation. The simulation results are shown below (Fig. 3).

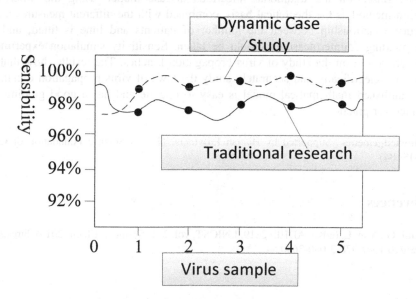

Fig. 3. Sensitivity simulation results

Through the above results, the following conclusions can be made, and the problems in the medical field can be transformed into the field of mathematics for analysis and discussion, and the development trend of infectious diseases and the prediction results for the future can be quantitatively obtained, which has strong theoretical and reliability. The parameters involved in the model have corresponding data sources. It can be easily calculated by combining certain data, and the relationship between the variables is clear, which is easy to solve the model. Since the mathematical model is based on continuous differential equations, an accurate analytical solution will not be obtained. We will fit the parameters under the premise of reasonable parameters. Considering the total number of people in the region, the population is divided into five categories: confirmed patients, suspected patients, healers, deaths and normal people, and these categories are divided into contagious and non-infectious. Countermeasures can be drawn based on the changes in the results. In addition, the sensitivity analysis of the incubation period and the healing period of the infectious disease was carried out. It was found that the change of the incubation period would have a greater impact on the results of the whole model, and the change of the healing period would only shorten the duration of the infectious disease, but has little effect on the cumulative number of patients.

5 Conclusions

With the continuous development of science and technology, in the context of current big data, the research on the law of traditional virus propagation dynamics has been developed to the bottleneck. The traditional law of virus propagation dynamics is less sensitive and the mathematical model is not easy to operate. Therefore, it is proposed to study the dynamics of viral propagation in combination with big data and kinetic models. Based on the traditional infectious disease model, using the differential equation method as the theoretical basis, combined with the different measures taken, the curve relationship between the number of patients and time is fitted, and the corresponding countermeasures should be taken. Sensitivity simulation experiments were carried out on the study of virus propagation kinetics. The results show that the study can scientifically and accurately study the laws of virus propagation dynamics. The established mathematical model is easy to operate and has a good guiding significance for practice.

Acknowledgements. Supported by Hainan Provincial Natural Science Foundation of China (118MS084)

References

1. Gui, G., Yun, L. (eds.): ADHIP 2019. LNICST, vol. 301. Springer, Cham (2019). https://doi.org/10.1007/978-3-030-36402-1

2. Reardon, F., Graham, D.A., Clegg, T.A., et al.: Quantifying the role of Trojan dams in the between-herd spread of bovine viral diarrhoea virus (BVDv) in Ireland. Prev. Vet. Med. **152**, 65–73 (2018)
3. Zhang, Y.: Kinetic model for active hepatitis and drug therapy to control hepatitis B virus transmission. Harbin Univ. Sci. Technol. **45**(6), 36–39 (2017)
4. Qi, L., Beaunée, G., Arnoux, S., et al.: Neighbourhood contacts and trade movements drive the regional spread of bovine viral diarrhoea virus (BVDV). Vet. Res. **50**(1), 158–169 (2019)
5. Fu, W., Liu, S., Srivastava, G.: Optimization of big data scheduling in social networks. Entropy **21**(9), 902 (2019)
6. Rouco, C., Aguayo-Adánn, J.A., Santoro, S., et al.: Worldwide rapid spread of the novel rabbit haemorrhagic disease virus (GI.2/RHDV2/b). Transbound. Emerg. Dis. **66**(4), 78–85 (2019)
7. Vincenti, G., Bucciero, A., Helfert, M., Glowatz, M. (eds.): E-Learning, E-Education, and Online Training. LNICST, vol. 180. Springer, Cham (2017). https://doi.org/10.1007/978-3-319-49625-2
8. Chris, O.: A possible role for domestic dogs in the spread of African horse sickness virus. Vet. Rec. **182**(25), 713–714 (2018)
9. Iwata-Yoshikawa, N., Okamura, T., Shimizu, Y., et al.: TMPRSS2 contributes to virus spread and immunopathology in the airways of murine models after coronavirus infection. J. Virol. **93**(6), 269–276 (2019)
10. Bello-Morales, R., Praena, B., Carmen, D.L.N., et al.: Role of microvesicles in the spread of Herpes simplex virus type 1 in oligodendrocytic cells. J. Virol. **88**(18), 155–263 (2018)
11. Bürli, C., Harbrecht, H., Odermatt, P., et al.: Mathematical analysis of the transmission dynamics of the liver fluke, Opisthorchis viverrini. J. Theor. Biol. **439**(42), 181–194 (2018)
12. Koca, I.: Modelling the spread of Ebola virus with Atangana-Baleanu fractional operators. Eur. Phys. J. Plus **133**(3), 100–109 (2018)

Research on Active Disturbance Rejection Method of Mobile Communication Network Nodes Based on Artificial Intelligence

Bing Li, Feng Jin, and Ying Li[✉]

Information and Communication College, National University of Defense Technology, Xi'an 710106, China
kmxttcll@sina.cn, xuennxe@163.com

Abstract. With the increasingly complex network environment and the interference of various other radio waves, the quality of mobile communication network is seriously affected. Aiming at the above problems, this paper studies an auto-disturbance rejection method for mobile communication network nodes based on artificial intelligence. According to artificial intelligence, an interference identification analysis model is constructed, which is used to identify and analyze the interference factors of mobile communication network nodes. Based on the recognition results, the characteristics of different interference types are summarized, and the interference problem is accurately judged. Then, the anti-interference work of mobile communication network nodes is completed by checking and processing the results. The experimental results show that the user is more satisfied with the quality of the mobile communication processed by this method than the traditional method of UAI participating in the identification and analysis of interference factors, which proves that this method is effective in anti-jamming and can meet the needs of users.

Keywords: Artificial intelligence · Mobile communication network · Anti-interference · Feature analysis

1 Introduction

In modern society, remote information exchange plays a key role, and remote information exchange can only be achieved by relying on mobile communication network. With the increasingly complex network environment of mobile communication industry and the interference of various other radio waves, problems such as increasing drop-out rate, reducing base station coverage, and sharply decreasing network indicators emerge one after another, seriously affecting the quality of calls. Under this background, how to reduce the interference in the network and improve network capacity and quality has become a problem that must be solved in the network operation and maintenance of operators [1, 2]. Reasonable and efficient analysis and solution of interference problems has become an important requirement in daily network operation and maintenance work.

Previous methods proposed by experts and scholars focused on the research of interference processing methods in mobile communication networks, but ignored the

S. Liu and L. Xia (Eds.): ADHIP 2020, LNICST 348, pp. 44–56, 2021.
https://doi.org/10.1007/978-3-030-67874-6_5

identification and detection of interference factors. Effective identification and detection will make mobile communication network interference processing more effective. Reference [3] proposes an efficient deployment method for optimizing coverage of key nodes in industrial mobile wireless networks. This method considers the industrial characteristics and mobility of wireless networks, and then simplifies the static and mobile node coverage problems into target coverage problems. A new cluster head deployment strategy is proposed based on the improved maximum cluster model and iterative calculation of new candidate cluster head positions. The maximum clique is obtained by double tabu search. Each cluster head updates its new position through an improved virtual force, and moves with full coverage to find the minimum inter cluster interference. The experimental results show that. This method can realize the location between interference nodes, but the anti-jamming effect is not good, and it takes a long time to identify interference nodes. In reference [4], an active interference suppression method for fully distributed communication delay is proposed. In this method, automatic generation controller is used instead of traditional control center, and automatic generation controller is embedded in each participating generating unit to avoid communication delay of control signal. Simulation results show that this method has good performance and robustness in the presence of communication delay, but when the number of attacks is large, the anti-jamming effect is not obvious, and it takes a long time to identify interference nodes.

Aiming at the weakness of traditional methods, this paper proposes an artificial intelligence-based ADRC method for mobile communication network nodes. The greatest advantage of this method is to introduce interference recognition into it. After verification and analysis, the communication quality of this method has been greatly improved, which promotes the development of mobile communication network.

2 Anti-jamming Method of Mobile Communication Network Node Based on AI

Mobile network is an important branch in the research of computer related fields. Nodes in mobile network communication system will enter and exit continuously, and then form a certain rule. Mobile network systems need to strictly request the bandwidth and timeliness of data transmission to ensure high-quality communication results. Through an efficient anti-jamming scheme, it can ensure that the mobile network system has stable information transmission and related functions under strong interference.

2.1 Disturbance Recognition and Analysis

Artificial Intelligence, abbreviated as AI. It is a branch of computer science. It produces an intelligent machine that can respond in a similar way to human intelligence. The research in this field includes robots, language recognition, image recognition, natural language processing and expert systems. Artificial intelligence can simulate the process of human consciousness and thinking. Recognition of interference factors in mobile communication networks is a weak part of traditional interference methods. Traditional methods focus on the research of anti-interference methods, but neglect the role of interference recognition. In this paper, pattern recognition in artificial intelligence is used to analyze the types of interference in order to deal with them pertinently in the future. Effective interference identification will greatly improve the efficiency of anti-interference and the speed of network operation.

(1) disturbance recognition analysis model

The interference spectrum data FAS collected by frequency sweeper is taken as the basis of the system, and the location function of interference cell is completed by pattern recognition. On the one hand, similar disturbance patterns are matched and located from the shape angle, on the other hand, from the FAS data feature point of view, and random forest pattern recognition method is used to locate similar disturbance characteristics of the cells, as shown in Fig. 1.

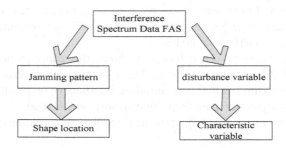

Fig. 1. Disturbance recognition analysis model.

Decision tree is not only a more understandable classifier, but also has advantages in training speed and memory usage. At the same time, considering the recognition of multiple possible classifications of target network elements, the system chooses the stochastic forest learning model based on decision tree. On the one hand, it has the advantages of decision tree, and at the same time, stochastic forest can improve the stability of training. It can count the votes of multiple classifications belonging to a single cell. It is a multi-classifier, i.e., multi-output model, which is the use of the system. The reason of disturbance in machine forest model learning [5].

Pattern recognition describes the characteristics of objects by extracting their features and phenomena. After learning and understanding the features, it establishes a learning rule model from objects to recognition results. The process is shown in Fig. 2.

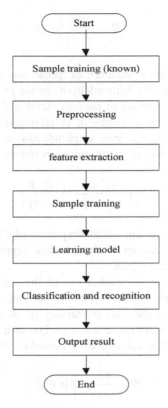

Fig. 2. Patterns recognition process.

As shown in Fig. 2 above, feature extraction and sample training are two key steps in the process of pattern recognition, which have a great impact on the recognition results, so feature extraction should be representative and the data of training set must be reliable. The extracted feature variables constitute the feature dimension of the pattern space, and the feature dimension is better representative for classification; the number of training samples should be enough, and the ratio of the sample number to the model space dimension should be at least greater than or equal to 3, preferably greater than or equal to 10. The main methods of sample training are decision tree, Bayesian classification training, random forest, neural network, support vector machine and so on. Stochastic forest method is used to build learning model and complete the learning process [6].

(2) feature extraction

Assuming that x represents a cluster of random variables consisting of n communication interference signals, y represents the frequency range of each cluster, and m represents a constant threshold of interference energy, the spectrum amplitude corresponding to the same frequency of each random variable in the random process is constituted into a new sequence, which is expressed by formula (1).

$$R = \frac{m \times y}{n \oplus \{p\}} \cdot x \tag{1}$$

In the formula, p represents a sequence of spread spectrum codes.

Assuming that a represents the finite additivity of the probability of communication interference signal, with the increase of random variable T' in the random process T, which consists of the spectrum sequence of communication interference signal, the probability of the corresponding spectrum amplitude also increases. Formula (2) is used to construct the spectrum matrix of communication interference signal.

$$G(f) = \frac{T \pm T'}{a \cdot f} \otimes \frac{E \cdot R}{h(2\pi f)} \tag{2}$$

In the formula, f represents the initial frequency of FM, E represents the linear frequency, R represents the number of time-domain segments, and h represents the sampling period of interference signal.

Assuming that v represents the symbol rate of the source, s represents the direct sequence pseudo-random code rate, b represents the corresponding amplitude of the same frequency, W represents the total probability of all amplitudes under each frequency, and z represents the spectrum distribution law of instantaneous interference of the fixed frequency signal, W is calculated by formula (3):

$$W = \frac{b \otimes z}{s} \otimes [v \oplus j] \tag{3}$$

In the formula, j represents the analysis window length of Fourier transform.

Assuming that d represents the spectrum statistical characteristics of full-time communication interference signal and k represents the change of signal position, formula (4) is used to calculate d:

$$d = \frac{R \oplus x(f)}{W \otimes r} \cdot ok \tag{4}$$

In the formulas, r and o represent the transformation parameters of the time domain and frequency domain of the communication interference signal.

In summary, in the process of optimum detection of communication interference signals in mobile networks, a new sequence of spectrum amplitudes corresponding to the same frequency of each random variable in the random process is formed by using probability and statistics method, and the spectrum matrix of communication interference signals is formed. The spectrum distribution law of instantaneous interference of fixed frequency signals is obtained, and the characteristic spectrum of communication interference signals is extracted to realize the shift. Optimal detection of communication interference signals in mobile networks lays the foundation [7].

(3) sample training

The Stochastic Forest method is to build a forest consisting of a decision tree, and there is no correlation between each decision tree. The basic component of random forest is decision tree. In the process of building multiple trees, feature subsets are selected randomly to make each decision tree different. Selecting subsets from the extracted feature variables to decide how to split the data is the best. It has proxy splitting with missing values and can judge the closeness (i.e. similarity) of two samples. The number of variables to be split and the number of trees in random forest are two important parameters in the training process. The number of training samples is represented by N, the number of characteristic variables by M and the number of trees by K. The steps are as follows:

1) The k-group training subset is formed by repeated sampling from training sample N, and each training subset is the training sample set of each classification tree.

2) Each classification decision tree is trained, and m (m < M) feature variables are randomly selected at each node of the tree. One or more of the M feature variables are selected to split according to the minimum impurity of the nodes, so that the tree can grow continuously without pruning in the process of growth.

3) According to the generated random forest, the target data are predicted and the classification results of each tree are counted.

The specific steps of sample training are shown in Fig. 3.

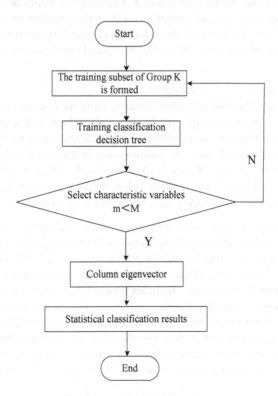

Fig. 3. sample training process.

2.2 Anti-jamming Processing of Mobile Communication Network Nodes

By summarizing the characteristics of different interference types, can distinguish the interference types from the spectrum performance of the received signals, make a rough judgment on the interference problems, and then make a survey and processing according to the results. The main interference in mobile communication system includes co-frequency interference, adjacent-frequency interference, inter modulation interference, repeater interference and other interference [8].

(1) anti-co-frequency interference

At present, mobile communication systems adopt cell structure to improve frequency utilization by using the same frequency multiplexing method, and the same frequency can be reused every certain distance. Under the condition of certain distance interval, the co-frequency interference in the system will not affect the normal communication too much. However, with the cell splitting and the increase of co-frequency multiplexing coefficient, a large number of co-frequency interference will seriously affect the normal operation of the system. When the carrier-to-interference ratio of the same frequency interference is less than a certain value, it will directly affect the communication quality of the mobile phone. Serious calls will be dropped or users can not establish normal calls.

Spread spectrum technology is used as information transmission mode. Spread spectrum code is used to modulate the transmitted information at the transmitter. The bandwidth of the original signal is broadened. The received information is decomposed coherently with the same spread spectrum code at the receiver to recover the information data. Through this process, the intensity of interference signal is reduced. In spread spectrum communication, the wider the spread spectrum, the stronger the anti-jamming ability.

(2) anti-adjacent frequency interference

Due to frequency planning and other reasons, there may be unreasonable design of adjacent frequency or coverage in adjacent cells, which will lead to adjacent frequency signals falling into the passband of adjacent frequency receivers and causing adjacent frequency interference. In addition, due to the near-far effect, the influence of adjacent frequency interference is also increased. When the adjacent carrier-to-interference ratio is less than a certain value, it will also affect the quality of communication, resulting in dropped calls or the inability to establish normal calls.

Aiming at the problem of adjacent frequency interference, power control technology is mainly used for anti-interference. Power control is to change the transmission power of mobile station or base station by radio within a certain range [9]. Power control can minimize transmission power and improve interference to other calls under the condition of good reception. Power control includes forward power control and reverse power control. Reverse power control is divided into open-loop power control which is only participated by mobile station and closed-loop power control which is jointly participated by mobile station and base station. For the adjacent frequency interference caused by far-near effect, the power control technology can be used to improve it. When the distance between the mobile station and the base station is closer, reducing the transmitting power of the mobile station can reduce the interference to other users. When the distance is longer, the transmitting power of the mobile station can be increased to overcome the increased path attenuation, so that the signal transmitted by the mobile station has the same signal strength as possible when it reaches the base station.

(3) anti-inter modulation interference

Because a large number of non-linear circuits are used in the communication system, when two or more different frequency signals enter at the same time, there will be inter modulation. If the frequency of modulated signal falls into the receiving frequency band, there will be inter modulation interference, and the direct consequence of interference is the waste of base station resources, but also the drop of words [10]. Frequency hopping is the main measure to solve intermodulation interference. Frequency hopping is to hop the carrier frequency of communication at several frequency points. Frequency hopping can play the role of frequency diversity and improve the error code characteristics caused by fading, but frequency hopping can also play the role of interference source diversity. Slow frequency hopping technology is adopted in mobile communication network. The frequency hopping rate is 217 hops. Frequency hopping is carried out between two slots. Fixed frequency is used in one slot and another frequency is used in the next slot to reduce the influence of interference.

(4) anti-repeater interference

The main reason for repeater interference is that the broadband repeater amplifies the useful upstream signal, and at the same time amplifies the noise, resulting in wideband interference. Mobile communication networks are gradually replacing existing broadband repeaters by using optical fiber frequency selective repeaters or by stretching RRU. However, due to the relative lag of station construction and site selection, many owners in weak coverage areas will install illegal broadband repeaters by themselves to solve their own coverage problems, while introducing a source of interference [11, 12].

The way to solve the interference of repeater is to start from two aspects: using optical fiber frequency selective repeater or stretching RRU to gradually replace the existing broadband repeater; Follow up the pace of urban construction, solve the problem of weak coverage through network planning and construction, and coordinate the elimination of private installed illegal repeaters through the radio management committee [13, 14]. In other words, to fundamentally solve the problem of repeater interference, it is necessary to solve the problem of user coverage through continuous network planning and construction, in order to suppress the emergence of repeater interference sources.

3 Mobile Communication Quality Testing and Analysis

In the above chapters, the quality of mobile communication will be affected after interference, so this simulation experiment analyses the quality of mobile communication with the participation of artificial intelligence, in order to judge the performance of the auto-disturbance rejection method of mobile communication network node based on artificial intelligence.

In the experiment, 200 users were selected for a one-month communication quality experience, and 100 mobile communication networks with artificial intelligence for anti-interference processing were selected as group A. Another 100 mobile communication networks using unattended artificial intelligence to participate in anti-interference processing, this group is group B. A month later, the communication satisfaction of 200 users (mainly in drop-off, connection, interference noise and other three aspects) was investigated. With the above data as an indicator, a mobile communication quality evaluation model was constructed, as shown in Fig. 4 below.

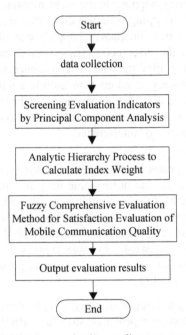

Fig. 4. Mobile communication quality assessment model.

After the establishment of the model, the model is used to evaluate the call quality. The evaluation results are shown in Table 1.

Table 1. Evaluation results (bits).

Name	Group A	Group B
Very satisfied	34	5
Satisfied	64	32
Commonly	2	45
Dissatisfied	0	10
Very dissatisfied	0	8

As can be seen from Table 1, 34 of the evaluations given by group A users are very satisfactory, 64 of them are satisfactory, and 2 of them are general. In the evaluation given by group B users, 5 are very satisfied, 32 are satisfied, 45 are general, 10 are unsatisfactory, and 8 are very unsatisfactory. Compared with group B, group A users are more satisfied with the quality of mobile communication than group B, which proves that the communication quality is higher after the anti-jamming treatment of this method.

On the basis of the above experiments, in order to further verify the performance of the mobile communication network node anti-interference method under this method, the attack data in KDD cup-99 data set is selected to test the anti-interference effect of different methods. Table 2 shows the test data set.

Table 2. Test data set.

Attack categories	Dataset type	Benign sample	Malicious samples
Dos	1	2157	2254
Probing	2	3965	3647
	3	2954	3015
U2L	4	3654	3519
R2L	5	1987	2024

The KDD CUP-99 data set contains real data sets generated by various user types, different network traffic and attack means. The whole data set contains about 5 million data records. The data exception types are divided into four categories, including DOS denial of service, R2L unauthorized remote host access, U2R unauthorized local super user privilege access, and probing port monitoring or scanning. Under normal access, the whole data set should contain 5 types, which are labeled as 1,2,3,4,5 respectively. Each data record contains 41 features, including 32 continuous features and 9 discrete features. Because the whole data set is very large, only 10% of the data are selected for experiment.

Based on the above data, the anti-interference effect of the traditional method of UAI participating in the identification and analysis of interaction factors is compared with that of the method in this paper. The results are shown in Fig. 5. Among them, Fig. 5(a) shows the anti-jamming effect comparison when the amount of attack data is small, and Fig. 5(b) shows the anti-jamming effect comparison when the amount of attack data is large.

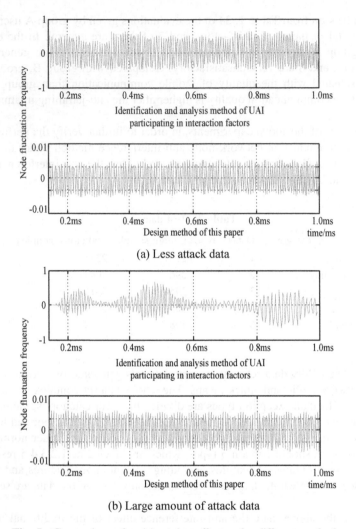

Fig. 5. Comparison of anti-jamming effect under different methods.

According to Fig. 5(a), in the case of less attack data, the anti-interference effect advantage of the design method in this paper is not obvious, which shows that in this case, both the traditional method and the design method in this paper have better anti-interference effect.

According to Fig. 5(b), when there are many attack data, the anti-interference effect of the design method in this paper is obviously better than that of UAI participating in interaction factor identification and analysis method, which shows that the design method in this paper can adapt to different attack data amount, and the anti-interference effect is better. This is because the method in this paper constructs an interference identification analysis model based on artificial intelligence technology. Using this model to identify and analyze the interference factors of mobile communication network nodes, can summarize the characteristics of different interference types, and accurately judge the interference types, so as to improve the anti-interference effect.

In order to further verify the effectiveness of the proposed method, the anti-jamming effect of different methods is compared with the interference node identification time. The comparison results are shown in Fig. 6.

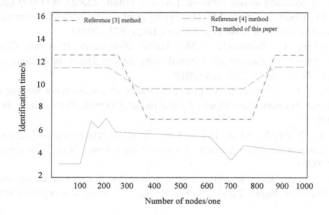

Fig. 6. Comparison of jamming node identification time.

It can be seen from Fig. 6 that the time consumed by the method of reference [3] and the method of reference [4] is much longer than that of the method in this paper, and the longest time of this method is 7S. It shows that the recognition efficiency of this method is higher, and it can realize the identification of interference nodes in a short time.

4 Conclusion

In summary, the mobile communication network is the carrier of modern people's long-distance communication and plays a key role in production and life, so it has important practical significance to ensure the quality of mobile communication. People are disturbed by various factors in the process of communication, such as dropping calls, switching, congestion, murmur and so on. Aiming at this phenomenon, an artificial intelligence based ADRC method for mobile communication network nodes is proposed. The innovation of this method and the application of artificial intelligence in interference recognition make up for the shortcomings of traditional anti-jamming methods.

References

1. Bai, X., Wang, Z., Sheng, L., et al.: Reliable data fusion of hierarchical wireless sensor networks with asynchronous measurement for greenhouse monitoring. IEEE Trans. Control Syst. Technol. **27**(3), 1036–1046 (2019)
2. Kizel, F., Etzion, Y., Shafran-Nathan, R., et al.: Node-to-node field calibration of wireless distributed air pollution sensor network. Environ. Pollut. **233**(2), 900–909 (2018)
3. Li, X., Li, D., Dong, Z., et al.: Efficient deployment of key nodes for optimal coverage of industrial mobile wireless networks. Sensors **18**(2), 545 (2018)
4. Han, W., Wang, G., Stankovic, A.M.: Active Disturbance Rejection Control in fully distributed Automatic Generation Control with co-simulation of communication delay. Control Eng. Pract. **85**(4), 225–234 (2019)
5. Balaji, S. Julie, EG., Robinson, Y.H., et al.: Design of a security-aware routing scheme in Mobile Ad-hoc Network using repeated game model. Comput. Stand. Inter. **66**(10), 103358–103358 (2019)
6. Guangjun, L., Nazir, S., Khan, H.U., et al.: Spam detection approach for secure mobile message communication using machine learning algorithms. Secur. Commun. Netw. **2020**(11), 1–6 (2020)
7. Hsieh, W.-B., Leu, J.-S.: Implementing a secure VoIP communication over SIP-based networks. Wireless Netw. **24**(8), 2915–2926 (2017). https://doi.org/10.1007/s11276-017-1512-3
8. Gupta, M., Chaudhari, N.S.: Anonymous two factor authentication protocol for roaming service in global mobility network with security beyond traditional limit. Ad Hoc Netw. **84**(3), 56–67 (2019)
9. Tingyu, X., Kai, K., Junjin, W., et al.: Research on the robustness of supply chain networks under the random and intention two different attack methods. Math. Practice Theory **48**(16), 40–47 (2018)
10. Fu, W., Liu, S., Srivastava, G.: Optimization of big data scheduling in social networks. Entropy **21**(9), 902 (2019)
11. Fengsheng, W.: A method for purifying abnormal intrusion signals in multimode fiber networks. Laser J. **40**(03), 120–124 (2019)
12. Wu, Y., Duong, T.Q., Swindlehurst, A.L.: Safeguarding 5G-and-beyond networks with physical layer security. IEEE Wirel. Commun. **26**(5), 4–5 (2019)
13. Li, B., Fei, Z., Zhang, Y., et al.: Secure UAV communication networks over 5G. IEEE Wirel. Commun. **26**(99), 114–120 (2019)
14. Yin, L., Haas, H.: Physical-layer security in multiuser visible light communication networks. IEEE J. Sel. Areas Commun. **36**(1), 162–174 (2018)

Research on Anonymous Reconstruction Method of Multi-serial Communication Information Flow Under Big Data

Ying Li$^{(\boxtimes)}$, Feng Jin, Xiao-xia Xie, and Bing Li

Information and Communcition College National University of Defense
Technology, Xi'an 710106, China
xuennxe@163.com

Abstract. The existing methods of dynamic reconfiguration of network information flow have some drawbacks, such as security, reliability and bad influence on the performance of the original network. Therefore, an anonymous reconfiguration method of multi-serial communication information flow under large data is proposed. Firstly, the original information flow is acquired in the communication network, and the cooperative filtering of multi-serial communication is carried out. After filtering, the notification information of relay nodes is obtained in the information flow, and the communication status of the information flow is extracted. The characteristic information of the information flow is reconstructed and anonymized. Finally, the anonymous reconstruction of multi-serial communication information flow is completed. By analyzing and comparing the experimental results, it can be seen that the method proposed in this paper is superior to the traditional method in terms of both the effect of anonymity and the efficiency of operation when reconstructing the anonymous information flow of multi-serial communication, it effectively solves the shortcomings of traditional methods, such as poor anonymous effect of information flow and slow speed of information flow reconstruction. It shows that the method has a high degree of anonymity and has a strong practicability.

Keywords: Large data · Information flow · Multi-serial communication · Reconstruction method · Anonymous

1 Introduction

Big data is a new processing mode, which has stronger decision-making power, insight and process optimization ability, and can effectively deal with massive, high growth rate and diversified information assets [1, 2]. With the continuous popularization of big data and Internet in personal and commercial communications, great changes have been brought to people's lives and work [3]. After a long time of use of computer internet and the development of big data industry, people have higher requirements for the network world in the use process [4]. Internet users hope to protect their privacy while enjoying the communication services provided by multi-serial network operators [5]. The emergence of serial transmission technology is an important condition to achieve the above requirements [6].

© ICST Institute for Computer Sciences, Social Informatics and Telecommunications Engineering 2021
Published by Springer Nature Switzerland AG 2021. All Rights Reserved
S. Liu and L. Xia (Eds.): ADHIP 2020, LNICST 348, pp. 57–68, 2021.
https://doi.org/10.1007/978-3-030-67874-6_6

Traditional multi-serial-port parallel communication data transmission system can not acquire serial passwords independently, so it needs to select and open serial ports manually, and users need to know serial passwords beforehand, which greatly reduces the efficiency of the system. Reference [7] proposed a method of ACNS collision information reconstruction based on compressed sensing. The method developed ACNS terminal based on compressed sensing theory, and proposed the process of acceleration acquisition, compression and restoration in ACNS. Taking 20 km/ h as the critical speed triggered by ACNS, the collision acceleration data at this speed can be obtained through sled collision test; based on orthogonal matching pursuit algorithm, using discrete cosine transform matrix, the collision acceleration data of ACNS are obtained, The results show that the collision information can be reconstructed accurately, but the effect of anonymity is not good. Reference [8] proposes an automatic detection method for implicit information leakage in Business Process Execution Language (BPEL) based on information flow. This method constructs a BPEL representation meta model based on Petri net for transformation and analysis. Based on the concept of position noninterference of Petri net, Petri net reachability graph is used to estimate the interference of Petri net, so as to detect the components of hidden information leakage in Web services. This method can detect the hidden information accurately, but it takes a long time to reconstruct the information. Therefore, a multi-serial parallel communication data transmission system which can independently identify serial numbers has been developed, and has been widely used in real life and achieved good results.

XON/OFF is used to complete the data transmission control of multi-serial parallel communication based on software flow. When the input data of the software in the serial port receiver is higher than the threshold value, XOFF characters are transmitted to the serial port data sender. After the sender collects the XOFF spontaneously, the sending data is terminated. Otherwise, when the amount of data at the receiving end is below the threshold, XOF characters are sent to the serial data sender and data is sent. This is the working principle and mode of multi-export communication. In this way, the transmission efficiency of user information can be guaranteed, but in addition, the user also needs to protect the identity information and privacy, and also needs to gradually realize anonymity in the transmission of data. An important goal of anonymous communication is to hide the identity or communication relationship between the two sides, so that the eavesdropper can not directly know or infer the communication relationship between the two sides. In general, information flow re engineering is also called information flow re-engineering. It is a process of optimizing the combination of information flows in business processes according to the strategic objectives of enterprises and customer needs. It can be approximated as the early planning stage of business process re-engineering. In the network data transmission and communication, the user's data information is more anonymous by reorganizing the information flow, which makes the user's information more secure in the transmission process.

2 Design of Anonymous Reconstruction Method for Information Flow

The whole method of anonymity reconstruction of information flow is realized by two steps, namely, anonymity of information flow and reconstruction of information flow. The original information flow in the communication system is reconstructed and processed. Finally, the reconstructed information flow is encrypted. Finally, the function of anonymous reconfiguration of information flow is realized, and the processed information flow is output.

2.1 Raw Information Flow Acquisition

Firstly, the data flow analyzer takes the defined information flow rules as the basis, and takes the middle of the source code output by the static security checking tool as the input to get the information flow among the variables in the source code. Then the information flow is filtered to get the information flow among the hidden channel variables. On this basis, the information flow graph is constructed. Combined with the characteristics of the hidden channel information flow and the external security rules, the hidden channel in the system source code is detected by reverse iteration traversing the information flow graph. The information flow generation method is shown in Fig. 1.

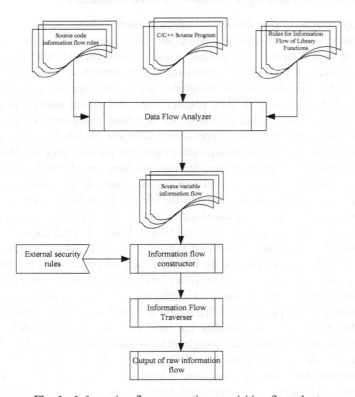

Fig. 1. Information flow generation acquisition flow chart.

The main function of the information flow generation part in the flow chart is to generate the information flow between source variables. Its main modules are as follows:

Source code information flow rule module. This module is based on the simple information flow rules given by Tsai and Gligor.

Library function information flow rule module. Functions that do not have function implementations in source code are called Library functions, including commonly known C/C ++ runtime libraries and third-party function libraries. According to the interface description of Library function, the module specifies the information flow between parameters and parameters, and between parameters and return values. When the data flow analyzer encounters a library function in the analysis process, the information flow generation rules of the matching function are extracted directly from the module, and the information flow is generated. As a supplement to the source code information flow rule module, this module ensures the integrity of the information flow.

Information flow graph constructor module. This module takes the information flow between source variables as input. Firstly, it filters out the information flow among the hidden channel variables, then organizes the information flow into an information flow graph. Finally, according to the implementation process of each module in the graph, the original information flow data can be obtained.

2.2   Collaborative Filtering of Multi-serial Communication

The data transmission of multi-serial communication includes serial data receiving module, parallel-serial conversion module and serial output selection module. The main work of serial data receiving module is level conversion and data transceiver. The original data stream is processed by multi-serial communication, which includes level conversion, UART IP characteristic parameter analysis and register control data baud rate design. The processing of data parallel-serial conversion module is mainly to analyze external channel signal, parallel-serial switching mode and timing simulation design. The processing of the serial port output selection module is mainly to analyze the serial port data of the target channel and to simulate the time series of ModelSim. It can be seen that the designed system realizes the function of multi-serial port data transmission, and the serial port baud rate can be adjusted. The multi-serial data is transmitted to the DSP processor through the UART IP of FPGA. The 8 UART IP implements the receiving of 8 kinds of serial data and the real-time adjustment of the baud rate of communication data. The data from 8 kinds of serial channels are fused into one channel and transmitted to the DSP processor serially. The expansion of multi-serial port of DSP is accomplished by FPGA, which simplifies the data transmission process of system communication and reduces the operation cost of the system. Serial parallel communication data is converted to data signal by level conversion circuit, and then transmitted to the pin of the FPGA. The designed serial data receiving module uses four MAX232 chips to complete the level conversion of 8 UART. The level conversion diagram of multi-serial communication circuit is shown in Fig. 2.

According to the conversion mode in the figure, the FPGA processor in multi-serial parallel communication data transmission is used to collect data and send it to the computer port. The display and control software is responsible for receiving data and

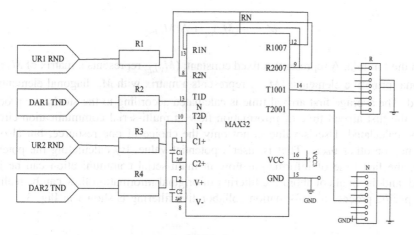

Fig. 2. Level conversion diagram of multi-serial communication circuit.

storing it. The processor transmits 8 kinds of serial channels via multiple UART IP, mainly transferring the communication data from PIO output to the parallel-serial conversion module. In this multi-serial communication architecture, collaborative filtering of the original input information flow is divided into two steps: computing the first arrival time of information, computing the potential value of users, and finally realizing the collaborative algorithm. Set up the first arrival time matrix M of each state in the information flow model. Its element m_{ij} takes i as the initial state and j as the expectation for the first time. Suppose $R = \lim_{t \to 0} Q(i, j, t)$, element r_{ij} in matrix R is the rate at which the semi-Markov process moves from state i to j. Taking constant $v \geq \max(r_i)$ and replacing the diagonal element of R with $1 - \frac{r_i}{v}$, discrete Markov chains equivalent to continuous-time Markov chains are extracted. Thus, discrete Markov chains equivalent to continuous-time Markov chains are extracted. The first arrival time matrix of discrete Markov chain is calculated and then converted to continuous time Markov chain first arrival time matrix.

If the discrete Markov chain transfer matrix and the first arrival time matrix are P_v and M_v respectively, the concrete expression formulas are shown in formulas 1 and 2.

$$P_v = I + \frac{1}{v}Q \tag{1}$$

$$M_v = \left[1 - Z_v + E(Z_v)_{dg} \right] D \tag{2}$$

In the formula, I represents the unit matrix of the original information flow; E is the matrix of all elements 1; D is a diagonal matrix with diagonal element $d_{ij} = \frac{1}{\pi(i)}$, and $\pi(i)$ represents the static distribution of discrete Markov chain state i; $(Z_v)_{dg}$ represents the matrix obtained by setting Z_v non-diagonal elements to 0, so the matrix M can be expressed as:

$$M = \frac{1}{v}(M_v)_{of} + \Lambda(M_v)_{dg} \tag{3}$$

In the formula, Λ represents a fixed constant; $(M_v)_{dg}$ represents a matrix of M_v non-diagonal primitive elements; $(M_v)_{of}$ represents a matrix with M_v diagonal elements all zeroed. The average first arrival time is calculated according to the stochastic process. Thus, the first arrival time of information flow in multi-serial communication circuits can be calculated. User's value is not only the choice of one resource, but also the influence on other users. That is, user's potential value. By calculating the potential value, the first time of information flow in multi-serial communication can be integrated, and the final collaborative filtering of original information flow can be realized. The specific process of information collaborative filtering is shown in Fig. 3.

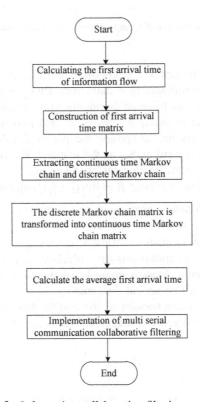

Fig. 3. Information collaborative filtering process.

2.3 Communication Information Known by Relay Nodes

Select the node T in the multi-serial communication network to receive information slices $\left(I_{Y_1}^*, I_{U_1}^*, I_{T_1}^*\right)$ and $\left(I_{Z_2}^*, I_{D_2}^*, I_{T_2}^*\right)$ from nodes S and S'. Node T grouped according to stream ID, and decoded data packets with the same stream ID number jointly.

Packets belonging to information slices T_1 and T_2 have the same ID number. Node T decodes the first piece of information received with the same stream ID number packet to get its own information. The number of information fragments contained in the data package is equal to the length of the established path, and the number of information fragments divided by the source information node is equal to the number of sub-nodes of the source information node. Node T decodes the information slice graph according to the path establishment stage, and regards the information slice received by the direct precursor node as an information slice of the data package sent to the corresponding direct successor node in the table, sorting out and sorting strictly according to the corresponding relationship in the chart. Node T takes the second piece of information received from node S as the first piece of information sent to node U. After the information slices are ordered by node T, the ordered information slices are coded by the same method as the source, multiplied by a random reversible matrix 8. The information slices received by node T are completely different from the information slices forwarded by node T, which is equivalent to further confusing and confusing the forwarded information. After T node coding, $\left(I_{U_1}^*, I_{Y_2}^*\right)$ becomes $\left(I_{U_1}^*, I_{Y_2}^*\right)'$, and its expression is:

$$\left(I_{U_1}^{*\prime}, I_{Y_2}^{*\prime}\right) = \begin{pmatrix} B_1 \\ B_2 \end{pmatrix} \left(I_{U_1}^*, I_{Y_2}^*\right) = \left(I_{U_1}^*, I_{Y_2}^*\right)' \tag{4}$$

In the formula, $\begin{pmatrix} B_1 \\ B_2 \end{pmatrix}$ is a 2 * 2 random reversible matrix, which is equivalent to a local coding coefficient matrix. The coefficient matrix obtained by the above multiplication must be reversible so as to ensure that the information received by the next relay node can be decoded. After processing, the information flow direction under relay forwarding is determined. When the direct communication mode between two nodes fails, different relay nodes are selected to complete relay forwarding of different levels of information by the relay node. On the basis of the above, extract the main features and reconstruct the state space of multi-serial communication information flow.

Let $H(x, y)$ denote two different data interference feature points of multi-serial communication information flow terminal, and the distance between the main feature vectors x and y of multi-serial communication information flow is expressed as data mining dimension. The data set is divided into $2n$ subsets. Based on sequential resampling method and principal feature matching, the expression of vector space set of feature points distribution for data to be mined is obtained. The phase space reconstruction of clustering features is achieved by defining a fuzzy clustering center and searching for multiple trajectories. When the disturbance of the state space of data storage is large, the fusion subset of state space features of data set is obtained by decomposing the semantic pheromone based on the fuzzy C-means clustering [9–11]. The reconstructed results of multi-serial communication information flow characteristics are obtained and output. According to the above method, feature extraction and state space reconstruction of large-scale multi-serial communication information flow are realized. Assuming that the density of the input data is a priori information random variable, a large-scale data mining feature model in the cloud environment after state space reconstruction is obtained. Based on this model, data mining is carried out to improve the matching ability and balance of data mining.

2.4   Establishing an Anonymous Path to Output the Result of Anonymous Reconstruction

Source node S needs to establish anonymous paths before sending messages to destination node. Source node S needs to send the next hop IP address of each relay node to the corresponding node separately. Node S' is a reliable pseudo-source owned by source S, and destination D is randomly allocated to a certain location. Source S divides the IP addresses of all forwarding relay nodes except the successor nodes into two pieces. These two pieces of messages are sent to the corresponding relay node through two different paths. In order to prevent wiretappers from getting relevant information from a single message, the IP address information is multiplied by the random reversible matrix A before sending. At this time, the reversible matrix A is 2 * 2, which is equivalent to confusing encryption of the message. Source S sends the first half and the second half of node U's IP address to two direct forward nodes T and W along two different paths. Any successor node of its direct successor node can be known, because T receives only the first half of the IP address of Y and Z nodes, and does not know that s is the source information node and D node is the destination node. It only knows that node S is its predecessor node, and node D may be another forwarder. At this time, the information transmitted on the path is still the address information of the direct successor node of T node [12]. But information symbols have undergone tremendous changes. This ensures that even if an attacker intercepts all packets received and forwarded by all r nodes, it is difficult to visually find the relationship between data streams. In order to ensure the same size of data packets, the relay node should fill the data packets in the two information slices randomly before forwarding. How to effectively anonymize the reconstructed information flow is to achieve the best anonymity effect, the highest data availability and the least time and space cost. Generalization and suppression techniques are usually used to achieve anonymity of information flow, which makes information flow data more general and abstract. Bottom-up local re-coding anonymization steps are as follows:

Input: To be published table T

Step 1: PT is the result table for re-encoding identity attributes

Step 2: Check PT and add group labels to tuples that satisfy the anonymity requirement of identity preservation

Step 3: While (Number of tuples in PT with no grouping label > 0) and do (Quasi-identifying attribute groups are not generalized to the highest level)

Select a quasi-identifying property:

The selected attributes of the remaining tuples are generalized:

Adding group labels to tuples that satisfy anonymity requirements

Step 4: If (Number of tuples in PT with no grouping label > 0)

Remove tuples from Q1 groups that can be moved out of tuples and add them to the remaining tuples.

Adding group labels to tuples that satisfy anonymity requirements

Step 5: Return PT
Output: Publish table PT

3   Experimental Analysis

In order to test the application performance of anonymous reconfiguration method in multi-serial communication information flow under large data, simulation experiments are carried out. Firstly, the corresponding simulation scene needs to be built, and the simulation results of the fusion network built by the simulation scene are shown in the Fig. 4.

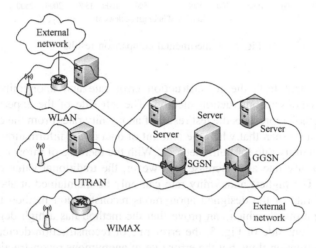

Fig. 4. Simulation experiment scene.

Because of the influence of various factors in the process of experiment, the experimental data will produce errors. In order to avoid the influence of errors on the experimental results, repeated measurement will be carried out to get the average value, so as to reduce the impact of errors.

Therefore, the designed anonymous reconstruction method is compared with the traditional reconstruction method for evaluation. In order to comprehensively evaluate and test the performance of the two methods, a set of artificial data with both uniform and Gaussian distributions is generated. Suppose that there are two attributes, and the range of each attribute is an integer in the interval [13]. The mean square deviation of the Gauss distribution is 1, and the default weight of each attribute is 1. The quality of anonymity and reconstruction error are used as experimental indicators for evaluation [14]. With the change of information flow radix, the quality of anonymity and reconstruction error of this method will also show different trends. The experimental results show that the comparison results are shown in Fig. 5.

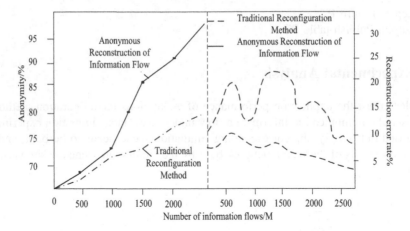

Fig. 5. Experimental comparison results

This experiment tests the reconstruction error rate and anonymity rate of the designed anonymous reconstruction method. The left side of the experimental comparison result graph represents the test result of anonymity rate. From the curve trend in the graph, it can be seen that when the amount of data in the information flow is zero, there is no anonymity rate for both methods. With the increase of information flow, the value of anonymity rate also increases. However, the traditional value of anonymity rate increases. The highest anonymity rate can only be maintained at about 80%, and the anonymity rate of the designed anonymous reconstruction method has exceeded 95% in the experiment, which can prove that the method has a high degree of anonymity. On the right side of Fig. 5, the error rate of reconstruction decreases with the increase of information flow, but the error rate of anonymous reconstruction method is more stable than that of traditional reconstruction method, which shows that the reconstruction accuracy of this method is higher.

Experimental data show that this method mainly realizes the function of anonymity and reconstruction of information flow. This method is superior to the traditional method in terms of both the effect of anonymity and the efficiency of operation.

In order to further verify the effectiveness of the design method, compare the anonymous reconfiguration time of multi serial communication information flow under different methods, and the results are shown in Fig. 6.

It can be seen from the analysis of Fig. 6 that the anonymous reconstruction time of the multi serial communication information flow in this design method is far lower than that of the traditional method, and its maximum time is only 3.9 s, while the reconstruction time of the traditional method is 7.9 s, which shows that the reconstruction efficiency of this design method is higher, and it can realize the real-time anonymous reconstruction of the communication information flow. This is because the method in this paper first obtains the original information flow in the communication network, and then cooperatively filters the multi serial communication to reduce the influence of redundant data on the speed of information reconstruction. After filtering, the

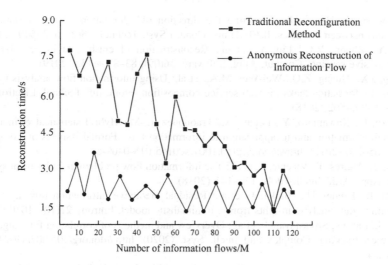

Fig. 6. Time comparison of information flow reconstruction.

notification information of relay node is obtained in the information flow, and the communication state of information flow is extracted, so as to improve the efficiency of information flow reconstruction.

4 Conclusion

Privacy protection has attracted more and more attention in various data applications. While enjoying various conveniences brought by information technology, people hope that their privacy information will be protected. Dynamic reconfiguration through anonymous reconfiguration of information flow in multi-serial communication can ensure important information security to reach the destination, and can also ensure the transmission performance of the overall network to a large extent through the diversion of information flow.

References

1. Su, Wei, Yu, Yongguang: Free information flow benefits truth seeking. J. Syst. Sci. Complexity **31**(4), 964–974 (2017). https://doi.org/10.1007/s11424-017-7078-4
2. Bijani, S., Robertson, D., Aspinall, D.: Secure information sharing in social agent interactions using information flow analysis. Eng. Appl. Artif. Intell. **70**(4), 52–66 (2018)
3. Bingwen, T.: Classified mining and optimizing technology for big data. Modern Electron. Technol. **40**(24), 34–36 (2017)
4. Ming, L., Ren, Z., Mei, H., et al.: An improved Bayesian network structure learning algorithm based on information flow. Syst. Eng. Electron. Technol. **40**(6), 25–28 (2018)
5. Wei, Q., Courtney, K.: Nursing information flow in long-term care facilities. Appl. Clinical Inform. **09**(02), 275–284 (2018)

6. Naghoosi, E., Huang, B.: Detecting the direction of information flow in instantaneous relations between variables. IEEE Trans. Control Syst. Technol. **28**(2), 542–549 (2020)
7. Lu, Y., Zhang, Y.C., Liu, Y., et al.: Reconstruction of crash data in acns based on compressive sensing. Comput. Appl. Software, **36**(09), 83–87 + 133 (2019)
8. Jiang, J.X., Huang, Z.Q., Wei-Wei, M.A., et al.: Using information flow analysis to detect implicit information leaks for web service composition. Front. Inf. Technol. Electron. Eng. **19**(04), 494–502 (2018)
9. Biondi, F., Kawamoto, Y., Legay, A., Traonouez, L.-M.: Hybrid statistical estimation of mutual information and its application to information flow. Formal Aspects of Comput. **31** (2), 165–206 (2018). https://doi.org/10.1007/s00165-018-0469-z
10. Luis, B., Andrés, T., Vicente, J., et al.: The information flow problem in multi-agent systems. Eng. Appl. Artif. Intell. **70**(4), 130–141 (2018)
11. Wahl, B., Feudel, U., Hlinka, J., et al.: Residual predictive information flow in the tight coupling limit: analytic insights from a minimalistic model. Entropy **21**(10), 1010 (2019)
12. Liu, S., Liu, D., Srivastava, G., et al.: Overview and methods of correlation filter algorithms in object tracking. Complex and Intell. Syst. (2020). http://doi.org/10.1007/s40747-020-00161-4
13. Fu, W., Liu, S., Srivastava, G.: Optimization of big data scheduling in social networks. Entropy **21**(9), 902 (2019)
14. Liu, S., Glowatz, M., Zappatore, M., Gao, H., Gao, B., Bucciero, A.: E-Learning, E-Education, and Online Training, pp. 1–374. Springer, USA (2020). http://doi.org/10.1007/978-3-319-49625-2

Mobile Communication Network Channel Allocation Method Based on Big Data Technology

Feng Jin[1], Bing Li[1], Ying Li[1], and Shi Wang[2(✉)]

[1] Information and Communcition College National University of Defense
Technology, Xi'an 710106, China
kmxttc@sina.com
[2] CRRC Qingdao Sifang Co., Ltd., Qingdao 266000, China
xuenxuen20@163.com

Abstract. In mobile communication systems, the purpose of channel allocation is to maximize the use of spectrum resources. The existing channel allocation is at the cost of frequent channel reallocation, so its practical application is not strong. Aiming at the problems of inaccurate allocation results and high bit error rate of traditional channel allocation methods, a channel allocation method based on big data technology is proposed and designed. This method makes use of the advantages of big data technology to discretize the channel data of mobile communication network. According to the requirements of the channel discretization standard and allocation algorithm of mobile communication network, it optimizes the channel allocation algorithm and realizes the effective channel allocation of mobile communication network. The validity of big data channel allocation method is confirmed by experimental demonstration and analysis. In the mobile communication network channel allocation, the allocation accuracy is high, and the allocation error is almost zero, which is better than the traditional method, and the allocation time is much lower than the traditional method. It shows that this method can realize the effective allocation of mobile network channel, and the allocation result is very reliable, which can guarantee certain network security.

Keywords: Big data technology · Mobile communication network · Channel allocation · Discretization processing · Allocation algorithm

1 Introduction

In the mobile communication system, the problem of rational allocation and optimal utilization of resources is collectively referred to as channel allocation problem [1], and its purpose is to make maximum use of spectrum resources. There are three existing channel allocation schemes: channel fixed allocation algorithm, dynamic channel allocation algorithm and hybrid channel allocation algorithm [2]. These algorithms require more or less network-wide and system-wide information, often at the expense of frequent channel redistribution, and are often not very practical.

S. Liu and L. Xia (Eds.): ADHIP 2020, LNICST 348, pp. 69–79, 2021.
https://doi.org/10.1007/978-3-030-67874-6_7

In recent years, big data technology is widely used in the solution of combinatorial optimization problems. Big data technology is to introduce a big self-feedback item [3], introduce big data technology in channel allocation to construct a big data channel allocation network, and draw on the data distribution strategy to carry out the channel allocation process of the mobile communication network [4]. However, there have been more or less deficiencies in the research process, resulting in the channel optimization technology not being effectively optimized and innovated. In reference [5], a dynamic subcarrier allocation algorithm for PLC channel based on adaptive genetic algorithm is proposed. The algorithm combines genetic algorithm and water injection algorithm to allocate the dynamic subcarriers of OFDM system by using the better global search ability of genetic algorithm. The new algorithm first cross operates the individuals in the population to obtain two new subgroups, The simulation results show that the performance of the improved genetic algorithm is greatly improved, the system transmission rate is faster, and the channel capacity is larger, but this method has the problem of inaccurate distribution results. In reference [6], an improved tree structure is proposed to solve the channel conflict in laser communication. Firstly, the mechanism of data transmission and channel switching in the network is analyzed, and the probability of channel conflict is identified by using the continuous idle time slots. Then, the average number of time slots and the conflict matrix are derived according to the tree decomposition method. Finally, the channels in the laser communication are reasonably allocated to each link based on the conflict matrix to improve the channel conflict and network performance. The experimental results show that the time slot simulation value of this method is closer to the theoretical value, which can improve the data throughput of laser communication, but this method has the problem of high bit error rate.

Based on this, this paper proposes and designs a channel allocation method for mobile communication networks based on big data technology.

2   Design of Big Data Channel Allocation Method

When designing the big data channel allocation method, the channels of the mobile communication network are first classified, and the mobile communication network with the same channel characteristics is discretized. Secondly, the channel allocation algorithm is introduced, and the calculation process of the algorithm is optimized to simplify the calculation steps of the algorithm. Finally, according to the requirements of the mobile communication network channel discretization standard and the allocation algorithm, the channel allocation process of the mobile communication network is performed. The channel allocation process of mobile communication network based on big data technology is shown in Fig. 1.

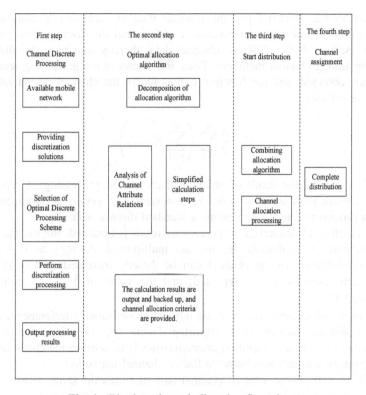

Fig. 1. Big data channel allocation flow chart.

2.1   Channel Discretization

In any cellular mobile communication network, the first step is to divide a given wireless spectrum into a set of discrete, interference-free channels. All of these channels can be used simultaneously in the service area and can guarantee the reception quality of certain signals. In order to divide a limited wireless spectrum into such channels, the big data-based channel allocation method employs various techniques such as frequency division, time division, and code division. A good distribution scheme can increase the capacity of the mobile communication network, reduce the overall overhead of the mobile communication network, and attract users by providing better services. Big data technology divides the spectrum into a number of discrete frequency bands, and splits the channel into several discrete time intervals called time slots to achieve the purpose of separating channels [7]. For a given mobile communication network, the determinant of determining the total number of channels available to it is the level of signal quality received from each channel. Some of these channels are used as signaling channels for establishing communication between the mobile station and the base station, and another portion is used to carry the provided services, that is, as a traffic channel.

Inspired by the literature [8], the channel data in each mobile communication network is divided into one classification member, and the discretization processing method is used to divide different channel data, thereby obtaining the distribution relationship among channel members. Thus, the discretization processing results of m channels are obtained, and the function expression of the channel discretization processing is as follows:

$$E = \frac{A}{2}\sum_{i=1}^{n}\left(\sum_{q=1}^{m}V_i - d_i\right) \tag{1}$$

Where, E represents the result of channel discretization processing; A represents a mobile communication network node; V_i represents a proportional parameter of discretization processing; and d_i represents a standard discrete value.

After the channel discretization processing result is obtained, the mobile communication networks with discrete features are multiplexed. As long as $V_i - d_i \geq A$ is satisfied, the influence of the channel can be directly increased; If $V_i - d_i < A$, the above discretization process is repeated, and the value of $\sum_{q=1}^{m}$ is continuously adjusted until $V_i - d_i \geq A$.

The quality of channel received signals and co-channel interference caused by channel multiplexing are the most important factors [9]. Therefore, the purpose of improving the channel discretization characteristics is to achieve the characteristics of wireless propagation path loss between the co-channel networks.

Let $\frac{A}{2}\sum_{i=1}^{n} \geq 0$ denote a set of channel data that uses the same network to communicate with each other. Because of the wireless signal propagation loss, when the value of i increases, $\frac{A}{2}\sum_{i=1}^{n} < 0$ or $\frac{A}{2}\sum_{i=1}^{n} = 0$ occurs. Therefore, it is necessary to reasonably control the value range of i to minimize the propagation loss of the wireless signal to ensure the accuracy of the channel discretization processing result.

The channel discretization processing result of the mobile communication network is output and saved, so as to improve the accurate data foundation for the design and optimization process of the next channel allocation algorithm, thereby ensuring the superiority of the channel allocation algorithm.

2.2   Channel Allocation Algorithm Design

The mobile communication network channel algorithm based on big data technology firstly defines the service as priority and non-priority. The real-time service takes precedence over the non-real-time service, and the total number of different types of service channels will be different. The big data based channel allocation algorithm has the function of channel borrowing.

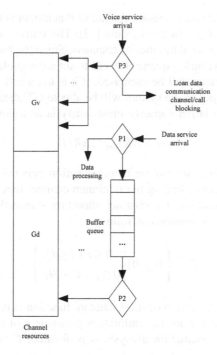

Fig. 2. Structure diagram of the big data channel allocation algorithm.

For example, when a voice service arrives but there is no voice channel allocation temporarily, the service will borrow data channel resources, thus reducing the blocking rate of the voice service. The structure of the channel allocation algorithm based on big data technology is shown in Fig. 2.

According to the operation structure of the big data channel allocation algorithm shown in Fig. 2, and compared with the general dynamic channel allocation strategy, the operation rules of the channel borrowing function are obtained, and the specific calculation process is as follows:

(1) The initial channel resources of the voice and data services are allocated by the system [10]. When the channel is allocated to the access users, the channel number assigned to the voice starts from the two ends, the channel number assigned to the voice is from small to large, and the data service is reversed.

(2) When the arriving voice service has no channel resources available, if the data channel resource is idle, the data channel numbered to the minimum value may be borrowed; If there is no idle channel, the voice traffic is blocked for communication [11, 12].

(3) When the data service arrives and the buffer queue is non-empty and there is free space, that is, $0 \leq k \leq B$, the buffer queue waits. If the buffer queue is full, the data service can borrow idle voice channel resources.

For the convenience of research, the transmission of the source when studying the mobile communication network adopts the *on/off* model. The voice and data services adopt a similar model. The transmission feature of this model is to generate the service data frame and the rate $R \geq 0$ in the T_0 time [13]. The transmission rate $R = 0$ in the T_{off} time, N represents the total number of channels. Since the buffer queue is set in the channel allocation policy, buffer queue B needs to set a threshold. If the B value is too large, the delay of the queue will be increased. If it is too small, the data loss rate will increase and the throughput of the system will be affected. Therefore, the condition that the single-column queue buffer capacity must satisfy is as follows:

$$B_T \geq NRT_{on} - NR/E \tag{2}$$

The problem of allocating this mobile communication network channel is then converted to a simple problem of finding the minimum channel function. According to the Markov state transfer principle, the channel allocation standard matrix of the mobile communication network is obtained as follows:

$$S = \left\{ (i,j,0) \left| \begin{array}{l} 0 \leq i+j \leq C, \\ 0 \leq j \leq C \leq B_T \end{array} \right. \right\} \tag{3}$$

Where, S represents the channel allocation standard function matrix range of the mobile communication network; i, j are the calibration parameters of the Markov state transition principle, and no orientation analysis is performed; C represents the buffered queue channel value.

The allocation of each channel characteristic in the mobile communication network system is identical, that is, the total number of channels of each mobile communication network is the same and the service type is the same. Then, when the new call arrival rate and the packet arrival rate respectively meet the parameters, the calculation process of the channel allocation of the mobile communication network is realized [14].

So far, the design process of the channel allocation algorithm of the mobile communication network is completed.

2.3   Mobile Communication Network Channel Allocation Implementation

According to the channel discretization processing result of the above mobile communication network, referring to the allocation algorithm in the literature, combined with the big data technology, the design and optimization process of the channel allocation algorithm of the big data mobile communication network is carried out based on the allocation algorithm [15].

First, the initial allocation of channels is performed according to the vertex coordinates of the channel discretization process, that is, $f(i,j) = ia + Sj$ is satisfied, where $j = (2x - 1)a$, and $x \geq n+1, a \geq 1$. The initial allocation results of the network channel are shown in Table 1.

Table 1. Channel initial allocation results.

i/j	−2	−1	0	1	2
−5					a+2b
−4				(x−4)a+2bS	a+b
−3			(2x−4)a+(2S+1)b	(x−3)a+bS	(2x−3)a+2 b S
−2		(x−3)a+(S+1)b	(2x−3)a+(2S+1)b	(x−2)a+ bS	(2x−2)a+2b S
−1	(2x−3)a+(S+1)b	(x−2)a+(S+1)b	2(x−2)a+(2S+1)b	(x−1)a+ bS	(2x−1)a+2 bS
0	(2x−2)a+(S+1)b	(x−1)a+(S+1)b		Xa+ bS	a−b
1	b	xa+(S+1)b	a	x+1)a+bS	2a−b
2	a+b	(x+1)a+ bS	2a	(x+2)a+bS	
3	2a+b	(x+2)a+(S+1)b	3a		
4	(2a+3)+Sb	(x+3)a+Sb			
5	(2x+2)a+(S−1)b				

According to the initial allocation result shown in Table 1, the big data allocation algorithm is introduced, which simplifies a series of insignificant calculation steps. The optimal allocation process of multiple channels in the field of cellular mobile communication networks is realized directly by allocating the channel data arriving at the current time in the mobile communication network [16, 17]. In the allocation result matrix, there must be a channel P, which has an allocation criterion in any mobile communication network, so the channel is used as an optimal allocation channel. After all the channels are optimized to the optimal standard, the allocation is performed again to realize the optimal initial allocation process of the channels in the mobile communication network.

Based on the optimal initial allocation, the channel allocation in each mobile communication network in the algorithm model is set. A mobile communication network has S_1 downlink channels and is divided into two parts: a voice channel and a data protection channel. The S_n channels are reserved as data channels for the protection service to compensate for the data packet loss rate. The $S^1 \sim S_n$ channels are then used as voice channels. When the voice channel is idle, the data service can borrow the S_n channel for data transmission. Once a voice call request arrives and the available voice channel set is found to be empty, the data service should immediately release the borrowed voice channel, stop the transmission in the voice channel, and continue to queue in the data buffer. In addition, a queuing buffer is set for the voice service, so that the handover call preferentially occupies the voice buffer to ensure handover priority.

So far, the design of the algorithm is completed. Further consider the threshold of the voice queue buffer, set to S_p. When the number of queues in the buffer exceeds S_v, the system will unconditionally block a new call. For handover call queuing, the threshold is set to $S_n (S_p \leq S_v)$. If the number of handover call queues in the buffer exceeds S_h, the algorithm will refuse to accept the new mobile communication network channel assignment.

3   Simulation Experiment Demonstration and Analysis

In order to ensure the effectiveness of the channel allocation method of mobile communication network based on big data technology designed in this paper, the simulation experiment demonstration analysis is carried out. Set the experimental object to the channel data of a mobile communication network, and perform a distribution demonstration experiment. During the experiment, the initial allocation results of the channel are shown in Table 1.

In order to ensure the validity of the experiment, the traditional channel allocation method and the big data-based channel allocation method are used to compare and experiment, and the channel allocation accuracy of the two methods is statistically analyzed. The experimental results are shown in Fig. 3.

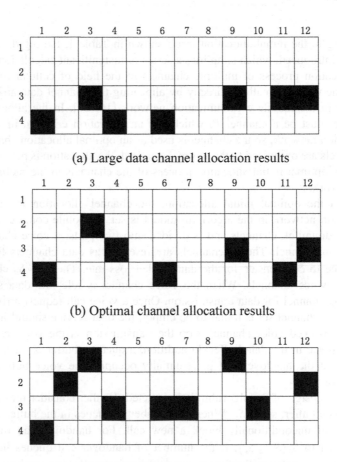

(a) Large data channel allocation results

(b) Optimal channel allocation results

(c) Traditional channel allocation results

Fig. 3. Comparison of experimental argumentation results.

According to the analysis of Fig. 3, the channel allocation method based on big data is more suitable for the allocation standard of the optimal allocation result in the process of allocating the channel of the mobile communication network; There is only a distribution bias in the (3, 3), (3, 4), (4, 11) network, and the allocation types of (3, 3), (3, 4) are consistent, and the final allocation result is not affected. Therefore, the channel assignment result of only one network element is not optimal. However, the distribution result of the traditional channel allocation method is quite different from the optimal allocation result. There is obvious distribution difference between the allocation type and the network structure, and the allocation type is almost uncon-nected. Therefore, it can be concluded that the big data-based channel allocation method designed in this paper not only improves the allocation type of channel data in mobile communication networks, it also improves the stability and accuracy of the distribution process, and gradually aligns the distribution results with the optimal distribution results, and has extremely high effectiveness and practical promotion significance.

In order to further verify the effectiveness of the method in this paper, the bit error rate is taken as the index to compare the reference [5] method, the reference [6] method and the method in this paper. The results are shown in Fig. 4.

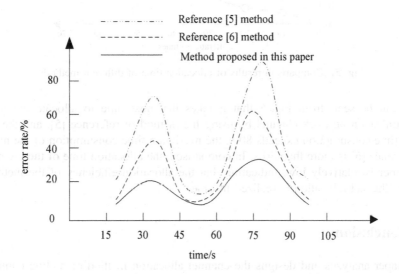

Fig. 4. Error rate comparison results of different methods.

It can be seen from Fig. 4 that the maximum BER of the method in this paper is less than 30%, while maximum BER of the method in reference [5] method is 80% and that of the method in reference [6] method is about 60%. It can be seen from the analysis of Fig. 4 that the error rate of channel allocation in this method is significantly lower than that in reference [5] method and reference [6] method, which shows that the allocation result obtained by this method is more reliable. This is because this method uses the advantages of big data technology to discretize the channel data of mobile communication network, so as to optimize the channel allocation process of mobile communication network.

In order to further verify the application effect of this method, the channel allocation time is taken as the index to compare the allocation effect of different methods. The results are shown in Fig. 5.

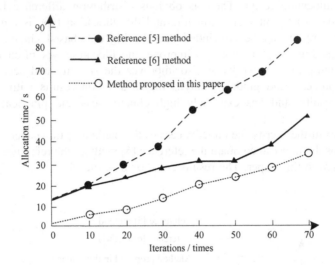

Fig. 5. Comparison results of allocation time of different methods

It can be seen from Fig. 5 that it takes the most time to allocate the mobile communication network channel by using the method of reference [5], and the maximum time consumption exceeds 80 s; the maximum time consumption of the method in reference [6] is more than 50 s. In comparison, the allocation time of the method in this paper is relatively low, indicating that the allocation efficiency of the method is higher Channel allocation is realized in time.

4   Conclusion

This paper analyzes and designs the channel allocation method of mobile communication network based on big data technology. Relying on the advantages of big data technology, the mobile communication network channel is discretized, and the allocation steps are simplified, and the big data channel allocation method is designed. The experimental results show that the big data channel allocation method designed in this paper has extremely high efficiency. When the channel is allocated, the allocation accuracy is greatly improved, and the allocation error can be effectively reduced, and the allocation time can be saved. It is hoped that the research in this paper can provide theoretical basis and reference for the channel allocation method in China's mobile communication network.

References

1. Han, C., Dianati, M., Cao, Y., et al.: adaptive network segmentation and channel allocation in large-scale V2X communication networks. IEEE Trans. Commun. **67**(1), 405–416 (2019)
2. Xu, J., Guo, C., Yang, J.: Bio-inspired power control and channel allocation for cellular networks with D2D communications. Wireless Netw. **25**(3), 1273–1288 (2018). https://doi.org/10.1007/s11276-018-1728-x
3. Choi, J., Jung, S.Y., Kim, S.L. et al.: User-centric resource allocation with two-dimensional reverse pricing in mobile communication services. J. Commun. Netw. **21**(2), 148–157 (2019)
4. Hong, S G. Park, J. Bahk, S.: Subchannel and power allocation for D2D communication in mmWave cellular networks. J. Commun. Netw. **22**(2), 118–129 (2020)
5. Rong, Y., Wangbin, C., Chengqun, Y.: Dynamic subcarrier allocation of PLC channel based on adaptive genetic algorithm. Power Syst. Protect. Control **47**(12), 111–116 (2019)
6. Kai, F., Nian, C.: Research on channel conflict resolution of laser communication. Laser J. **39**(05), 107–110 (2018)
7. Hu, J., Yang, Q.: Dynamic energy-efficient resource allocation in wireless powered communication network. Wireless Netw. **25**(6), 3005–3018 (2018). https://doi.org/10.1007/s11276-018-1699-y
8. Shuai, L., Gelan, Y.: Advanced Hybrid Information Processing, pp. 1–594. Springer, USA (2015)
9. Liu, S., Lu, M., Li, H., et al.: Prediction of gene expression patterns with generalized linear regression model. Front. Genetics, **10**, 120 (2019)
10. Chien, H.T., Lin, Y.D., Lai, C.L., et al.: End-to-end slicing with optimized communication and computing resource allocation in multi-tenant 5G systems. IEEE Trans. Vehicular Technol. **69**(2), 2079–2091 (2020)
11. Zewde, T.A., Gursoy, M.C.: Optimal resource allocation for energy-harvesting communication networks under statistical QoS constraints. IEEE J. Selected Areas Commun. **37**(2), 313–326 (2019)
12. Xianghong, S.: The role of big data analysis in mobile communication network. Commun. World **46**(10), 52–53 (2017)
13. Liang, Y.-J.: Dynamic resource allocation in mobile heterogeneous cellular networks. Wireless Netw. **25**(4), 1605–1617 (2017). https://doi.org/10.1007/s11276-017-1617-8
14. Liu, Y., Zhang, H., Long, K., et al.: Energy-efficient subchannel matching and power allocation in NOMA autonomous driving vehicular networks. IEEE Wireless Commun. **26**(4), 88–93 (2019)
15. Yan, Z., Pan, C., Wang, K., et al.: Energy efficient resource allocation in uav-enabled mobile edge computing networks. IEEE Trans. Wireless Commun. **18**(9), 4576–4589 (2019)
16. Fu, W., Liu, S., Srivastava, G.: Optimization of big data scheduling in social networks. Entropy **21**(9), 902 (2019)
17. Liu, S., Fu, W., He, L., Zhou, Jiantao, Ma, M.: Distribution of primary additional errors in fractal encoding method. Multimedia Tools Appl. **76**(4), 5787–5802 (2014). https://doi.org/10.1007/s11042-014-2408-1

Intelligent Optimization Design of Reactive Voltage Sensitivity Parameters for Large-Scale Distributed Wind Farms

Hai Hong Bian[1], Jian-shuo Sun[1], and Xu Yang[2(✉)]

[1] Nanjing Institute of Technology, Nanjing 210000, China
yhbgv690250@sina.com
[2] State Grid Jiangsu Electric Power Company Yangzhou Power Supply Company, Yangzhou 225000, China
zxg560020@sina.com

Abstract. Aiming at the problem that the reactive voltage sensitivity parameter of large-scale distributed wind farm is low overall, the parameter intelligent optimization design of the reactive voltage sensitivity of large-scale distributed wind farm is carried out. Firstly, design a wind farm equivalent circuit and optimize the parameters of the traditional reactive voltage sensitivity optimization model, and set the objective function to adjust the model weight coefficient. Then the bat algorithm is improved according to the parameter intelligent optimization model, and the reactive volt sensitivity parameter of the scaled distributed wind farm is intelligently optimized according to the improved bat algorithm. Finally, a simulation experiment is carried out to test the performance of intelligent optimization of reactive voltage sensitivity parameters of large-scale distributed wind farms. It is concluded that the reactive power sensitivity parameter of the large-scale distributed wind farm reactive voltage sensitivity parameter optimization is significantly higher than that of the reactive voltage sensitivity parameter optimized by the traditional reactive voltage sensitivity parameter optimization method.

Keywords: Scale · Decentralized · Wind farm · Reactive voltage · Sensitivity · Parameters · Intelligent optimization

1 Introduction

Reactive voltage sensitivity is no stranger to large-scale distributed wind farm electric power workers. With the increase of voltage level of transmission system, the problem of reactive voltage sensitivity has been paid more and more attention [1]. However, there is no unified conclusion on the concept of reactive voltage sensitivity [2]. In the relevant report, the IEEE believes that if the large-scale distributed wind farm power system can maintain the voltage, so that when the load increases, the power consumed by it will also increase. At this time, the wind farm power system is in a state where the reactive voltage sensitivity is high. On the contrary, the wind farm power system is in a

S. Liu and L. Xia (Eds.): ADHIP 2020, LNICST 348, pp. 80–90, 2021.
https://doi.org/10.1007/978-3-030-67874-6_8

state where the reactive voltage sensitivity is low [3]. In general, high reactive voltage sensitivity means that when the wind farm power system is operating under given initial conditions, the sudden increase in load or the change in system structure still has the ability to maintain all nodes operating sensitively. The essence is that the system has sufficient safety margin when facing these emergencies [4]. The reactive voltage sensitivity reduction refers to the process in which the reactive voltage sensitivity value is lower than the specified range when the wind farm power system is operating, and the reactive voltage sensitivity value is gradually attenuated [5]. Most of the performance of the reactive voltage sensitivity reduction phenomenon is the continuous decline of the reactive voltage sensitivity, but the instability of the reactive voltage sensitivity rise also exists and occurs [6]. The collapse of reactive voltage sensitivity means that when the wind farm power system is in a state of large voltage instability, the load continues to try to increase the current to obtain more power. This reduces the reactive voltage sensitivity to an unacceptable range [7]. Intelligent optimization design of reactive voltage sensitivity parameters can effectively improve the reactive voltage sensitivity of large-scale distributed wind farms [8].

2 Intelligent Optimization Model Design of Reactive Voltage Sensitivity Parameter

2.1 Wind Farm Equivalent Circuit Design

Because of the randomness and uncontrollability of wind speed and the difference of wind speed in the wind farm, the wind turbine group division is different from the traditional synchronous generator group. Wind farms are generally built in remote areas, and the environmental conditions are relatively poor. Wind turbines run in dynamic environment with changing wind speed and wind direction. The traditional methods of calculating the output power of wind farms depend on the standard power characteristic curve of wind turbines provided by the manufacturers. There may be differences between the actual operation conditions and the design conditions of the wind turbine; in addition, the factors that change the structural parameters or operation mode of the wind turbine in the long-term operation process will also have an impact on the performance of the wind turbine, resulting in the difference between the actual operation characteristics and the design of the wind turbine. The actual operation of the wind turbine does not necessarily follow the static wind speed power characteristic curve given in the technical manual under the specific operation environment.

For large wind farms with complex terrain and irregular layout, the principle of grouping wind turbines with the same or similar operating points is adopted, and the equivalent wind farm of multi typhoon generator model is used. Considering the active and reactive power output and voltage of each wind turbine, these important output characteristics integrate the operation information of wind turbine, such as the operation environment, the fluctuation of wind speed, speed and parameters of wind turbine, including the dynamic information of wind turbine when the wind speed fluctuates, and can also directly and accurately reflect the operation point of wind turbine. The wind speed of wind turbines in different locations may be close to or greatly different, and even the wind speed measured by two wind turbines with similar geographical locations may be quite different. It can be concluded that the wind speed can not only reflect the size of the wind force acting on the blades of the wind turbine, but also reflect the topography of the installation location of each wind turbine unit and the mutual influence of adjacent wind turbines in the wind speed. Therefore, the actual measured wind speed data can be used for cluster division. The measured active power of each wind turbine unit is the final result of the physical process that its wind speed is transformed into electric energy through complex wind energy. It is the final comprehensive feedback of the wind speed, geographical location, terrain and actual operation performance of each wind turbine unit to the grid,

Grid connected operation of wind power generation is an effective way to realize large-scale wind energy development and utilization. However, unlike conventional energy, wind energy is a kind of random energy with small energy density, which has the characteristics of "intermittence" and "randomness", which leads to the output power of wind farms fluctuating with the change of wind speed being uncontrollable and unpredictable. When the wind speed is large, the power generation of the wind farm is large; when there is no wind or the wind speed is very small, the power generation of the wind farm is zero. However, the wind speed depends on the natural conditions, which will change every moment. In recent years, with the increasing number of wind farms and the increasing scale of wind farm construction, wind farms have become an important part of power grid. The fluctuation of wind farm output power will bring many adverse effects on the safety, stability and economic operation of the power system. Reactive power reflects the reactive power compensation of the wind farm; electricity and current reflect the operation of each connecting line of the wind farm; various temperatures of the wind turbine reflect the operation of the wind turbine itself. But compared with these data, wind speed and active power are more closely related to the fluctuation of wind farm output power, which can more reflect the actual fluctuation of wind farm output power. Therefore, the wind speed measurement data or active power measurement is selected as the basis for cluster classification. If the wind farm is composed of several different types of wind turbines, first of all, it is necessary to divide the wind turbines according to the type and capacity of the wind turbines, and then take the actual wind speed or active power in a certain period of time as the input of the data sample based on the spectral clustering algorithm to obtain Wind turbine group division results.

The wind farm equivalent circuit used in the wind farm electric field sensitivity parameter intelligent optimization model is shown in Fig. 1.

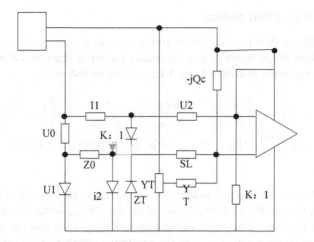

Fig. 1. Wind farm equivalent circuit

U1 and U2 are the high and low voltage side voltages of the substation, ZT and YT are the transformer impedance and the excitation admittance, k is the transformer ratio, and SL is the total load on the low voltage side of the transformer. QC is the reactive power compensation capacity of the low-voltage side of the transformer, Z0 is the equivalent impedance of the upper system of the substation, and U0 is the equivalent voltage of the upper system. It is calculated by the measured high-voltage side voltage U1 of the wind farm and the voltage loss of the upper system line. Regardless of the voltage characteristics of the load, when the capacitor bank is switched or the main transformer tap position is adjusted, it is considered that the upper system equivalent voltage U0 and the load power SL remain unchanged [9].

2.2 Parameter Optimization

In the optimization of reactive voltage sensitivity parameters of large-scale distributed wind farms, the primary criterion to be met is the voltage pass rate [10]. Then, each time the transformer tap and the capacitor are operated, it takes cost and time, and the cost of the tapping operation is high. The number of times of tapping and capacitor operation is minimized under the condition of ensuring the voltage qualification rate. Then, the power supply bureau also has operational requirements for the high-voltage side power factor of the large-scale distributed wind farm, but it is not a strict assessment index. For a 110 kV large-scale distributed wind farm, its power factor is required to meet the operational requirements as much as possible, and too much reactive power cannot be reversed to the higher-level system. Finally, the low-voltage side voltage of the large-scale distributed wind farm must meet the reverse voltage control requirements as much as possible to ensure that the high-voltage side power factor can be adapted to the reactive power requirements of the superior system.

2.3   Objective Function Setting

Under the condition that the number of times of control equipment is not given, the objective function of the intelligent optimization model of reactive voltage sensitivity parameter of large-scale distributed wind farm is set as follows:

$$
\min J_1 = \alpha_1 \left(1 - \frac{N_{Uqua}}{N} \right) + \alpha_2 \frac{K_{Tap}}{K_{T\max}} + \alpha_3 \frac{K_C}{K_{C\max}} + \alpha_4 \left(1 - \frac{N_{\cos \varphi qua}}{N} \right)
$$
$$
+ \alpha_5 \sum\nolimits_{t-1}^{N} |U_{2t} - U_{2t,aim}| + \alpha_6 \sum\nolimits_{t-1}^{N} |COS\varphi_{1t} - COS\varphi_{1t,aim}|
\tag{1}
$$

(1)   Where, J1 is the objective function, the optimization target is the minimum value of J1, N is the number of time periods of the day, N_{Uqua} is the number of qualified periods of the low-voltage side of the day, and K_{Tap} is the number of times of the transformer tapping day. K_{Tmax} is the maximum number of movements allowed by the transformer tap, K_C is the number of movements of the capacitor group in one day, K_{Cmax} is the maximum number of movements allowed by the capacitor, and $N_{\cos\varphi qua}$ is the number of qualified periods of the high-voltage side of the day. U_{2t} is the low-voltage side voltage of the t-th period, $U_{2t,\ aim}$ is the low-voltage side voltage control target of the t-th period, and $\cos\varphi 1t$ is the high-voltage side power factor of the t-th period. $\cos\varphi 1t$, aim is the high-voltage side power factor control target for the t-th period, and $\alpha 1 \sim \alpha 6$ are the weight coefficients.

The first term of the above objective function J1 represents the voltage failure rate, and the calculation formula of the voltage yield is the sum of the time of the monitoring point voltage within the acceptable range and the percentage of the total time N of the voltage monitoring [11, 12]. Similarly, the fourth term of the objective function represents the high-voltage side power factor failure rate of a large-scale distributed wind farm for one day. The fifth term of the objective function indicates the degree of deviation of the low-voltage side voltage from the voltage control target, and the voltage control target is set according to the inverse voltage regulation principle.

The sixth term of the objective function indicates the degree of deviation of the high-voltage side power factor from the power factor control target, and the power factor control target is matched with the reactive demand of the superior system. According to the importance of the power system to each index, the corresponding weight coefficient is selected. In this model, $\alpha 1 = 105$, $\alpha 2 = 104$, $\alpha 3 = 103$, $\alpha 4 = 102$, $\alpha 5 = 10$, and $\alpha 6 = 1$ are set. The intelligent optimization model of reactive voltage sensitivity parameter does not treat voltage amplitude safety and power factor safety as hard constraints, but modifies it to voltage yield rate and power factor yield as indicators in the objective function. The reason is that due to the limitation of the number of times of control equipment operation and the fluctuation of the scaled distributed wind farm load and voltage [13, 14]. If the optimal operation of the control equipment cannot meet the requirements of the voltage-reliable all-day qualification of the large-scale distributed wind farm, it is necessary to ensure the highest pass rate and power factor pass rate of the large-scale distributed wind farm throughout the day. Therefore, the voltage pass rate and the power factor pass rate index are added to the objective function.

2.4 Adjustment Weight Coefficient

Through the solution of the model, the optimality of the objective function corresponding to different weight coefficients is found under the constraint condition. The change of the number of times of control equipment has a certain influence on the control effect of reactive voltage sensitivity. According to the objective function, the comprehensive optimal result of each index can be obtained theoretically [15–17]. However, due to the influence of the algorithm or the number of iterations, not every calculation can obtain the global optimal solution. At the same time, in order to explore the influence of the change of the number of times of control equipment on the control effect of the reactive voltage sensitivity of the scaled distributed wind farm, if the decrease in the number of times the control device is operated has little effect on the control effect, the number of actions can be continuously reduced until the important indicator fails. The grid dispatcher of large-scale distributed wind farms is provided with various control effects optimization schemes, and the dispatcher can select the next-day optimization scheme according to the operation experience [18, 19]. The objective function of the intelligent optimization model of reactive voltage sensitivity parameter is affected by the change of the number of actions of the control device:

$$
\min J_2 = \alpha_1 \left(1 - \frac{N_{Uqua}}{N} \right) + \alpha_2 \left(1 - \frac{N_{\cos \varphi qua}}{N} \right) + \alpha_3
$$
$$
\sum_{t-1}^{N} |U_{2t} - U_{2t,aim}| + \alpha_4 \sum_{t-1}^{N} |COS\varphi_{1t} - COS\varphi_{1t,aim}| \tag{2}
$$

(2)   In the formula, J_2 is the objective function, the optimization target is the minimum value of J_2, the meanings of other variables are the same as the first formula, and $\alpha 1 \sim \alpha 4$ are the weight coefficients, taking $\alpha 1 = 105$, $\alpha 2 = 102$, $\alpha 3 = 10$, $\alpha 4 = 1$. In addition, it is necessary to improve the constraint of the maximum number of action times of the transformer daily maximum number of times of the capacitor, and the specific limit values of the number of times of operation of the control device are as follows:

$$
\sum_{t-1}^{N} C(t-1) + C(t) = n_{Qg} \left(n_{Qg} \leq n_Q \right)
$$
$$
\sum_{t-1}^{N} Tap(t-1) + Tap(t) = n_{Tg} \left(n_{Tg} \leq T \right) \tag{3}
$$

(3)   Where N is the total number of time periods in a day, $C(t-1) + C(t)$ represents the number of change groups of the capacitor bank from the t-1 period to the t period, and n_{Qg} is the number of times the capacitor bank is given. n_Q is the maximum number of operations allowed for the capacitor bank. $Tap(t-1) + Tap(t)$ indicates the value of the change of the tap from the $t-1$ period to the t period, and n_{Tg} is the number of times the tap is given, n_T is the maximum number of actions allowed by the tap [20].

The above objective function J2 reduces the index of the number of times of tap and capacitor group daily action with respect to J1, and gives a specific limit value of the number of times of operation of the control device in the constraint condition. The purpose of J2 is to obtain a variety of optimization schemes by continuously reducing the number of times of tapping and capacitor operation, and to explore the influence degree of the change of the number of times of control equipment on the reactive voltage sensitivity control effect of large-scale distributed wind farms. The grid dispatcher of large-scale distributed wind farms is provided with various control effects optimization schemes, and the dispatcher can select the next-day optimization scheme according to the operation experience. After the design of the intelligent optimization model of reactive voltage and sensitivity parameters of large-scale distributed wind farms is completed, the optimization parameters of reactive voltage sensitivity need to be calculated to complete the intelligent optimization design of reactive voltage sensitivity parameters of large-scale distributed wind farms.

3 Improved Bat Algorithm Based on Parameter Intelligent Optimization Model

1) Initialization. Set large-scale distributed wind farm parameters and algorithm parameters, including population size popsize, maximum iteration number K_{max}, inertia factor ω, maximum speed limit V_{max} and other parameters.
2) The initial population is produced. Under the condition that the variation range of the control variable is satisfied, the initial position X_i and the initial velocity V_i of each bat in the population are randomly assigned.
3) Position correction. For a bat that does not satisfy the constraint of the minimum time interval between two adjacent actions of the control device, the bat is improved according to the "position correction strategy that considers the minimum time interval of the two adjacent actions of the control device". The bat that does not satisfy the constraint condition of the maximum number of movements of the control device is improved according to the "position correction strategy that takes into account the maximum number of movements of the control device."
4) Power flow calculation, fitness evaluation. The Newton-Raphson iterative method is used to calculate the power flow. The fitness value f(Xi) of each bat is evaluated according to the power flow calculation results, and the optimal solution of each bat and the optimal solution of the population are found and stored. The formula for calculating the fitness value is:

$$f(X_i) = 1/J_i \tag{4}$$

Where J_i is the objective function of the i-th bat.

5) The bat algorithm starts iterating, updating the iteration number k and the inertia factor ω.
6) Global search. According to the frequency of each bat updated F_i and speed V_i.
7) Guided by a feasible direction. First, the bats with unqualified power factor are improved according to the "speed correction strategy considering power factor

safety". Then, the bat with unqualified reactive voltage sensitivity parameters is improved according to the "speed correction strategy considering the safety of the reactive voltage sensitivity parameter". If the speed exceeds the maximum speed limit V_{max}, it is limited to V_{max}.

8) Location update. Update the position Xi of the bat, if the position exceeds the defined range of the control variable, it is limited to the boundary of the defined range. And follow step 3 to re-position the position.

9) Neighborhood search. When a random number is greater than the pulse emissivity r_i, a random walk is performed; if a random number is not greater than the pulse emissivity r_i, no random walk is performed.

10) Fitness evaluation and convergence judgment. The Newton-Raphson iteration method is used to calculate the power flow, and the fitness value $f(X_i)$ of each bat is re-evaluated, and the individual optimal solution and the optimal population solution are updated and stored. Then it is judged whether the number of iterations k reaches the maximum number of iterations K_{max}, or the optimal solution of the population remains unchanged for 10 consecutive generations. If so, the program ends and the optimal individual is output; otherwise, the process proceeds to step 5 to continue the iteration.

The bat algorithm is improved by the intelligent optimization model of reactive voltage sensitivity parameter, the optimization parameters of reactive voltage sensitivity are calculated, and the sensitivity parameters of reactive voltage of large-scale distributed wind farm are intelligently optimized.

4   Intelligent Optimization Based on Improved Bat Algorithm

The bat algorithm is a new type of heuristic intelligent algorithm. Compared with traditional intelligent algorithms such as genetic algorithm and particle swarm algorithm, it has more advantages in global search ability and calculation speed. Of course, similar to the traditional intelligent algorithm, there are also a large number of invalid searches in the random search process of the bat algorithm. The expert experience of guiding the feasible direction in the dynamic reactive power optimization calculation also plays an important role in improving the random search efficiency of the bat algorithm. Therefore, a feasible direction guiding strategy for dealing with the adaptive optimization model of reactive voltage and sensitivity parameters of large-scale distributed wind farms is proposed to improve the random search efficiency of the bat algorithm. The improvement strategy of the bat algorithm mainly has the following four points: a speed correction strategy considering the safety of the reactive voltage sensitivity parameter. When the reactive voltage sensitivity parameter is unqualified, the feasible direction guiding strategy is used to correct the speed direction and step size of the algorithm; considering the power factor safety speed correction strategy. When the power factor is unqualified, the feasible direction guiding strategy is used to correct the speed direction and step size of the algorithm; considering the position correction strategy of controlling the minimum time interval constraint of the adjacent two actions of the device. When the minimum time interval constraint condition of the adjacent two actions is not satisfied, the position correction strategy is used to correct the position of the algorithm; and the position correction strategy of controlling the maximum action number of the device is considered. When the maximum action number constraint condition is not satisfied on the day, the "peak clipping and valley filling" strategy is adopted to ensure that the equipment daily allowable action number constraint is satisfied.

5   Simulation Experiments

In order to ensure the effectiveness of the experiment, the traditional reactive voltage sensitivity parameter optimization method is compared with the intelligent optimization method of the reactive voltage sensitivity parameter of the scaled distributed wind farm, and the test results are observed. The bat algorithm parameters are set to: population size popsize = 20, maximum iteration number K_{max} = 100, maximum limit speed of taps and capacitors $V_{i,max}$ = 1, the constant α = 0.9, the constant γ = 2, the pulse emissivity ri^0 = 0.6, the minimum frequency F_{min} = 0, and the maximum frequency F_{max} = 2. The simulation conditions are: hardware platform: desktop computer, processor Intel(R) Core(TM)2 Duo CPU E7200 @2.53 GHz, memory 1.99 GB, hard disk 120 GB, Windows XP 32-bit operating system. Software platform: Matlab 7.10.0 (R2010a). The experimental results are shown in Fig. 2.

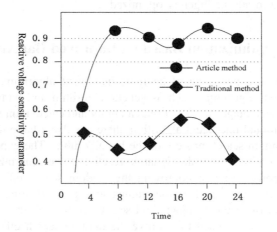

Fig. 2. Reactive voltage sensitivity parameters under intelligent optimization method

By comparing the reactive power and voltage sensitivity parameters of large-scale distributed wind farms after the intelligent optimization method, the traditional reactive power and voltage sensitivity parameter optimization method is adopted, and it can be found that after the intelligent optimization model is established, the reactive power and voltage sensitivity parameters of the whole day can be obtained The parameters of reactive power and voltage sensitivity are studied, and the intelligent optimization method of reactive power and voltage sensitivity parameters based on bat algorithm is improved. Most of them are kept above 0.9, which shows that it improves the sensitivity of reactive power and voltage sensitivity. The parameters of large-scale distributed wind farms are obviously higher than the traditional optimization methods of reactive power and voltage sensitivity parameters

6 Conclusion

Most of the wind power bases are located in remote areas of the land or at sea, especially large-scale distributed wind farms. Because the wind farms are far away from the load center, their long-distance transmission power has fluctuations. Therefore, the reactive voltage sensitivity parameter is not stable enough. The intelligent optimization method for reactive voltage sensitivity parameters of large scale distributed wind farm combining the intelligent optimization model of reactive voltage sensitivity parameter and the bat algorithm based on the intelligent parameter optimization and optimization of the sensitivity parameter can intelligently optimize the reactive voltage sensitivity parameter of the scaled distributed wind farm to significantly improve the sensitivity of reactive voltage sensitivity parameters of large scale distributed wind farms.

References

1. Dan, W., Jingjing, T., Wenxia, L., et al.: Coordination reactive power optimal dispatching method for decentralized wind farms and regional power grids considering uncertainty. Modern Electric Power **33**(34), 130–137 (2016)
2. Chen, D., Wang, C., Li, C.K., et al.: Study on the influence of decentralized wind power access on reactive power and voltage characteristics of 110 kV regional power grid. Rural Water Resour. Hydropower China **22**(23), 188–191 (2015)
3. Yang, J., Cui, J., Xing, Z., et al.: Decentralized reactive power control strategy for wind farms considering wind power prediction. Power Syst. Autom. **33**(35), 1118–1115 (2015)
4. Junhong, W., Yanjun, Z., Zuoxia, X., et al.: Multi-objective reactive power optimization of decentralized wind farms based on adaptive genetic algorithm. Northeast Power Technol. **37** (33), 111–116 (2016)
5. Xing, Z., Yanning, Xiao, W., et al.: Multi-objective reactive power optimization control of decentralized wind farms at different time levels. J. Electric. Mach. Control **20**(11), 246–252 (2016)
6. Liu, S., Liu, G., Zhou, H.: A robust parallel object tracking method for illumination variations. Mobile Netw. Appl. **24**(1), 5–17 (2019). https://doi.org/10.1007/s11036-018-1134-8
7. Zhicheng, X., Bo, Z., Ming, D., et al.: Random scene simulation of photovoltaic absorptive capacity of distribution network based on voltage sensitivity and optimization of control parameters of inverters. J. Electric. Eng. China **36**(26), 1578–1587 (2016)
8. Liu, S., Fu, W., He, L., Zhou, J., Ma, M.: Distribution of primary additional errors in fractal encoding method. Multimedia Tools Appl. **76**(4), 5787–5802 (2014). https://doi.org/10.1007/s11042-014-2408-1
9. Cao, W., Wang, H., Xue, H., et al.: Study on optimization of voltage attributes on Z. oscilloscope on GRO arrester. J. Xi'an Polytechn. Univ. **2018**(06), 686–690 (2018)

10. Liu, S., Lu, M., Li, H., et al.: Prediction of gene expression patterns with generalized linear regression model. Front. Genetics **10**, 120 (2019)
11. Warwick-Evans, V., Atkinson, P.W., Walkington, I., et al.: Predicting the impacts of wind farms on seabirds: an individual-based model. J. Appl. Ecol. **55**(2), 60–66 (2018)
12. González, J.S., Payán, M.B., Santos, J.M.R.: Optimal design of neighbouring offshore wind farms: a co-evolutionary approach. Appl. Energy **12**(1), 140–152 (2019)
13. Ko, Y., Andresen, M., Buticchi, G., et al.: Discontinuous modulation based active thermal control of power electronic modules in wind farms. IEEE Trans. Power Electron. **3**(99), 1 (2018)
14. Abdolrasol, M.G.M., Hannan, M.A., Mohamed, A., et al.: An optimal scheduling controller for virtual power plant and microgrid integration using binary backtracking search algorithm. IEEE Trans. Ind. Appl. **4**(54), 2834–2844 (2018)
15. Garcia-Ruiz, A., Dominguez-Lopez, A., Pastor-Graells, J., et al.: Long-range distributed optical fiber hot-wire anemometer based on chirped-pulse ΦOTDR. Opt. Express **26**(1), 463 (2018)
16. Corcovilos, T.A., Mittal, J.: Two-dimensional optical quasicrystal potentials for ultracold atom experiments. Appl. Opt. **58**(9), 2256 (2019)
17. Liu, S., Li, Z., Zhang, Y., et al.: Introduction of key problems in long-distance learning and training. Mobile Netw. Appl. **24**(1), 1–4 (2019). https://doi.org/10.1007/s11036-018-1136-6
18. Vlahakis, E., Dritsas, L., Halikias, G.: Distributed LQR design for a class of large-scale multi-area power systems. Energies **12**(14), 2664 (2019)
19. Vincenti, G., Bucciero, A., Helfert, M., Glowatz, M. (eds.): E-Learning, E-Education, and Online Training. LNICST, vol. 180. Springer, Cham (2017). https://doi.org/10.1007/978-3-319-49625-2
20. Corbellini, A., Godoy, D., Mateos, C., et al.: An analysis of distributed programming models and frameworks for large-scale graph processing. IETE J. Res. **10**(5), 1–9 (2020)

Distributed Reactive Energy Storage Structure Voltage Reactive Power Control Algorithm Based on Big Data Analysis

Yang Xu[1(✉)], Jie Gao[1], Yong-biao Yang[2], and Hai-hong Bian[3]

[1] State Grid Jiangsu Electric Power Company Yangzhou Power Supply Company, Yangzhou 225000, China
zxg560020@Sina.com
[2] Southeast University, Nanjing 210096, China
[3] Nanjing Institute of Technology, Nanjing 210000, China

Abstract. In the traditional distributed hybrid energy storage structure, there are security, island operation and capacity blocking. For this reason, a distributed hybrid energy storage structure voltage reactive power control algorithm is proposed based on big data analysis. First, establish a voltage reactive power control model, using the control or manual operation of the communication system to achieve automatic operation of the transformer tap or capacitor input capacity. Secondly, the Nash theorem is used to calculate the physical quantity obtained by the secondary user under certain transmission power conditions. Based on the game theory, the operating data of the power grid is obtained, and then the voltage reactive power control algorithm is realized. Finally, experiments show that the distributed reactive energy storage structure voltage reactive power control algorithm based on big data analysis has certain advantages compared with the traditional voltage reactive power control algorithm.

Keywords: Active power · Equilibrium node · Control variable · Nash theorem

1 Introduction

The AC-DC relay protection of the traditional distributed hybrid energy storage structure already has a large amount of tuning. Among them, the calculation index and formula are based on years of operation experience. Now, the distributed hybrid energy storage structure under big data analysis is connected to adjust the voltage, but there are still new problems in relay protection. In order to adapt to a variety of operation modes, each device in the voltage needs communication and corresponding control strategies, especially when the operation mode is converted or isolated network, the control mode determines whether the voltage frequency of the whole system can be stable. At present, there are two main control modes of voltage and reactive power control algorithm:

S. Liu and L. Xia (Eds.): ADHIP 2020, LNICST 348, pp. 91–102, 2021.
https://doi.org/10.1007/978-3-030-67874-6_9

1) The master-slave control mode refers to that there is a master controller in the voltage, which is responsible for maintaining the voltage frequency stability of the whole micro-grid during the operation of the isolated grid, and therefore adopts Vf control. As for other DGs, PQ control can be used. For those that can not be controlled, such as wind power without energy storage, photovoltaic maximum power tracking, these DGs are collectively known as the slave controller. In the voltage of master-slave control structure, some slave control units output constant power according to instructions, some slave control units have intermittent fluctuations, so once the load changes, it can only be tracked through the master control units, only when the power achieves a balance, the voltage and frequency can be stable, so the energy storage system that can achieve two-way rapid power flow is the best choice for the master controller. The master-slave control mode can realize the no-difference control of voltage and frequency. At present, there are corresponding demonstration projects in Holland, Greece and Japan. In this structure, the MU is not always controlled by Vf. When the MU is connected to the grid, the large power grid is responsible for maintaining the frequency stability. Otherwise, the system will be disturbed. That is to say, the main control unit has two control loops. When the voltage enters the isolated operation mode, the control loop switches quickly from PQ to Vf.

2) In the peer-to-peer control mode, all DG and energy storage systems have no master-slave relationship, they have the same status, and generally adopt droop control. The so-called droop control refers to the real-time acquisition of the output voltage and frequency, and the difference with the reference value, control feedback, local power compensation. In this structure, all the peer-to-peer control units are responsible for maintaining power balance, voltage stability and frequency stability, and work together to reach a new balance point when the load changes. Different from the master-slave control, even when the voltage operation mode changes, the control loop is still unchanged, which can realize the fast switching between grid connected and isolated network. In droop control, the coefficient directly affects the stability of the whole voltage. If the coefficient is not reasonable, it may cause voltage and frequency oscillation. Otherwise, each DG can automatically distribute power according to its own droop coefficient.

Due to the limitations of the traditional algorithm control device, once the grid has a large fault, it can not continue to provide support to the grid as much as the traditional thermal power. Considering the constraints of the branch flow, voltage access may cause other loads on the same line to be cut off, and frequent switching on the other hand will also affect the quality of the power supply. Due to the intermittent and forecasting errors of wind power, photovoltaic and other energy sources, the distributed hybrid energy storage structure may not be put into operation on time according to the requirements of the local main network controllers, which may result in high economic losses. To this end, based on big data analysis, the distributed hybrid energy storage structure voltage reactive power control algorithm is proposed to solve the problem of security, island operation and capacity blocking in the traditional distributed hybrid energy storage structure [1].

2 Distributed Hybrid Energy Storage Structure Voltage Reactive Power Control Algorithm

2.1 Establish Voltage Reactive Power Control Model

The voltage reactive power optimization compensation is to optimize the reactive power flow of the computing system and optimize the distribution of the reactive power compensation position and size in the system. Its real intention is to use the results of the reactive power calculation of the power system to achieve reactive power [2]. In fact, the optimal control of reactive power can be understood as a sub-process in the reactive power optimization process of the power system. That is to say, the control or manual operation of the communication system is used to achieve automatic operation of the transformer tap or capacitor input capacity. This is the most practical, simple, fast and safe operation method, which is also a direction for the development of power system control in the future [3]. In the process of research, the voltage reactive power control model is established based on this, and the voltage reactive power model structure is as follows:

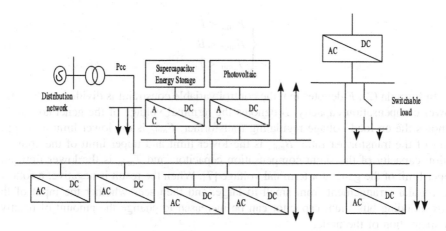

Fig. 1. Voltage reactive model structure

Figure 1 shows the voltage reactive power model structure. The voltage reactive power optimization of the distributed hybrid energy storage structure can be defined as the voltage reactive power control model that minimizes the active power network loss at a certain moment in the system operation. In the entire power grid, except for the balance node, the active power of the generator sets on all remaining nodes has been given. Moreover, the power level of the given node is always kept constant, so if you want to minimize the loss of the active power network of the system, you can turn this problem into the minimum active power injected by the balanced node. The objective function can also be written as the problem of the least active injection of the balanced nodes of the system [4]. The specific expression is as follows:

$$\begin{cases} m_e = b \sum_{ei} b_e (k_e \cos \theta + k_e) \\ n_e = b \sum_{ei} b_e (k_e \sin \theta - k_e) \end{cases} \tag{1}$$

In formula (1), m denotes the active voltage injected by the node, b denotes the reactive voltage injected by the node, k denotes conductance, n denotes susceptance, and e denotes phase angle difference [5]. The above is a node power equation expressed in polar coordinates. The constraint equation is also a mathematical model mainly used in the load flow calculation method such as Newton method.

Because the operation of the power system needs to meet certain economic and technical requirements, this requires certain constraints to be met in the calculation process. The constraints in the voltage reactive optimization problem contain state variable constraints and control variable constraints. The control variable constraint is divided into reactive power compensation capacity, the ratio of the load regulating transformer and the terminal voltage of the generator; the reactive power injected by the generator ensures that the voltage of each node is a state variable [6]. The constraint formula for all control variables in the system is as follows:

$$\begin{cases} F_{imin} < F_i \\ B_{imin} < B_i \\ T_{imin} < T_i \end{cases} \tag{2}$$

In formula (2), F denotes that the control variable constraint is divided into reactive power compensation capacity, B denotes the terminal voltage of the generator, and T denotes the on-load voltage regulating transformer, F_{imin} is the lower limit and upper limit of the transformer ratio, B_{imin} is the lower limit and upper limit of the compensation capacity of the shunt compensation capacitor, and T_{imin} is the lower limit and upper limit of the generator terminal voltage [7]. When the result is negative, it follows the regular arrangement from small to large, and then sees whether the input of the corresponding bus shunt capacitor can be increased to change the amount of reactive compensation of the node.

Distributed hybrid energy storage structure voltage reactive power control has its own characteristics: the three-level coordinated control system can be compatible with a variety of communication protocols, and its data acquisition program has multiple acquisition methods [8]. This is because the three-level coordinated control is performed between different levels, which inevitably has the problem that the data interface of different voltage reactive devices is incompatible. In addition, the communication protocol and acquisition method are different for voltage reactive devices produced by different manufacturers or different types of device models produced by the same manufacturer. So far, the construction of the voltage reactive power control model is completed.

2.2   Realizing Voltage Reactive Power Control Algorithm

The voltage qualification rate is the most important quality indicator of the power grid, and the grid line loss rate is the most important economic indicator of the power grid. Effective voltage control and reasonable reactive power compensation can not only ensure voltage quality, but also improve the stability and safety of power system operation, reduce power loss of power grids, improve transmission capacity of power grid equipment, and give full play to the economic benefits of power grid operation. Unqualified voltage will bring huge losses to the safe operation and social and economic benefits of electrical equipment. Reducing the power loss of the grid will greatly reduce the current shortage of power, which is of great significance [9]. With the development of technology, energy storage types are divided into three categories according to the energy storage form, namely mechanical energy storage, electromagnetic energy storage and battery energy storage. Table 1 shows the performance comparison results of various energy storage technologies. Among them, pumped storage has been widely used because of its mature technology and large capacity, while super capacitor, superconducting magnet, liquid flow battery, flywheel energy storage and other technologies are more convenient to use and have their own advantages, but they are still in the development stage.

Table 1. Comparison of advantages and disadvantages of various energy storage technologies

Energy storage type		Advantage	Inferiority
Mechanical energy storage	Pumped storage	High power, large capacity and low operation cost	Special geographical environment
	Flywheel energy storage	High-capacity	Low energy density
	Compressed air energy storage	High power, large capacity and low operation cost	High environmental requirements
Electromagnetic energy storage	Superconducting magnetic energy storage	High-capacity	High investment cost, low energy density
	Super capacitor energy storage	Long life, high efficiency, fast charging and discharging	High investment cost, low energy density
Battery energy storage	Lead acid battery	Low investment	Short life
	Ni MH, Ni Cd batteries	Large capacity, long life	Low energy density
	All vanadium flow battery	High capacity, high energy density	Low power density

Energy storage technology can directly and effectively solve the problem of unbalanced power supply and demand. At present, the role of energy storage system in the power grid can be summarized as four points: first, it can be used to cut peak and fill valley in the power system; second, energy storage technology can improve the reliability of power supply, and play a role of temporary power supply in case of power failure due to system failure, that is, the role of uninterruptible power supply (UPS) to reduce Accident and economic loss caused by sudden power failure; once again, the energy storage device can charge and discharge instantaneously, with similar functions as SVC, statecom and other devices, and has an auxiliary supporting role in maintaining the stability of the power grid, which is helpful for the system to reach a new balance point under the large disturbance state such as short circuit fault of the power grid; finally, it is a necessary device for intermittent energy generation such as wind power, photovoltaic and other energy storage devices It can not only suppress the power fluctuation, but also maintain the voltage and frequency stability of microgrid as the main control element under the operation condition of isolated network. In one cycle, the change curve of battery state of charge is shown in Fig. 2.

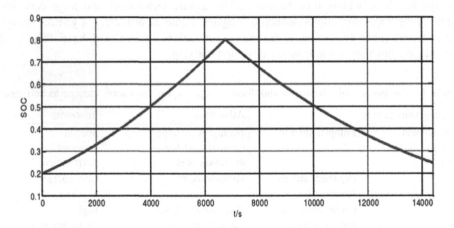

Fig. 2. Change curve of state of charge

The traditional voltage reactive power control strategy is that the control devices at each level first collect the local voltage value and the reactive power value, and then take corresponding control measures according to the control principles of the equipment at all levels. It can only implement voltage and reactive power control for this level, but it does not affect each other and is not related to each other. In the case where there is no mutual communication between the equipments at all levels, the regulation strategy of the distribution station area has been exhausted, the voltage at the end of the user is still low, and the substation side and the line part are regulated but cannot participate in the regulation. This causes a waste of resources used by the equipment. Therefore, the traditional voltage reactive power control method only relies on the independent voltage and reactive power control performed by the devices at all levels, and cannot realize the synchronous adjustment of the operating states of the devices at all levels. This makes the substation, line, and station equipment unable to complement each other effectively, and the equipment control margins of all levels are not fully utilized.

In order to effectively control the voltage and reactive power of the distribution network, according to the voltage reactive power control model, a voltage reactive power control algorithm is proposed for the distributed hybrid energy storage structure. The distributed reactive energy storage structure voltage reactive power control algorithm mainly relies on reactive power compensation devices at all levels. Through the wireless communication public data exchange platform to realize the data exchange between the control center and each reactive compensation control device, not only the operation data of the power grid is obtained, but also provides the analysis basis for the control [10, 11]. At the same time, it can also provide a good solution for remote control of the compensation device, greatly reducing the number of equipment actions and extending the life of the equipment.

Based on the voltage reactive power control model, the research of voltage reactive power control algorithm is realized. In the voltage reactive power control algorithm, the Nash theorem is used to optimize the voltage reactive power control algorithm, and the calculation formula is as follows;

$$h = \arg \max \eta (j_i - j_{i\min}) \qquad (3)$$

In formula (3), h represents the utility function, j represents the maximum utility in the cooperative game, and i represents the cognitive user. Where $\arg \max$ represents the minimum utility function value of user i. The utility function is the physical quantity obtained by the secondary user under certain transmit power conditions. In the voltage reactive control model, the utility function definition formula is as follows:

$$u_i(p_i, y_i) = \eta (y_i - y_i^{\min}) \qquad (4)$$

In formula (4), u denotes a cognitive user, y denotes a maximum utility function, and p denotes a policy space. In the process of calculation, the initial transmission power of a group of users needs to be selected, the power meets the minimum normal communication condition, and the initial interference power of the primary user is set to zero. The transmission power is obtained by using the Nash theorem.

Nash equilibrium is an important term in game theory. Nash equilibrium means that in a strategy combination, there are n participants. Under certain circumstances, if any player participant changes the strategy alone, then all users participating in the game will not increase the revenue [12, 13]. Before giving the definition of Nash equilibrium, we first need to give a mathematical description of the three elements of the game. We usually use $S_1, S_2.S_3, \ldots \ldots, S_N$ to represent the set of strategies that all participants can choose in a game with n participants. The i-th strategy of game participant j is denoted by $S_{ij} \in S_i$. $u_i, u_2.u_3, \ldots \ldots, u_N$ represents the revenue of each game participant, and u is the income of the i-th participant. The parameter setting values of the Nash theorem optimized voltage reactive power control algorithm are as follows (Table 2):

Table 1 is the parameter setting value of the Nash theorem optimized voltage reactive power control algorithm. At the same time, in the voltage control reactive power optimization, the high and low voltage reactive power compensation is planned,

Table 2. Parameter setting of Nash theorem optimization voltage reactive power control algorithm

Node 1	Node 2	Resistance	Reactance
1	2	0.2344	0.6578
2	3	0.3244	0.7654
3	4	0.2565	0.4356
4	5	0.3454	0.6458
5	6	0.3455	0.6238
6	7	0.7678	0.4738
7	8	0.8788	0.7939
8	9	0.5467	0.2375
9	10	0.4566	0.3896
10	11	0.7864	0.3893
11	12	0.4676	0.9046
12	13	0.1256	0.5693
13	14	0.1266	0.5486
14	15	0.2652	0.5623
15	16	0.1653	0.2653

mainly including two types. First, a fixed capacitor bank with a corresponding capacity of 15% capacity is configured. The high voltage network is arranged on the voltage load bus, and the low voltage network is arranged on the distribution side. The second is to install a fixed compensation capacitor bank at the voltage feeder load center or somewhere according to the reactive demand at low load. The qualified voltage values of the power grid are not only one, but the range of the upper and lower limits. If the three-level joint control is not carried out, there is no mutual communication between the distribution transformer and the user. This will cause the voltage value of the low-voltage user side to be smaller than the minimum value of the voltage interval, and the regulation side has the regulation margin but does not participate in the regulation. The reason why the transformer control device does not operate is because its own voltage value is normal, so it will not adjust and control the low voltage of the low voltage user. This results in a low voltage user side voltage deviation from the normal range of voltage is not improved.

If the voltage reactive three-level joint control is adopted, when the above situation occurs, if there is a control margin on the distribution side, the voltage reactive power control device at the distribution will immediately take effective measures to adjust. This is based on the analysis of the voltage on the distribution side. If the transformer, the line, and the distribution side are combined for analysis, the control strategy is more complicated and contains more levels of logic [14, 15]. This kind of coordinated control is a comprehensive consideration of the coordinated control of various voltage levels and various types of voltage regulation and reactive power equipment in rural power grids. It includes "adjacent coordination" and "neighborhood coordination" based on "adjacent coordination". "Adjacent coordination" is divided into three situations: mutual control of "feeder" and "substation", mutual control of "user" and "distribution", mutual control of "distribution" and "feeder"; "Neighboring coordination" refers to fully digging the control capabilities of equipment at all levels on the basis of the adjustment control of "adjacent coordination" to achieve comprehensive control of cross-level assisted voltage regulation [16].

After configuring the appropriate reactive power compensation capacity, the distribution network has a certain reactive power compensation capability. However, the non-intelligent and uncontrollable fixed compensation method cannot adapt to the dynamic requirements of the system load change, and the compensation effect is poor. At present, the power sector mainly uses a variety of reactive power compensation intelligent control components to perform a certain degree of intelligent control on the capacitor bank, so that the control effect has been correspondingly improved, and the power grid has a certain optimization ability. However, for a large-scale distribution network system, it is impossible to achieve the optimization goal of the system or even a distribution network feeder by relying on local optimization control alone. Therefore, based on the voltage reactive power control model, a distributed reactive energy storage structure voltage reactive power control algorithm based on big data analysis is proposed.

3 Experimental Results

In order to verify the validity of the distributed reactive energy storage structure voltage reactive power control algorithm based on big data analysis, the traditional voltage reactive power control algorithm and the distributed hybrid energy storage structure voltage reactive power control algorithm based on big data analysis are respectively carried out. In the experiment, the PI parameters of the voltage outer loop PI controller of the two algorithms are consistent. The test parameters are as follows:

Table 3 is the test environment. The comparison results of the two algorithms are as follows:

Table 3. Test environment

Experimental parameters	Numerical value
Net side power supply voltage	34
Given voltage on DC side	35
DC side resistance load	36
AC side equivalent inductance	63
DC side support capacitor	24
Control cycle	2
Switching period	2.3

As can be seen from Fig. 3, in the steady-state state, the sinusoidal current obtained by the distributed reactive energy storage structure voltage reactive power control

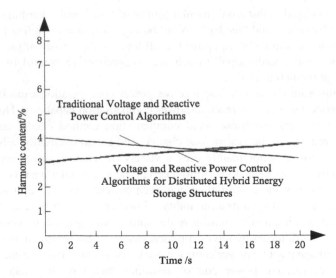

Fig. 3. Comparison of experimental results

algorithm is relatively large, while the sinusoidal current obtained by the conventional voltage reactive power control algorithm is relatively small. The analysis results of the power grid side current show that the harmonic content of the traditional voltage reactive power control algorithm is relatively high, which is 8.70%. Moreover, harmonics are mainly distributed in the low frequency band, and the switching frequency is not fixed. The current harmonic content on the network side of the distributed reactive energy storage structure voltage reactive power control algorithm is small, and the harmonic content is 3.38%. The high frequency harmonic components are mainly distributed around 2 times of the switching frequency, and the switching frequency is fixed. Compared with the traditional voltage control algorithm, the harmonic content of the distributed harmonic storage structure voltage reactive power control algorithm is between the two, the harmonic content is 4.25%. The harmonic high frequency component is mainly distributed around 2 times of the switching frequency, while the harmonic low frequency component is distributed between the two, which is a great improvement compared with the traditional voltage control algorithm.

4   Conclusion

With the development of distributed power technology, voltage and reactive power control is a new type of voltage structure. It can not only connect the traditional voltage, but also operate in isolation. Because of the intermittent energy access, the safe and stable operation of voltage and reactive power control can not be separated from the energy storage system. The hybrid energy storage system of liquid flow battery and super capacitor can give full play to their advantages, increase the growth space of their own advantages, and significantly extend the cycle life of energy storage elements. In the safe and reliable operation of distributed hybrid energy storage structure under big data, energy storage technology is the key, which plays an important role in uninterrupted power supply, improvement of power quality and improvement of photovoltaic power supply performance. The voltage reactive power control algorithm combines the characteristics of energy and power storage, which reduces operating costs and greatly improves the utilization of energy storage. In this paper, the distributed hybrid energy storage structure is used and its characteristics are analyzed, and the voltage reactive power control model is established. On this basis, the voltage reactive power control algorithm is implemented. The shift of voltage and frequency is also regarded as the key point of optimization. Taking the minimum square sum of shift as the objective function, on the basis of stability optimization results, the droop coefficients KPF and kqv are updated through multiple simulation calculations, and finally an optimization program with internal and external cycles is formed. The intermittent energy such as fan can be regarded as the load with negative power, and its transient characteristics are ignored and replaced by load. Then the small signal model is derived in the simplified microgrid. In the future research, in order to get more accurate parameter optimization results, we should study the small signal model derivation method for complex networks, and then expand the application scope of this method.

References

1. Yufei, Chen: Research on power control of distributed hybrid energy storage system with DC bus access. Yanshan University **12**(3), 1234–1324 (2017)
2. Fan, Q., Zheng, X., Wang, P., et al.: Distributed coordination control of independent hybrid microgrid based on dynamic regulation of hybrid energy storage. Power Syst. Protection and Control, **12**(7), 2314–2341 (2018)
3. Liu, S., Liu, D., Srivastava, G., et al.: Overview and methods of correlation filter algorithms in object tracking. Complex Intell. Syst. (2020). https://doi.org/10.1007/s40747-020-00161-4
4. Li, P., Duan, L., Dong, Y., et al.: Energy management strategy of photovoltaic DC microgrid with distributed hybrid energy storage system. Power System Protect. Control **45**(13), 1142–1148 (2017)
5. Yang, L., Yan, X., Yu, Z.: Bus voltage control strategy of low voltage DC distribution network based on hybrid energy storage. Electric. Electric. **12**(1), 1130–1134 (2017)
6. Wu, C., Chen, D., Liu, X., et al.: Stratified coordinated control strategy for AC/DC hybrid energy storage system. Commun. Power Supply Technol. **34**(3), 1144–1146 (2017)
7. Yang, L., Hao, S., Yan, X., et al.: Improved voltage zoning control strategy for DC distribution network based on hybrid energy storage. Zhejiang Electric. Power **36**(6), 2120–2125 (2017)
8. Yan, Y.: Improvement of massive multimedia information filtering technology in big data environment. J. Xi'an Polytechnic Univ. **12**(04), 569–575 (2017)
9. Pan, F., Cheng, F., Luo, C., et al.: Study on charging and discharging control strategy of microgrid hybrid energy storage system. Northeast Power Technol. **34**(3), 1567–1768 (2018)
10. Yao, X., Zhong, L., Wang, W.: Channel capacity analysis of energy capture wireless communication based on hybrid energy storage structure. Comput. Sci. **45**(2), 453–467 (2018)
11. Li, H., Huang, Y., Ma, F.: Coordination control strategy of hybrid energy storage system based on charged state. China Electric. Power **50**(1), 158–163 (2017)

12. Shuai, L., Gelan, Y.: Advanced Hybrid Information Processing, pp. 1–594. Springer, USA (2015)
13. Dam, S.K., John, V.: A Soft-switched Fast Cell-to-Cell Voltage Equalizer for Electrochemical Energy Storage. arXiv, **15**(12), 25–31 (2019)
14. Liu, S., Bai, W., Liu, G., et al.: Parallel fractal compression method for big video data. Complexity **2018**, 1–6 (2018)
15. Zhang, Delong., Li, Jianlin, Hui, Dong: Coordinated control for voltage regulation of distribution network voltage regulation by distributed energy storage systems. Protection Control Modern Power Syst. **3**(1), 3–9 (2018)
16. Lai, C.-H., Cheng, Y.-H., Hsieh, M.-H., et al.: Development of a bidirectional DC/DC converter with dual-battery energy storage for hybrid electric vehicle system. IEEE Trans. Vehicular Technol. **67**(2), 1036–1052 (2018)

Performance Optimization Analysis of Carbon Nanotube Composites Based on Fuzzy Logic

Tian-hui Wang[1] and Wen-chao Zheng[2(✉)]

[1] Zhuhai College of Jilin University, Zhuhai 519041, China
[2] Institute of Data Science, City University of Macau, Macao 999078, China
guanny06@163.com

Abstract. Materials have always been a hot issue in people's eyes. With the increasing demand for materials, the performance of various carbon nanotube composites is insufficient to meet people's needs. Therefore, the performance of carbon nanotube composites based on fuzzy logic is proposed. Optimization Analysis. Firstly, the performance equivalent parameters are calculated. On this basis, the material ratio and the standard geometry are refined. Finally, the performance of the carbon nanotube composite is optimized by the fuzzy relation matrix. The experimental results show that the optimization method can effectively improve the stability, conductivity and bearing capacity of composite materials, and prove that the optimization method can improve the performance of composite materials.

Keywords: Fuzzy logic · Carbon nanotubes · Performance optimization · Composite

1 Introduction

Carbon nanotube composite material is a kind of nanocomposite material. The special structure and superior properties of carbon nanotubes have attracted the attention of many people and have become a hot research topic in nanocomposites [1]. Through the development of technology, carbon nanotube composite materials have gradually developed into various types, including: carbon nanotube-metal composite materials, carbon nanotube-ceramic composite materials, and polymer-based carbon nanotube composite materials, even so There are still some problems to be solved in the field of composite research of carbon nanotubes.

At present, relevant experts in this field have also obtained some good research results. In literature [2], flexible poly (3, 4-ethylene dioxyethiophene)/single-walled carbon nanotube (PEDOT:SWCNT) thermoelectric composites were prepared by dynamic three-phase interfacial polymerization and physical mixing. Conclusion: The content of SWCNT has great influence on the thermoelectric property of composite. The maximum power factor reaches 253.7 ± 10.4, which is one of the highest power values of polymer-based thermoelectric composite materials. In literature [3], a simple electrophoretic deposition method was proposed to deposit copper and carbon

S. Liu and L. Xia (Eds.): ADHIP 2020, LNICST 348, pp. 103–114, 2021.
https://doi.org/10.1007/978-3-030-67874-6_10

nanotubes on the surface of carbon fiber to improve the thermal conductivity and interfacial properties of carbon fiber reinforced composites. Surface morphology, crystallization property, thermal conductivity, interlaminar shear strength (ILSS) and element distribution of the composite were characterized by scanning electron microscopy (SEM), X-ray diffraction (XRD), thermal constant analysis, short-beam bending test and SEM energy dispersive X-ray diffraction (SEM - EDX). However, the above traditional methods fail to calculate the equivalent parameters of material properties and ignore the refinement of material ratio and standard geometric shape, which results in the unsatisfactory application effect.

In order to solve the defects of its performance, it is improved based on fuzzy logic. Based on multi-valued logic, fuzzy logic is used to study the science of fuzzy thinking, language form and its laws. Through this reasoning method, the performance of carbon nanotube composites can be improved, so that it can be more effectively utilized in various fields.

2 Performance Optimization Model Design of Carbon Nanotube Composites

2.1 Performance Equivalent Parameter Calculation

In order to accurately optimize the performance of carbon nanotube composites, it is necessary to calculate various properties, convert various performance into equivalent parameters, and convert some performance into intuitive data so that the performance status can be calculated by calculation. First, the various properties should be equivalently converted to reflect the conductivity, thermal conductivity, stability, and mechanical properties of the carbon nanotube composite. First, the conductivity is equivalently converted into an electromagnetic parameter, and the electromagnetic parameters are calculated. The macroscopic electromagnetic parameters characterizing the electromagnetic properties of materials have different physical quantity representations in applications without application. In the study of absorbing properties of materials, complex magnetic permeability and complex permittivity are used. The absorbing properties of materials depend on their complex permittivity and complex permeability, so accurate measurement of the electromagnetic parameters of materials is a prerequisite for material absorbing properties. At present, the measurement method of electromagnetic parameters of materials mainly uses the cavity method, and the prepared ring sample is placed in a coaxial line or a rectangular waveguide sampler, and then the C-scattering parameter of the sample to be tested is automatically detected by a microwave vector network analyzer. The measurement process is shown in Fig. 1.

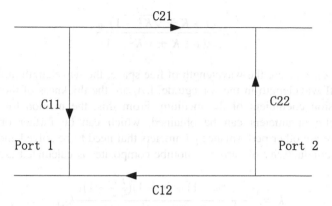

Fig. 1. Electromagnetic parameter C measurement flow chart

The meaning of the C parameters in the figure is: S11 is defined as the ratio of the reflected energy of the port 1 to the energy of the input signal of the port 1; S21 is defined as the ratio of the energy transmitted to the port 2 through the sample to be tested and the energy of the input signal of the port 1; S22 is defined as the ratio of the reflected energy of port 2 to the energy of the input signal of port 2; S12 is defined as the ratio of the energy transmitted by port 2 through the sample to be tested to the energy of port 2; according to the measured S11, S21, S12, S22 scattering parameters, using the formula to calculate electromagnetic parameters.

$$K = \frac{(C_{11}^2 - C_{21}^2) + 1}{2C_{11}} \tag{1}$$

$$T = \frac{(C_{11} + C_{21}) - (K \pm \sqrt{K^2 - 1})}{1 - (C_{11} + C_{21})(K \pm \sqrt{K^2 - 1})} \tag{2}$$

$$\frac{1}{\lambda^2} = -[\frac{1}{2\pi l} ln(\frac{1}{A_t})]^2 \tag{3}$$

$$\mu = \frac{1 + K \pm \sqrt{K^2 - 1}}{\lambda(1 - K \pm \sqrt{K^2 - 1})\sqrt{\frac{1}{\lambda_0^2} - \frac{1}{\lambda_c^2}}} \tag{4}$$

$$\varepsilon = \frac{(\frac{1}{\pi^2} + \frac{1}{\lambda_0^2})\lambda^2}{\mu} \tag{5}$$

For coaxial systems, the electromagnetic wave is a TEM wave, and the formula for calculating the complex permittivity and complex permeability is:

$$\mu = \frac{(1 + K \pm \sqrt{K^2 - 1})\lambda_0}{\lambda(1 - K \pm \sqrt{K^2 - 1})} \tag{6}$$

$$\varepsilon = \frac{(1 + K \pm \sqrt{K^2 - 1})\lambda_0}{\lambda(1 + K \pm \sqrt{K^2 - 1})} \qquad (7)$$

In Eqs. 1–7, $\lambda_0, \lambda, \lambda_c$ are the wavelength of free space, the wavelength in the medium, and the cutoff wavelength in the waveguide. $1, A_t$ are the thickness of the sample and the transmission coefficient of the medium. From this, the method for solving the electromagnetic parameters can be obtained, which can be further optimized. In addition, there are other performance parameters that need to be solved and calculated. The thermal conductivity of carbon nanotube composites is calculated as:

$$k = \frac{\frac{k_{c,f}}{k_{c,ao}} + (n - 1) + (n - 1)(\frac{k_{c,f}}{k_{c,ao}} - 1)v}{\frac{k_{c,f}}{k_{c,ao}} + (n - 1) + (\frac{k_{c,f}}{k_{c,ao}} - 1)v} k_{c,ao} \qquad (8)$$

Where: $k_{c,ao}$ is the composite thermal conductivity, n is the fiber shape factor, v and v_0 are the volume ratio of the composite, respectively; $k_{c,f}$ is the thermal conductivity of the thermal conductivity factor, so that the thermal conductivity can be calculated. The mechanical properties mainly refer to the compressive properties and are also important parameters for the mechanical properties of carbon nanotube composites. The mechanical properties of carbon nanotube composites were obtained by fitting the scaling law:

$$E = a\rho_c^b \qquad (9)$$

Where: E is the target value for characterizing mechanical properties, and a and b are the fitting coefficients, which are the density of the carbon nanotube composite.

2.2   Refinement of the Material Ratio

When preparing the carbon nanotube composite material, the carbon nanotube as the filler can reduce the impurity doping amount, thereby improving the performance of the composite material; In the composite material in which the carbon nanotubes are combined in a loosely combined manner, the loading of adjacent carbon nanotubes is not caused by the failure of a small amount of fibers, thereby realizing material reinforcement; Carbon nanotubes have the typical stability and affinity of carbon materials, but the difference is that the outer layer of carbon nanotubes has high chemical activity and can form stable chemical bonds with matrix materials. The material thus prepared is thus enhanced in stability. According to the ratio of ferric chloride: citric acid = 1:2, nitric acid drill: citric acid = 3:2, the corresponding drugs were weighed, placed in two beakers, dissolved in a small amount of distilled water, and then placed in a water bath at 80 °C for stirring. During this process, HCl gas and gas are continuously released. The solutions in the two beakers were then mixed and then placed in a water bath at 80 °C until the liquid was a viscous gel. The obtained sol was placed in an oven at 120 °C for 3 h to obtain a dried gel. The obtained gel was calcined in a muffle furnace at 500 °C for 2 h, and the sample was naturally cooled in a furnace to obtain a drill ferrite [4].

Then, the ferrite and the carbon nanotubes of different masses are uniformly mixed, 20 mL of anhydrous alcohol is added, and then ultrasonically dispersed for 40 min. After drying, the product is uniformly mixed with paraffin, and composite samples with different mass fractions of carbon nanotubes are obtained. The mass fraction of each raw material in the sample is shown in Table 1.

Table 1. Material quality score ratio

Sample	Mass fraction		
	$FeCl_3 \cdot 6H_2O$	$Co(NO_3)_2 \cdot 6H_2O$	Citric acid
1	23.65%	46.28%	0.36%
2	27.31%	43.95%	0.56%
3	24.58%	41.97%	0.74%
4	21.97%	47.55%	0.81%

In order to obtain better results, a protective film can be applied on the surface of the prepared carbon nanotube composite material, and the coating can affect the electrical conductivity of the carbon nanotube composite material, that is, the electromagnetic parameter, that is, the absorbing coating. The absorbing properties of absorbing materials depend on two important factors: one is the impedance matching of the material to the air, and the other is the ability of the material to attenuate the propagation of electromagnetic waves, both of which depend on the electromagnetic parameters of the material. As the content of carbon nanotubes increases, the ability of the material to transmit microwaves into the coating increases rapidly, and the impedance matching of the coating with air gradually decreases. Therefore, when the content of carbon nanotubes is small, as the content of carbon nanotubes increases, the total loss of the coating increases with microwaves; when the content of carbon nanotubes in the coating increases to a certain extent, the total amount of materials to microwaves The loss reaches the maximum; continue to increase the content of carbon nanotubes in the coating, because the microwave energy entering the coating is rapidly reduced, even if the loss of the material to the microwave is enhanced, the microwave transmitted into the coating is greatly weakened, so the coating The total loss of the layer to the microwave is rapidly weakened [5]. In order to achieve the desired absorbing performance, the ratio of the coating needs to be very precise, and the index of "thin, wide, light and strong" should be achieved. In addition to selecting suitable absorbing materials, it must have a perfect structural design.

2.3 Refinement of Standard Geometric Shapes

Carbon nanotubes are a one-dimensional quantum material with a distinctive structure, also known as a bucky tube. The reason why its structure is different is that it is a coaxial tube with hexagonal carbon atoms stacked. A number of layers or dozens of layers, each layer has a fixed distance between them, the value is generally 0.34 nm, the diameter is generally 2–20 nm. Because its structure is not necessarily a single hexagon, it may also contain pentagons and heptagons, so in the process of superimposition and weaving, it may appear uneven in some places, so that the carbon nanotubes are not always straight. If the pentagon is located at the top of the carbon nanotube, a carbon nanotube seal is formed. Conversely, when the heptagon appears, the nanotube is recessed, as shown in Fig. 2.

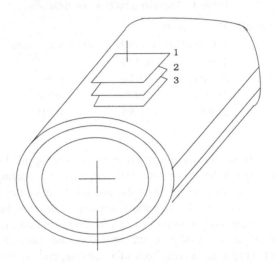

Fig. 2. Carbon nanotube geometry

Its structural speciality is also reflected in the fact that two adjacent carbon nanotubes are not connected to each other, but have a certain distance. Assuming that the displacement of the upper and lower surface layers at the interface is continuous, the lower surface of the first layer has the following displacement relationship with the upper surface of the second layer:

$$x_1 + \left(\frac{h_1}{2}\right)\theta_1 = x_2 + \left(\frac{h_2}{2}\right)\theta_2 \tag{10}$$

In the formula, h_1, h_2 respectively represent the thicknesses of the respective layers, x_1, x_2 respectively represent the mid-plane displacement of the respective layers, and θ_1, θ_2 respectively represent the corners of the plane normal axis and the Y-axis of the respective layers.

After specifying the movable displacement of the carbon nanotube composite material, the selection of the center of gravity in the material plays a stable role in the entire geometric structure. If the selection of the re-point is slightly biased, the stability of the structure will be affected [6]. In the actual operation, if you want to get a more accurate result, you can choose a higher center of gravity sampling, thus reducing the correction of the limited base set. In the calculation of various materials, carbon nanotubes need special attention, and its energy band calculation is different from other materials. In addition to the general precision and the setting of the base group, the structure is also considered. The difference caused. When selecting the center of gravity point, you cannot use the method of setting the center of gravity of the usual material, but should consider its unique characteristics, using only one center of gravity point in the XY direction and more center of gravity point in the Z direction. First, you must check the band structure to manually customize the center of gravity to meet the needs. The initial selection start point is (0, 0, 0), and the end point is (0, 0, 0.5), and is continuously optimized according to the experimental results. T he process of normalizing the geometry finds the structure with the lowest energy, that is, the structure that can be relatively stable, by calculating the optimized structure. In addition, in order to prevent the energy from being unreasonably accounted for in the change of the total energy, the calculated total energy error is increased, so that the error of other attribute calculations is further increased, and the energy is minimized to adjust the structure. It combines the advantages of high specific strength, specific stiffness and stability of the composite.

2.4   Fuzzy Relation Matrix Implementation Performance Optimization

The properties of carbon nanotube composites depend on various factors such as the properties, proportions and geometry of the components of the composite. After the various performance factors in the material are formulated and calculated, the fuzzy relation matrix is applied to finally optimize the performance. Let the set of factors be $A = \{a_1, a_2, \ldots a_m\}$, the composite performance set be $B = \{b_1, b_2, \ldots b_j\}$, and the influence of set A on set B:

Where R is the fuzzy relationship, each element in the set A is the amount, performance, phase geometry and other factors of each component in the composite; each element in the set B is the performance parameter of the composite, including the process performance and interface properties of the composite., mechanical properties, physical properties, stability, etc., then ARB means that set A has a fuzzy influence on set B [7, 8]. If there is no resistance and synergistic effect between the components in the composite, it can be called the orthogonal component factor. At this time, the composite property is written as the algebraic sum of the individual factors.

$$b_j = \sum_{d=1}^{l} g_{jd}(a_{d1}, a_{d2}\ldots, a_{di})$$ (11)

Where: l is the number of components; a_{di} is the i-th factor of the d-th component. In general, composite properties can be written as a relational expression of component content [9]. The formulation test of materials is often to find out the relationship between performance and content. If the component addition amount is set to:

$$X = \{x_1, x_2, \ldots, x_j\}$$ (12)

Then the relationship between performance and factors can be written as:

$$b_j = \sum g_{jd}(x_d | x_d \in X) \tag{13}$$

In the formula, x_d is the amount of component d, X is the amount of component added; $g_{j,d}$ is the function of the degree of influence of the amount of component D (relative) on the properties of composite material J. Its form is determined by the properties and phase geometry of component D. That is to say, C includes all factors except the amount of component d, such as the performance and phase geometry [10–12]. The subordinate degree vector of the properties of composite material type J to the performance level is H, and the performance value is at this time:

$$b = \sum hk \Big/ \sum h \tag{14}$$

Then the influence of adding vector on J performance is aRb [13]:

$$R = \left\{ \begin{array}{l} R_{11} \cap R_{12} \cap \ldots R_{1q} \\ R_{21} \cap R_{22} \cap \ldots R_{2q} \\ \ldots \ldots \\ R_{p1} \cap R_{p2} \cap \ldots R_{pq} \end{array} \right\} \tag{15}$$

The degree of influence of the amount added on the performance, in the graph is to move the performance curve $+(b_{j1} - b_{j0})$, that is, the number of intervals moved $+(b_{j1} - b_{j0})$. The performance of the material is optimized by the fuzzy relation matrix to obtain the required performance and required performance level of the material.

3 Experimental Results and Analysis

In order to verify the optimization effect of the performance of the carbon nanotube composite, the experimental verification was carried out, the carbon nanotube composite material was prepared according to the optimization scheme, the sample characterization was measured and analyzed, and the phase composition of the sample was determined by X-ray diffraction method, and observed by scanning electron microscopy. The particle size and morphology of the ferrite component in the sample were measured by microwave vector network analyzer to determine the electromagnetic parameters of the sample composite in the 2–18 GHz band [9]. After the initial parameters were determined, the performance test was carried out. In order to ensure the rigor of the test, a comparative experiment was set up in the test, and the unoptimized carbon nanotube composite material was the most contrasted, and the conductivity comparison result is shown in Fig. 3.

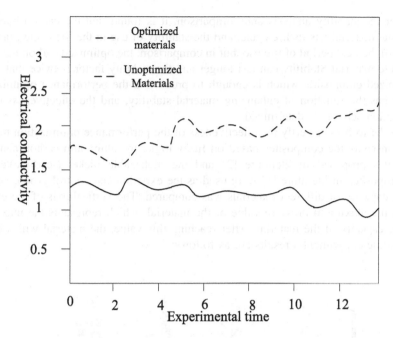

Fig. 3. Conductivity comparison results

It can be seen from the figure that the conductivity changes gradually as the material usage time increases, and the conductivity of the carbon nanotube composite material before and after optimization is between 1–2 when not in use, but with time The growth of the optimized carbon nanotube composites is on the rise, while the unoptimized carbon nanotube composites are declining. It can be seen that the optimization of the carbon nanotube composite can improve the conductivity (Fig. 4).

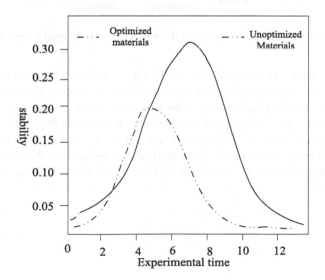

Fig. 4. Stability comparison results

After the stability analysis and comparison, it is found that the carbon nanotube composite material has its life cycle, and the stability of use in the life cycle gradually rises until the best period of life use, but in comparison, the optimized carbon nanotube composite material Stability can last longer and the stability factor is twice that of the unoptimized composite, which is enough to prove that the performance optimization scheme has the function of enhancing material stability, and the effectiveness of the optimization scheme is determined.

In order to further verify the effectiveness of the performance optimization method of carbon nanotube composites based on fuzzy logic, the aluminum carbon nanotube composites proposed in literature [2] and the metastable nickel (-carbide)/carbon nanocomposites in literature [3] were used as the experimental control group, and the bearing capacity of different materials was compared. The vertical axis of the experiment is the maximum pressure value of the material, which represents the maximum pressure capacity of the material. After reaching this value, the material will fracture. The specific experimental results are as follows:

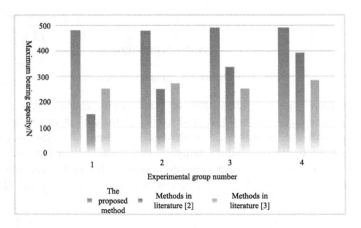

Fig. 5. The pressure properties of materials in different experimental groups were tested

According to the experimental results in Fig. 5, under the condition of constant pressure on the four groups of experimental materials, the carbon nanotube composites analyzed by methods in literature [2] and [3] broke under the pressure of 150 N to 400 N. In contrast, the performance optimization effect in this paper is more obvious. When the pressure level is nearly 500 N, the optimized carbon nanotube composite material in this paper will fracture. The experimental results show that the properties of carbon nanotube composites have been optimized effectively and the toughness has been enhanced.

4   Conclusion

The unique properties of carbon nanotubes give them many opportunities in the field of composites. Carbon nanotube composite materials have been widely used in life. However, there are still some problems with carbon nanotubes, such as the dispersion of the prepared mixture, the preparation cost and quality of the carbon nanotubes. However, the good performance exhibited by carbon nanotubes and their composites will definitely make them better applied to life.

Acknowledgments. 1. Teaching Quality and Teaching Reform Project of Guangdong Undergraduate Colleges and Universities: Construction Project of Experiment Demonstration Center (2017002);

2. Development and Management Research of "Designed Experiment" Project of Engineering Physics Experiment Course in Independent College (GDJ2016057).

References

1. Lu, X., Hu, Y., Li, W., et al.: Macroporous Carbon/nitrogen-doped carbon nanotubes/polyaniline nanocomposites and their application in supercapacitors. Electrochimica Acta **189**(12) 158–165 (2016)
2. Wusheng, F., Cunyue, G., Guangming, C.: Flexible films of poly(3,4-ethylenedioxythiophene)/carbon nanotube thermoelectric composites prepared by dynamic 3-phase interfacial electropolymerization and subsequent physical mixing. J. Mater. Chem. A **10**(6), 1039–1048 (2018)
3. Yan, F., Liu, L., Li, M., et al.: Preparation of carbon nanotube/copper/carbon fiber hierarchical composites by electrophoretic deposition for enhanced thermal conductivity and interfacial properties. J. Mater. Sci. **53**(3), 8108–8119 (2018)
4. Su, Y., Shi, B., Liao, S., et al.: Silver nanoparticles/n-doped carbon-dots nanocomposites derived from siraitia grosvenorii and its logic gate and surface-enhanced raman scattering Characteristics. ACS Sustain. Chem. Eng. **4**(3) 1728–1735 (2016)
5. Deng, W., Shen, L., Wang, X., et al.: Using carbon nanotubes-gold nanocomposites to quench energy from pinnate titanium dioxide nanorods array for signal-on photoelectrochemical aptasensing. Biosens. Bioelectron. **82**, 132–139 (2016)
6. Tang, J., Mu, B., Zong, L., et al.: Facile and green fabrication of magnetically recyclable carboxyl-functionalized attapulgite/carbon nanocomposites derived from spent bleaching earth for wastewater treatment. Chem. Eng. J. **322**, 102–114 (2017)
7. Liu, S., Bai, W., Srivastava, G., Machado, J.A.T.: Property of self-similarity between baseband and modulated signals. Mobile Netw. Appl. **25**(4), 1537–1547 (2019). https://doi.org/10.1007/s11036-019-01358-9
8. Karimi, P., Ostojastarzewski, M., Jasiuk, I.: Experimental and computational study of shielding effectiveness of polycarbonate carbon nanocomposites. J. Appl. Phys. **120**(14), 211–252 (2016)
9. Hoseini, A.H.A., Arjmand, M., Sundararaj, U., et al.: Significance of interfacial interaction and agglomerates on electrical properties of polymer-carbon nanotube nanocomposites. Mater. Des. **125**, 126–134 (2017)

10. Sanli, A., Benchirouf, A., Müller, C., et al.: Piezoresistive performance characterization of strain sensitive multi-walled carbon nanotube-epoxy nanocomposites. Sens. Actuators A **254**, 61–68 (2017)
11. Wang, Y., Zhao, X.: A theoretical model of effective electrical conductivity and piezoresistivity of carbon nanotube composites. Philos. Mag. Lett. **22**(3), 1–6 (2018)
12. Chen, H.L., Ju, S.P., Lin, C.Y., et al.: Investigation of microstructure and mechanical properties of polyvinylidene fluoride/carbon nanotube composites after electric field polarization: a molecular dynamics study. Comput. Mater. Sci. **149**(9), 217–229 (2018)
13. Awan, F.S., Fakhar, M.A., Khan, L.A., et al.: Interfacial mechanical properties of carbon nanotube-deposited carbon fiber epoxy matrix hierarchical composites[J]. Compos. Inter. **15**(5), 1–19 (2018)

Network Dynamic Bad Information Security Filtering Algorithms Based on Large Data Analysis

Wenchao Zheng[1]([✉]), Yin-zhu Cheng[2], Ze-yu Zhang[1], and Yong-qing Miao[1]

[1] Institute of Data Science, City University of Macau, Macao 999078, China
yanghong1911@163.com
[2] School of Economics and Management,
Xinjiang University, Xinjiang 830046, China

Abstract. In view of the low filtering accuracy of traditional bad information in the massive data environment, the security filtering algorithm of network dynamic bad information is innovated and improved in the big data environment. Combining the data set analysis algorithm with the grey statistics theory, this paper evaluates the dynamic information security status of the network structure, extracts the information security features in the evaluation results, compares the data features in the network structure, detects the dynamic time domain range of bad information, and filters and corrects the information in the time domain by nodes and channels, so as to realize the security of the dynamic bad information of the network The experimental results show that the dynamic bad network information security filtering algorithm based on big data analysis is more accurate and effective than the traditional algorithm, with high accuracy and the shortest time, and can be used in the network dynamic bad information security effectively, which meets the research requirements.

Keywords: Big data · Network dynamics · Bad information · Filtering

1 Introduction

With the arrival of the era of big data, the rapid development of network technology has gradually penetrated into various fields, and become an indispensable part of people's daily life and work. With the continuous expansion of network scale, the increasing number of network equipment, applications and services, the effective management of network dynamic bad information is gradually increasing difficulty. The information age brings not only science and technology, but also technology security issues such as science and technology. Network is the most potential new technology product in the 21st century. With its rapid development, network has penetrated into every field of people's daily life. Therefore, how to effectively diagnose the hidden dangers of network operation, effectively filter the bad network information and reduce the loss of information to the network has become the focus of current research [1, 2]. For this reason, researchers in related fields have done a lot of research.

© ICST Institute for Computer Sciences, Social Informatics and Telecommunications Engineering 2021
Published by Springer Nature Switzerland AG 2021. All Rights Reserved
S. Liu and L. Xia (Eds.): ADHIP 2020, LNICST 348, pp. 115–126, 2021.
https://doi.org/10.1007/978-3-030-67874-6_11

In reference [3], an intelligent prediction method of network abnormal failure rate based on big data analysis and fault spectrum feature extraction is proposed. The correlation spectrum feature detection method is used to collect network abnormal failure data, the collected network fault information feature is matched and filtered, and the adaptive beamforming method is used to focus the network fault big data, The extracted network fault big data is classified and identified by fuzzy clustering method, and the intelligent prediction of network abnormal fault rate is realized under the big data, so as to ensure the safe operation of network information. This method can effectively reduce the security of network information, but it focuses on the study of network fault, and seldom considers the network security information. In reference [4], a network information security monitoring system is designed to solve the problems of weak de joint analysis ability, response, processing ability and insufficient monitoring breadth and depth. The data layer of the system uses web service and Oracle10g database to collect and store network logs, and carries out data exchange, communication and encryption through soap and SSL; uses J2EE platform to realize data analysis service; uses flex to realize cross platform display. The proposed system can quickly respond to all kinds of network security attacks and improve the work efficiency and work level of network information security monitoring. However, the lack of consideration on the filtering of complete network information leads to poor filtering effect of bad security information. In reference [5], aiming at the deficiency of existing information security evaluation algorithms in dealing with subjective attitude deviation of multi experts, and the traditional sequential machine learning model method to deal with the problem of deviation accumulation in time period, a deep sequential information security evaluation algorithm based on depth fuzzy correction is proposed. Firstly, the expert fuzzy evaluation index is constructed by trigonometric fuzzy function, and then the modified weighted DS evidence reasoning correction index is used, then the loss and possibility matrix features are created, and finally the information security is evaluated by deep time series network. The simulation experiments are carried out on MIT data set. The experiments analyze whether the features can cope with multi expert conflicts, and evaluate the accuracy, robustness and time efficiency of the algorithm. The proposed algorithm has stronger fuzzy evaluation ability, stronger ability to deal with conflict opinions among experts, and more accurate information security evaluation in time sequence, but the efficiency of the algorithm is low.

In the complex Internet environment, the messages sent and received by the network are affected by bad information, which leads to the impact of network security. An intelligent filtering method of bad information in mobile network environment based on mode matching is proposed. This method analyzes the weight of all the information in the network. Based on this, the network information feature database is designed. According to the ontology element weight of positive information, Bayesian classification algorithm is introduced to classify the network information in the database, and the classified network information data are screened for bad information to realize the filtering of bad information. This method can effectively classify information, but there are many positive information in the filtered information, which is not conducive to universal application. In reference [6], an efficient filtering method for spam information in double buffered communication networks is proposed. In this method, the existing information in the network is processed by dimension, and then the network information is partitioned according to the different dimensions. The principal component analysis method is introduced to gain the network information of each area. The boosting algorithm is used to construct the spam information filter, and the extracted information features are input into the spam information filter to filter the spam information existing in the double buffer communication network, To achieve efficient filtering of spam information in double buffer communication network. This method can realize the filtering of bad information quickly, but the accuracy of filtering information needs to be improved.

Based on the above problems, this paper proposes a dynamic bad information security filtering algorithm based on large amount of data. By using the nonlinear time series location algorithm, we can quickly extract and filter the bad information features, ensure the normal operation of the network, and avoid the loss caused by the failure of network security information. The experimental results show that the proposed algorithm can effectively filter the network dynamic bad information security and has high efficiency.

2 Network Dynamic Bad Information Security Filtering Algorithms

2.1 Network Information Security Situation Assessment

By investigating a large amount of data, a method of evaluating network security situation based on data set analysis algorithm is proposed. The design of network information security situation assessment algorithm is mainly based on grey statistical theory. Assuming that the characteristic quantity of network information's security behavior is S_0, the acquisition factor's behavior characteristic is $S_i - S_0$, the data security is observed on the network information sequence t, and defined as $S_0(t)$, $(t = 1, 2, \ldots, n)$, the sequence of network security information's characteristic behavior can be deduced by combining grey statistical theory, and the sequence of network security information's characteristic behavior is $S_0(t) = [S_0(1), S_0(2) \ldots S_0(n)]$. From this, we can deduce that the sequence of network bad information characteristic factors is as follows: $S_i(t') = [S_i(1), S_i(2), \ldots, S_i(n)]$. Among them, t' is the indefinite value of information detection time, bad information index, object number, etc. It can be concluded that the correlation coefficients of sequence Si and S0 at t point are defined as:

$$\delta_{0i}(t) = \frac{\min_i \min_t |S_0(t) - S_i(p)|^k + \mu \max_i \max_t |S_0(t) + S_i(p)|}{|S_0(t) + S_i(p)| + \mu \max_i \max_t |S_0(t) - S_i(p)|^a} \tag{1}$$

Among them, μ is the evaluation value of the attack that may occur during the operation of network bad information, k is the operation time of network, and a is the number of kinds of bad information that network structure suffers. If n is the degree of harmfulness of bad information to network security attacks, the above algorithms are combined to collect and process the characteristics of bad information in large data environment. The algorithms are as follows:

$$\delta(S_0, S_i) = \frac{1}{n} \sum_{i=1}^{t} 10\mu_{0i}(t), \mu \in (0, 1) \tag{2}$$

Combining the above formulas, we can conclude that the evaluation algorithm of the t security situation evaluation index of network structure R under the influence of bad information is as follows:

$$Rs_i(t) = \delta(S_0, S_i) \sum_{i=1}^{t} 10^{\mu_{0i}} v \tag{3}$$

Among them, v is the number of network structure information supply types. Combined with the above algorithm, the hidden dangers of bad information security in cloud computing network are extracted, and the situation of network information security is evaluated.

2.2 Dynamic Time Domain Location of Bad Information

In the context of the current large data network environment, the scale of data is relatively large, resulting in complex bad information, relatively strong perturbation to the network structure, and easy to lead to network information failure. Therefore, combined with data feature mining algorithm to locate the bad information in the network today, in order to mine the bad information accurately and avoid the interference of bad information (Fig. 1).

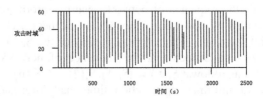

Fig. 1. Time domain analysis model of bad information attacks

Aiming at the attack degree analysis model of bad information in the figure above, the time-domain location analysis and processing are carried out. The scalar time-domain location sequence of network bad information dynamic data is set as follows: $a(t), t = 0, 1, .., n - 1$. The basic screening function algorithm of bad information location signal can be described as:

$$x = [x_1, x_2, ..., x_N] \in Q^{mN} \tag{4}$$

The time-domain feature analysis method is used to decompose the characteristic data of bad information in the network in two gradient directions in the horizontal and vertical direction. The maximum gradient difference algorithm of data mining can be obtained as follows:

$$f = \frac{1}{i \times j} \sum_{x=1}^{j} Rs_i(t) \sum_{j=1}^{i} |f(x)| a(t) \tag{5}$$

Formula (5): f is the correlation coefficient of bad light signal feature mining. According to the above formulas, the directional feature range algorithm of bad information location area is as follows:

$$p = a(t) \frac{Q^{mN}}{\sqrt{(x)^m} \sqrt{f(x)^N}} \tag{6}$$

Assuming that the channel of bad information communication data transmission channel in network structure is continuous, the obtained frequency domain model is:

$$G_n = \zeta_{ij}[a(t_0)]^i + \delta \prod f * Rs_i(\bar{t}) \tag{7}$$

In the formula, $\bar{t}$ is the range of difference of information feature ranking and ζ_{ij} is the error parameter of bad information filtering. Through the above steps, it can provide an effective reference and data basis for the filtering of bad information security by investigating the bad information domain of the network [7].

2.3   Network Dynamic Bad Information Filtering Method

According to the demand of bad information filtering in network, the information security filtering model is constructed. By establishing the data security set, the security data eigenvector is set as: $F=(X, Y)$, where X is the set of limited non-empty security activity information terminal objects e_n in the network structure, and Y is the set of all bad information elements in X. If the direction of security information transmission is: $(e_n, e_m) \in Y$. Among them, e_n is the impact of bad information on network information e_m. Each item e_n has a designated security weight of z_n, which fully considers the failure probability of any bad information active terminal object depending on e_n, so as to filter bad information effectively. Its filtering algorithm is as follows:

$$L = \prod F(X, Y) * z_n[e_n \rightarrow e_m]' G_n \tag{8}$$

Combining with the above algorithm, the bad information is filtered, and the information transmission path weighting more than z_n is found in the network structure, as well as the shortest path from the bad information feature point to other feature points, so as to avoid the attack of bad information and potential information security risks through the route checking and control [8]. Connect all effective paths in the process of information transmission in time, and form bad information filtering and early warning location domain whose structure is shown in the Fig. 2.

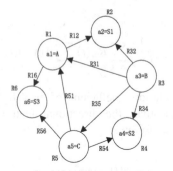

Fig. 2. Time-domain location of bad information filtering

It can be seen from the graph that different kinds of network nodes are universal in network structure. Impression needs to filter bad information in time by modifying network parameters to provide support for fast fault location of network communication [9, 10]. Combined with the above algorithm, the steps of network bad information security filtering in large data environment are improved and perfected. The network dynamic bad information filtering process is shown in the following Fig. 3.

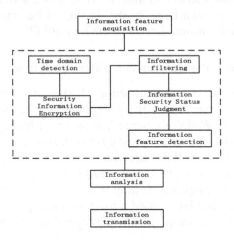

Fig. 3. Network dynamic bad information filtering process

In order to accurately complete the information filtering process and avoid the harmful information influence, damage and other phenomena in the process of information use, the security encryption steps are added. By calculating the time domain range of information security, the network security structure is encrypted and strengthened, and the dynamic bad information of the network is intelligently filtered and intercepted to effectively protect the network information system. Data security [11, 12].

2.4 Implementation of Dynamic Information Security Filtering

After determining the network topology, the coverage rate of network nodes and the occupancy rate of network information transmission channels are calculated by judging the bad information, so as to search and filter the bad information pertinently. The filtering steps are as follows:

All bad information received within a fixed time interval forms a set of bad information feature sources;

Step 1: Relevant information of all bad information received has N sets of entity objects.

Step 2: For any object in the entity's bad information set, it can accurately locate and screen the bad information.

Step 3: In the process of information filtering, bad information and security information can be divided into two subsets, namely, left subset U (bad information set) and right subset N (new security set). Each information feature in the set has the greatest interdependence.

Step 4: The iteration process is implemented until the iteration result is a single point source set. The common features of left and right subsets of information are sorted from bottom to top or from top to bottom in order to reduce the complexity of filtering the information. By calculating the time-domain location of bad information, we can accurately detect all possible factors of bad information in all locations, and ensure the normal operation of network communication.

Step 5: Randomly extract two common characteristics of bad information in any non-left subset;

Step 6: Find the nodes that can explain the common characteristics of two bad information at the same time;

Step 7: Use these nodes to replace the common characteristics of the original security information, and record and compare them.

Step 8: Repeated acquisition of bad information characteristics until there is no intersection between the two bad information.

Step 9: Determine the propagation node of bad information in the network structure.

In order to accurately obtain the feature interpretation of each bad information, it is necessary to find the bad information transmission nodes in the network structure, and calculate the bad information factor between each node through the cross chain storage method. The maximum common factor method is used to construct the filtering step formula of dynamic bad information in network structure, which is as follows:

$$(U_1 \cap N_1) \cup (U_2 \cap N_2) \cup ... \cup (U_{n-1} \cap N_{n-1}) \cup (U_n \cap N_n) \qquad (9)$$

According to the results of the above steps, the best feature interpretation of all bad information in the network structure can be accurately obtained, thus effectively completing the security filtering of dynamic bad information in large data networks.

3 Analysis of Experimental Results

3.1 Experimental Environment

In order to verify the accuracy of network dynamic bad information security filtering algorithm under large data analysis, experimental verification and analysis are carried out. The experiment is based on the Matlab environment. The specific experimental environment is shown in Fig. 4:

Fig. 4. Experimental environment

3.2 Experimental Parameter Setting

Assuming that the network nodes are distributed in 2000 m × 2000 m uniform array area in large data environment, the experimental parameters are set as shown in Table 1.

Table 1. Settings of experimental parameters

Parameter	Remarks
Network frequency band	3 kHz–8 kHz
Information carrier frequency and time domain	3 ms
Initial frequency of information	0.15 Hz
Number of sampling points	256个
Bad information domain scope	−15 dB–15 dB

According to the experimental environment and the results of parameter setting, the experimental contents were analyzed. In the same environment, the accuracy of the network dynamic bad information security filtering algorithm is compared and tested under the large data analysis, and the calculation results are recorded.

3.3   Analysis of Experimental Results

3.3.1   Analysis of Filtering Accuracy of Network Information with Different Algorithms

In order to verify the effectiveness of the proposed method, the experiment analyzes the accuracy of the proposed algorithm and the traditional algorithm for network bad information filtering, and the experimental results are shown in the table below (Tables 2 and 3):

Table 2. Bad information security filtering algorithm in this paper

Practical value	Forecast value	Error value
0.24	0.22	0.02
0.45	0.45	0.00
0.41	0.40	0.01
1.02	1.00	0.02
1.45	1.45	0.00
0.84	0.86	0.02
0.54	0.55	0.01

Table 3. Traditional bad information security filtering algorithms

Practical value	Forecast value	Error value
0.24	0.22	0.12
0.45	0.45	0.08
0.41	0.40	0.23
1.02	1.00	0.14
1.45	1.45	0.32
0.84	0.86	0.12
0.54	0.55	0.23

Comparing with the above results, it is not difficult to find that compared with the traditional bad information security filtering algorithm, the error rate of the proposed dynamic bad information security filtering algorithm is relatively lower, which shows that the accuracy of this method is relatively higher. The experiment analyzes the prediction value, the actual value and the error value. It can be seen from Table 1 that the maximum error value of the proposed algorithm is 0.02, and it is found in the experiment that the difference between the actual value and the predicted value of the proposed algorithm is relatively small; while the minimum error value of the traditional method is 0.08, and the difference between the international value and the predicted value is relatively large. Through the comparison, it can be seen that the proposed algorithm has a high accuracy when filtering the bad information of the network, which is verified The effectiveness of the proposed algorithm.

3.3.2   Gradient Analysis of Network Information Filtering Based on Different Algorithms

In order to verify the practicability of the algorithm, the gradient in the process of information filtering is detected. In the experimental process, the higher the gradient is, the worse the practicability is. The test results are plotted as follows (Fig. 5):

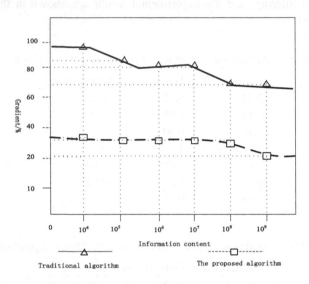

Fig. 5. Comparisons of experimental results

Through the above test results, it is not difficult to find that compared with the traditional bad information filtering algorithm, the proposed algorithm has lower data gradient and significantly improved stability, so it is proved that the proposed algorithm has better practicability and fully meets the research requirements.

3.3.3   Time Consuming Analysis of Network Information Filtering with Different Algorithms

In order to further verify the feasibility of the proposed algorithm, the time-consuming of the proposed algorithm and the traditional algorithm in filtering network bad information is analyzed in the experiment, and the experimental results are shown in Fig. 6:

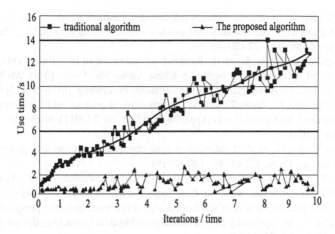

Fig. 6. Comparison of network information filtering time of different algorithms

It can be seen from the analysis of Fig. 6 that there is a certain gap in the time taken by the two methods for filtering the bad network information. Among them, the proposed algorithm takes the shortest time and the traditional method takes longer time. The validity and feasibility of the proposed algorithm are verified.

4   Conclusion

In the complex large data network environment, the accurate perception and prediction of information security situation has become a key issue of current research in various fields. Aiming at the problems of low precision and low efficiency of traditional bad information filtering algorithm, a dynamic bad information filtering algorithm based on data mining and large data information processing is proposed. Experiments show that the accuracy and practicability of the proposed method are significantly improved compared with traditional methods.

References

1. Qiang, J., Wei, Z., Song, L., et al.: Research on Personalized Adaptive Learning - A New Normal of Digital Learning in Big Data Age. China Audio Vis. Educ. **12**(2), 24–32 (2016)
2. Qingzhou, Z., Yong, L., Shiming, T., et al.: State monitoring and fault handling based on big data analysis of intelligent distribution network. Grid Technol. **40**(3), 774–780 (2016)
3. Lei, Y., Bin, L., Zhuoyu, W.: Research on the new model of teachers' information technology ability training - based on "internet plus" and "big data" thinking. China Audio Vis. Educ. **15**(8), 61–66 (2016)
4. Haijuan, Y.: Research on influencing factors of reform and innovation of personnel training mode of information management specialty in the background of big data - taking Hubei university as an example. Libr. Inf. Knowl. **24**(2), 21–29 (2016)

 5. Xianmin, Y., Si, T., Jihong, L.: Framework and development trend of educational big data: the overall framework of "research and practice column of educational big data." Modern Educ. Technol. **26**(1), 5–12 (2016)
 6. Si, Z., Qingtang, L., Shijie, L., et al.: Research on learner input in online learning space - big data analysis of online learning behavior. China Audio Vis. Educ. **13**(4), 24–30 (2017)
 7. Shuai, L., Gelan, Y.: Advanced Hybrid Information Processing, pp. 1–594. Springer, USA
 8. Liu, S., Fu, W., He, L., Zhou, J., Ma, M.: Distribution of primary additional errors in fractal encoding method. Multimed. Tools Appl. **76**(4), 5787–5802 (2014). https://doi.org/10.1007/s11042-014-2408-1
 9. Liu, S., Lu, M., Li, H., et al.: Prediction of gene expression patterns with generalized linear regression model. Front. Genet. **10**, 120 (2019)
10. Minghua, W., Jingui, Z.: An evaluation algorithm of information security based on fuzzy adjustment feature matrices and deep time sequence model. **46**(5), 464–470 (2018)
11. Nanzhong, W.: Reconstruction of teaching design framework from the perspective of mixed learning - also on the supporting role of educational big data in teaching design. China Audio Vis. Educ. **59**(5), 18–24 (2016)
12. Liu, S., Glowatz, M., Zappatore, M., Gao, H., Gao, B., Bucciero, A.: E-Learning, E-Education, and Online Training, pp. 1–374. Springer International Publishing, USA

Analysis of Intelligent Monitoring Model of Network Security Situation Based on Grid Power Flow

Shang Gao, Shou-ming Chen, Yun-de Liang, Yan-qian Lu,
and Jie-sheng Zheng[(✉)]

Guangdong Power Grid Corporation Information Center,
Guangzhou 510062, China
gaoshang527@outlook.com, zhengjiesheng857@outlook.com

Abstract. In order to introduce the grid power flow model to intelligently monitor the network security situation, a model based on grid power flow is established. In the construction of the network security situation intelligent monitoring system, the hierarchical database is protected and managed, the attacks brought by the network security situation are changed, and the network security situation level protection system is improved and improved. On this basis, the network trend correction factor is introduced, and the network security situation is normalized according to the network security situation value. The network information flow is processed uniformly, and the intelligent monitoring model of network security situation based on power flow is built. Compared with the traditional network security situation intelligent monitoring model, the application of network security situation intelligent monitoring model can effectively solve the uncertainty and fuzziness of information provided by various network security devices.

Keywords: Power flow · Network evaluation index · Network security situation · Risk assessment

1 Introduction

The power flow of power grid is based on the determination of the network structure and the equivalent load power of distribution network. By controlling the output power of large-scale thermal power and hydropower, as well as flexible adjustment of transformer adapters and reactive power compensation equipment, under the condition of network security constraints, the goal of minimizing power generation cost, minimizing network loss and minimizing environmental pollution can be achieved. Compared with economic dispatch, the optimal power flow model is more complex, and the solution method becomes the most important problem to be studied [1]. The optimization model is decomposed into active sub-problems and reactive sub-problems by utilizing the weak coupling relationship between "active-phase angle" and "reactive-voltage" of transmission network. The overall optimization is realized by alternating iterations, which reduces the scale of the problem and improves the calculation efficiency. In addition, with the advancement of optimization mathematical theory and

S. Liu and L. Xia (Eds.): ADHIP 2020, LNICST 348, pp. 127–136, 2021.
https://doi.org/10.1007/978-3-030-67874-6_12

computer technology, a series of effective solutions such as simplified gradient method, linear programming method, quadratic programming method and Newton method have emerged, especially the application of linear and non-linear interior point method with polynomial time complexity in optimal power flow, which makes it possible to quickly solve optimal power flow problems.

Network security situation is a macro response to the network operation status. It reflects the past and current status of a network, and monitors the possible network status in the next stage. Its original information comes from network management equipment, network security equipment, network monitoring equipment [2]. Through the mathematical processing and integration of these data, we can generate numerical values and charts that can reflect the operation of the network. By analyzing the relationship between network attack and network situation, an intelligent monitoring model of network security situation is proposed. The model provides knowledge support for network security situational awareness, understanding and decision-making. At the same time, through the combination of network security situation knowledge base and power flow theory, network security situation awareness and situation understanding can solve the uncertainty and fuzziness of information provided by various network security devices. By accurately synthesizing the information obtained by various network security devices into a unified description of the environment, the correct decision-making ability of the intelligent monitoring model of network security situation can be enhanced. Aiming at the problem that the existing network security technology can not monitor the future security situation of the network, a model of intelligent situation monitoring based on power flow is proposed, which takes advantage of the characteristics of network security situation value with non-linear time series and the advantages of neural network in dealing with chaotic and non-linear data. The experimental results show that. The model can accurately obtain the monitoring results of situation values and help network managers make security decisions.

2   Design of Intelligent Monitoring Model for Network Security Situation Based on Power Flow

Through the establishment of intelligent monitoring system for network security situation, the protection and management of hierarchical database, and the improvement of power flow coupling, the improvement of power flow coupling can be achieved. The specific operation method can be carried out in the following steps.

2.1   Construction of Intelligent Monitoring System for Network Security Situation

When evaluating the network security situation, we can evaluate it qualitatively or quantitatively. For qualitative evaluation, it means only describing the existing risks qualitatively. Quantitative evaluation method expresses the evaluation factors with specific numerical values, and then all these factors are included in the algorithm to calculate the risk value. If the calculated risk value is greater, it indicates that the network is insecure. In the qualitative evaluation of network security, in order to perceive the network security situation, independent attack recurrence, network traffic state change and network connection state evolution will be used. The qualitative risk assessment does not quantify the risk, but presents the state of network security in the form of recurrence. In the demonstration of network security situation, experts' knowledge in the field will be referred to [3]. Nowadays, many security products can easily get network evaluation index, SCYLLARUS system and Honeypot system.

Whether qualitative or quantitative evaluation index system, because of its different emphasis, in order to fully reflect the security situation of the network, it is necessary to consider them comprehensively. The combination of the two risk assessment methods can ensure the integrity and effectiveness of the assessment (Fig. 1).

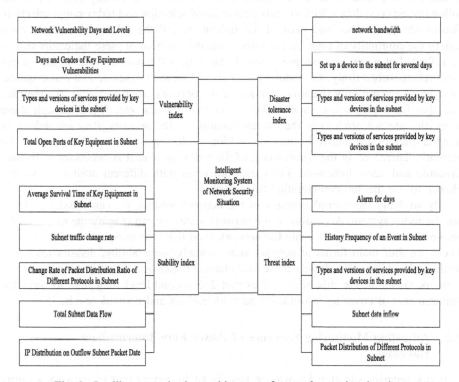

Fig. 1. Intelligent monitoring architecture of network security situation

As shown in the figure, the model uses multiple data sources and composite evaluation indicators to accurately perceive the whole network state in order to achieve the purpose of comprehensive evaluation of network security. Therefore, in order to evaluate the whole network entity (network, host, router, etc.) and all levels (network layer, operating system layer, application layer, etc.), On the premise of the following principles, an intelligent monitoring system of network security situation is established.

First, integrity and independence. In large-scale networks, the index system is used, because the index system can reflect the impact of all factors, so as to achieve a comprehensive assessment of the network situation. And because the index system will take into account various factors, so the number of indicators will be very complex, if all indicators are not different, completely integrated, it is difficult to make a comprehensive and effective evaluation [4]. Therefore, in the evaluation of network situation, several elements should be kept relatively independent, and these relatively independent properties will be reflected by the calculation of the index set. Secondly, it is systematic and hierarchical. In the network situation index system, systematicness and hierarchy are its main characteristics. Because the network itself is a hierarchical structure, it is necessary to ensure the hierarchy when establishing the index system, so as to reflect the rationality and regularity of the index system. Thirdly, it is scientific. The selection of network situation indicators must be based on science, in order to truly reflect the true state of network situation, its scientificity is mainly reflected in the following aspects: data selection, statistical method selection and index range selection. Fourth, the persistence and goal of the indicators. All the selected indicators must satisfy the continuity of time. At the same time, the function of these indicators should be reflected at a certain time point, which can reflect the overall state of the system comprehensively Fifthly, the index system should include dynamic and static indicators. There are not only dynamic changes in the network, such as traffic, change rate, but also fixed properties, such as hosts, routing devices, etc. [5]. Although these properties are not obvious in the overall situation of the network, they are indeed an integral part of the network situation, which is the basis for the existence of the network. Therefore, in the construction of the index system, it is necessary to include dynamic and static indicators. For these indicators with different attributes, we can choose to use manual configuration to achieve.

By studying the overall structure of the network system, we can classify and sort out the index system. According to the network state, we can classify the nature of the network, so that we can evaluate the network state from the perspective of the nature. There are four main forms of network state, namely vulnerability, disaster tolerance, weili, and stability. We classify these four characteristics as the first-level indicators of network situation. On this basis, we extract 2–5 second-level indicators to meet the requirements of covering information network entities and network levels.

2.2 Adjusting Monitoring Sequence of Power Flow Intermediate Database

Network security situation knowledge base is an important part of network security situation assessment system. It is responsible for storing network security situation information, providing knowledge support in network security situation awareness, understanding and decision-making methods. Referring to the relationship between network attack and network security, this paper uses attacker state and network state to construct network security situation knowledge base [6].

In the knowledge base of network security situation, the relationship among network status, attacker status and network security situation is as follows. Network security situation: The security-related information in the network is expressed in the form of P (X), where P is the predicate and X is the parameter set. Attack state: Describes the system security knowledge and resources that an attacker possesses, similar to the system state, in the form of P (X). Network security situation: the whole network and attacker's information set related to security, that is, the complete relationship between network state and attacker's state is shown in Fig. 2.

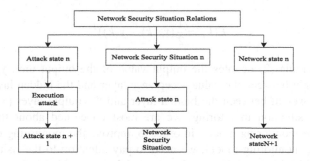

Fig. 2. Network security situation diagram

Network vulnerabilities are errors in the specific implementation and use of the network, but not all the errors in the network are vulnerabilities. Only errors that threaten network security are vulnerabilities. Many errors do not cause harm to network security under normal circumstances. Only when people intentionally use them under certain conditions can they affect network security. Although vulnerabilities may exist initially in the network, they do not appear on their own and must be found artificially. In actual use, users will find errors in the network, and attackers will intentionally use some of them and make them become a tool to threaten network security, then this error will be considered as a network vulnerability [7]. For an attacker, in order to attack a target network, it is necessary to discover the system vulnerabilities in the target system, and then consciously use the vulnerabilities to attack, that is, to find the premise of network attack.

At the same time, the network attack using network security vulnerabilities will bring dynamic state changes to the network security situation after the attack. The so-called dynamic network security state refers to the uncertainty of the attack results, such as some buffer overflow attacks, which may successfully obtain privileges, may also lead to process denial of service, and may not cause any results.

2.3 Introduction of Power Flow Correction Factor

Compared with the basic Elman neural network, the double feedback Elman network increases the feedback of the output layer nodes, which makes up for the deficiency of the basic Elman network. It takes the output layer feedback as the input of the hidden layer along with the input layer and the connection layer unit. This feature makes its information processing ability more powerful. Its mathematical model is expressed as formula (1):

$$x_c(k) = \alpha x_c(k-1) + x(k+1) \tag{1}$$

$x_c(k)$ in formula (1) represents the output of the receiving layer unit 1, αx_c represents hidden layer output. Double feedback Elman neural network still uses gradient descent idea to obtain the learning algorithm of the network. Formula (2) is used to represent the error function in the dynamic double feedback Elman neural network, i.e. the objective function:

$$E(k) = \frac{1}{2}(y_\partial(k) - y(k))^T \tag{2}$$

$E(k)$ in Formula (2) denotes the output value of the output unit. y_∂ denotes the connection weight between the value acceptance layer and the hidden layer. T denotes the connection weight between the hidden layer and the output layer [8].

In network situation monitoring, we are most concerned about the monitoring capability of the monitoring model. In order to improve the monitoring capability and accuracy of the monitoring model, we need to pay attention to the rising and falling direction of monitoring trend. If a trend correction factor is added to the model to reflect the monitoring trend, the monitoring trend can be effectively adjusted and the correct direction of monitoring can be guided, so that the monitoring accuracy can be improved [9]. Its core idea is in the process of monitoring. If the trend of monitoring value is different from the actual value or the direction of rise and fall is not consistent, the parameter of trend correction factor should be g, otherwise the parameter should be h. The trend correction factor can be expressed in formula (3):

$$f_{DP}(t) = if(y_d(t) - y_d(t-1)) \tag{3}$$

Formula (3) f_{DP} represents the trend correction factor, if represents the number of iterations, and y_d represents a relatively large value.

2.4   Normalization of Network Security Situation

The network security situation is estimated by analyzing the network security situation data, and the network model is shown in Fig. 3.

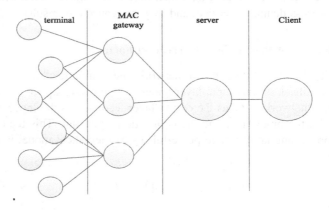

Fig. 3. Network model.

According to its linear frequency hopping spread spectrum technology, the network model is convenient to identify malicious attack information effectively, and can guarantee the network security situation.

Suppose that the data flow of m malicious attacks under the neural network is:

$$x(k) = [x_1(k), x_2(k), \cdots, x_m(k)] \tag{4}$$

In the formula, k represents the attribute value of the malicious network attack; $x_i(k)$ represents the feature vector of the malicious data of the communication network.

According to the process of network attack and the non-linear sequencing of alarms generated by security devices, the network security situation value x obtained by weighting all kinds of alarms can be abstracted as a function of time series t. Namely: $x = f(t)$, This situation value has the characteristics of nonlinearity [10]. Therefore, the network security situation value can be treated as a time series, so assuming the time series $x = \{x_i | x_i \in R, i = 1, 2, ...L\}$ with network security situation value, we now hope to monitor the following M situation values through the situation value of the first N time of the sequence [11].

In order to eliminate the possibility of large errors due to the use of data in the process without processing, this paper normalizes the situation values [12, 13]. The specific normalization formula is shown in (5):

$$x_i = \frac{x_i^t - x_{\min}}{x_{\max} - x_{\min}} x(k) \tag{5}$$

In formula (5), x_i is the calculated situation value, x_i^t is the normalized situation value, and $x_{\min}$ and $x_{\max}$ represent the maximum and minimum values in the network security situation value, respectively.

3 Model Effectiveness Verification

The experimental data select 120-day network status data of a university campus network to monitor the network security situation of the campus network. The evaluation indexes are carried out by using the index system established. There are four first-level indicators and 22 second-level indicators of risk, vulnerability, availability and reliability in the index system. We use Math to calculate these indicators according to the above calculation situation value method. Mathematica software is used to calculate the situation value, and 120 situation values are obtained.

According to the cyber security incident indicator system we developed in Sect. 3, we use the commonly used method of evaluating the grading scale of the Likert quantity. According to the semantic principle, we develop a five-level evaluation level standard: The levels correspond to an element of good, good, general, poor, and poor in the fuzzy set. For the convenience of calculation, we quantize the _5 kinds of fuzzy evaluations respectively, and assign their corresponding values to 5, 4, 3, 2, 1. The corresponding situational value table for the evaluation grading standards is shown in Table 1.

Table 1. Evaluation grading criteria corresponding situation value

Evaluation situation value	Comment	Grading
$x_i > 4.5$	Good	E1
$3.5 < x_i < 4.5$	Good	E2
$2.5 < x_i \leq 3.5$	Good	E3
$1.5 < x_i \leq 2.5$	Good	E4
$x_i \leq 2.5$	Good	E5

3.1 Preliminary Experiment Preparation

Using the basic Elman model and the dual feedback Elman neural network model with trend correction factor, in the experiment, the network security situation is monitored by these two models.

The basic Elman neural network and the dual feedback Elman neural network with trend correction factor are all based on a three-layer network structure, that is, there are _5 input layer nodes, 10 hidden layer nodes and 1 output layer node. The layer unit node, the former has one receiving layer unit node, and the latter has two receiving layer unit nodes. Because the input layer has _5 input nodes, the input is the continuous 5th network security situation value, and one output node of the output layer outputs the monitored value of the number of attacks on the 6th day.

3.2 Network Security Intelligent Monitoring Node Coverage Comparison

The network security trend monitoring model and the traditional network security situation intelligent monitoring model are used to monitor the network security situation. Monitor the cybersecurity situation values from day 91 to day 110 in the data set. That is, d86 (the situation value on the 86th day to the 90th day) was used to monitor d91 (the situation value on the 91st day), and d92 was monitored using d87–d91. And so on, monitor d110. The monitoring results and actual results are shown in Fig. 4:

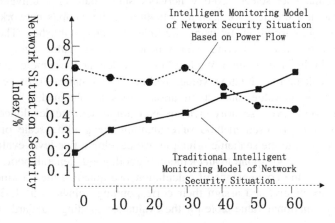

Fig. 4. Contrast experimental results

It can be seen from Fig. 4 that in the two models, the network traffic situation intelligent monitoring model of the power flow of the grid has three monitoring times, and the monitoring effect evaluation functions LS E and AAE are smaller than the basic Elman neural network. In Fig. 3, by comparing the results of the two models of the SYNFIood attack monitoring evaluation function, we can find that the AAE obtained by the SYNFIood attack of the network security situation intelligent monitoring model of the power grid is 0. 007462, and the AAE value of the basic Elman neural network monitoring is 0. 005552, the former increased by 34.2070 than the latter; this shows that the network security trend based intelligent monitoring model based on grid power is more capable of monitoring.

Compared with the traditional network security situation intelligent monitoring model, the network traffic situation intelligent monitoring model of the power grid not only increases the feedback of an output layer, but also increases the trend correction factor to control and adjust the monitoring trend of the rising and falling trend. Therefore, the model has Stronger information processing capability enhances monitoring performance, and its monitoring capability is significantly stronger than the traditional network security situation intelligent monitoring model.

In summary, the two monitoring models are feasible in the field of network security attack monitoring. However, through the monitoring and comparison of the two, the monitoring capability of the network security situation intelligent monitoring model based on the power grid trend is stronger, and the monitoring results obtained are quite satisfactory.

4   Conclusion

Network security management is not only a technical issue, but also a management issue. Therefore, how to find a solution from the security incidents that have occurred is the most concerned issue for researchers. Since the concept of network security situational awareness has been proposed, network administrators have begun to consider the security threat status of the overall network from multiple angles and macro perspectives, and to achieve the purpose of assisting decision-making based on the comprehensive evaluation index system of the network.

Acknowledgments. This research is funded by Jiangsu Tong Brand Professional Construction Project (Z215015002).

References

1. Ningbo, Peng, J., Changpeng, et al.: Analysis and research of power network monitoring signals based on equipment intelligent logic model modeling. Electron. Technol. Softw. Eng. (21), 216–217 (2017)
2. Xu, W., Dai, L.: Research on unified information model of monitoring integration platform based on smart grid. New Technol. Prod. China **23**(5), 41 (2017)

3. Jun, X., Lei, Z., et al.: Distribution system security region model based on power flow calculation. J. Electr. Eng. China, **37**(17), 334–336 (2017)
4. Anonymous. Construction method of large data application model for smart grid monitoring operation. Power Syst. Autom. **42**(20), 121–128 (2018)
5. Niu, W., Bao, P., Tang, H., et al.: Smart grid security vulnerability mining model based on data mining. Power Technol. **42**(4), 134–155 (2018)
6. European Network and Information Security Agency: Baseline guidelines for internet of things security in key infrastructure areas. Inf. Secur. Commun. Secr. **54**(1), 80–95 (2018)
7. Shuai, L., Weiling, B., Nianyin, Z., et al.: A fast fractal based compression for MRI images. IEEE Access **7**, 62412–62420 (2019)
8. Li, W., Wang, S., Li, X., et al.: Information security prevention and control system for power grid enterprises based on artificial intelligence. Power Inf. Commun. Technol. **17**(2), 105–109 (2017)
9. Chen, C., Tu, Z., Gu, L.: State grid corporation network and information security situation awareness practice . Power Inf. Commun. Technol. **45**(6), 3–8 (2017)
10. Jiang, C., Jiang, J.: Practice and innovation of network security internal audit in power grid enterprises. China Internal Audit **229**(7), 68–70 (2018)
11. Wang, H.: Application of information security situation analysis method and system in power informatization. Digit. Technol. Appl. **43**(2), 215–217 (2017)
12. Liu, S., Glowatz, M., Zappatore, M., Gao, H., Gao, B., Bucciero, A.: E-Learning, e-education, and online training, pp. 1–374. Springer, USA
13. Zheng, P., Shuai, L., Arun, S., Khan, M.: Visual attention feature (VAF): a novel strategy for visual tracking based on cloud platform in intelligent surveillance systems. J. Parallel Distrib. Comput. **120**, 182–194 (2018)

Online Monitoring Method for Hazard Source of Power System Network Based on Mobile Internet

Jie-sheng Zheng$^{(\boxtimes)}$, Bo-jian Wen, Wen-bin Liu, Guang-cai Wu,
and Gao Shang

Guangdong Power Grid Corporation Information Center,
Guangzhou 510062, China
zhengjiesheng857@outlook.com

Abstract. In the power system network, aiming at the low accuracy of traditional network hazard online monitoring method, an online monitoring method for hazard source of power system network based on mobile internet is proposed. Based on mobile internet, a power system network communication is constructed. The model uses this model to collect dangerous source data. After the hazard data is collected, the WAMS system is used to calculate the relative residuals of the hazard source data, and then the relative residuals are used to identify the hazard source parameters, and the branch with the hazard source parameters is present. The traveling wave positioning network is used to locate the dangerous source. After the hazard source is located, the hazard source is monitored online by the hazard source indicator. Under the condition that the experimental environment is the same, the method is compared with the online hazard source online monitoring method based on feature recognition technology and the online hazard source online monitoring method based on communication message parsing. The monitoring accuracy of these three methods is improved. The results are 41.1%, 68.8%, and 94.5%, respectively. The experimental results show that the monitoring accuracy of this method is higher than the traditional online hazard source online monitoring method, which proves the superiority of the method.

Keywords: Mobile internet · Power system · Network hazard source · On-line monitoring

1 Introduction

With the implementation of the national 1000 kV UHV networking strategy, a wide-area power system network with an installed capacity of more than 8 kW and spanning thousands of kilometers is being formed in China. Although the formation of a wide-area power system network will bring huge benefits to the national power system network, the security, stability and economy of the power system network will also face unprecedented challenges [1]. Through the research on the hazard source of power

S. Liu and L. Xia (Eds.): ADHIP 2020, LNICST 348, pp. 137–145, 2021.
https://doi.org/10.1007/978-3-030-67874-6_13

system network, people have a certain understanding of the cause of the hazard source: the traditional online monitoring method has slow information transmission speed, lacks wide-area synchronous measurement capability, and cannot monitor the dynamic process of the power system online in real time; The occurrence of potential chain accidents will lead to further expansion of the dangers. The main reason for the chain accidents is that the protection and stability control systems of traditional power systems can only rely on local information, and it is difficult to achieve global optimization and coordination control; The existing grid on-line monitoring method lacks robustness to changes in the operating state of the power system, especially in emergency situations such as cascading failures, which cannot provide accurate operational status information to dispatchers [2]. Therefore, an online monitoring method for hazard source of power system network based on mobile internet is proposed, which improves the monitoring accuracy of hazard source in power system network.

2   On-Line Monitoring of Hazard Source of Power System Network Based on Mobile Internet

2.1   Hazard Source Data Collection

A power system network communication model is constructed based on mobile internet, and the model is used to collect dangerous source data. The power system network communication model is mainly composed of five parts: local aggregator, building gateway, smart meter, user and trusted organization [3]. Local aggregators are affiliated with power grid companies; building gateways are generally affiliated with some outsourcing companies, such as mobile companies or China Unicom; trusted organizations are independent organizations, such as regional communication organizations or independent system operators. The communication architecture of the power system network communication model is shown in Fig. 1.

The power system network needs to know the user's power consumption information in real time, dynamically adjust the electricity price, and analyze and predict the power consumption in the next stage, which will generate a large amount of data transmission. In order to facilitate the acquisition of dangerous source data, the power system network communication architecture is divided into three hierarchical networks, namely regional regional networks, nearby regional networks and building regional networks. Among them, each building area network is composed of many users, and the information interaction interface between the user's smart meter and the smart grid is defined as the building gateway [4]. Each nearby area network consists of a number of building area networks, defining this layer of network nodes as local aggregators. The regional area network consists of a number of nearby regional networks, and the network nodes defining this layer are central aggregators. Information interaction between smart meters and building gateways can be transmitted via power line carriers, WiFi wireless broadband or Zigbee; information exchange between building gateways and local aggregators can be through broadband infrastructure such as WiMax Global Interoperability for Microwave Access or 3G/4G network, etc.; data communication and operational information control between the local aggregator and the central aggregator can be

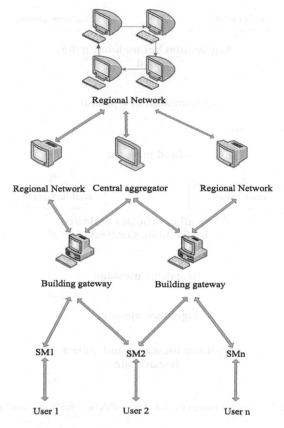

Fig. 1. Communication architecture of network communication model in power system

transmitted through optical fiber, and the security of communication can be guaranteed. Each user's home is equipped with a smart meter that can be used to periodically record the user's real-time power consumption data. And the smart meter periodically transmits the real-time power consumption data to the local aggregator through the building gateway. After receiving all the data, the local aggregator sends the data to the central aggregator in batches, and the central aggregator forwards the data to the control center. The control center adjusts the data transmission rate according to the power consumption information, so as to facilitate the collection of dangerous source data [5].

The trusted authority initializes according to the blind signature and generates initialization parameters, wherein the trusted authority is a pair of public and private keys issued by all entities such as smart meters, building gateways, and local aggregators [6]. Before accessing the power system network, the user presents the specific identity information to the local aggregator, and the local aggregator verifies whether it is a legitimate user [7]. After the verification is passed, the smart meter and the license book with the public and private keys and the unique secret number are embedded. To the user, finally, the local aggregator stores the data in its database, and then collects the dangerous source data [8]. The specific flow of hazard source data collection is shown in Fig. 2.

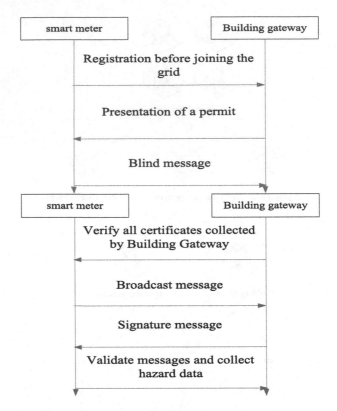

Fig. 2. Specific processes for data collection from risk sources

2.2 Hazard Source Parameter Identification

After the data collection of dangerous source is completed, the relative residual of dangerous source data is calculated by using WAMS system, and then the parameter identification of dangerous source is carried out by using the relative residual. Firstly, the WAMS system is used to estimate the linear state of the dangerous source data [9]. For a bus system with n busbars and m measurements, the linear measurement equation can be used to represent:

$$Z^{meas} = H \cdot X + e \tag{1}$$

Where Z^{meas} represents the m-dimensional PMU measurement; H is a $m \times (2n - 1)$ - dimensional measurement Jacobian matrix; X is a $2n - 1$-dimensional voltage state vector; e is a m-dimensional measurement noise vector. Then the linear state estimate with weighted least squares can be expressed as:

$$X^{est} = (H^T W^{-1} H)^{-1} (H^T W^{-1} Z^{meas}) \tag{2}$$

Where X^{est} is the voltage phasor matrix estimated by the linear state; W is the weight matrix. In order to measure the calculation error of the linear state estimation, the "measurement amount" calculated from the measured quantity and the voltage phasor obtained from the linear state estimation is compared here [10], and the difference between them is measured by using the relative residual, relative residual R^{mar} is calculated as:

$$R^{mar} = \frac{1}{n} \sum_{1}^{n} \left| \frac{Tn}{Z^{meas}} \right|$$

(3)

Where R^{mar} is the relative residual; T is the correlation coefficient between the measured quantity and the quantity to be determined; n is the number of measured quantities. Then use the relative residual R^{mar} to identify the source of the hazard and identify the branch with the hazard source parameter.

2.3 Hazard Source Location

For the branch with the dangerous source parameter, the traveling wave positioning network is used to locate the dangerous source. The traveling wave positioning network is mainly used to collect the initial traveling wave time information of the whole network, including the initial traveling wave arrival time and circuit breaker state information recorded by all traveling wave detecting devices in the power system network [11, 12]. First, determine the branch circuit with the dangerous source parameters, eliminate the invalid initial traveling wave arrival time, and then use all valid initial traveling wave arrival times to locate the dangerous source.

Analyze whether the initial traveling wave arrival time recorded by any substation in the power system network is valid, arrange all the initial traveling wave arrival times in a certain order, and then sequentially discriminate in order. The specific method is: the initial traveling wave arrival recorded for a substation. The time difference between the time, in turn, and the initial traveling wave arrival time calculation recorded by its neighboring substation. The dangerous source positioning master station arranges the effective arrival time of all effective initial traveling waves into two arrays on both sides of the dangerous source line, respectively, taking one line of arrival time from the two arrays, and calculating the dangerous source point according to the double-ended dangerous source traveling wave positioning algorithm. To the dangerous source distance of the substation, the traveling wave positioning steps of the hazard source of the power system network based on the above positioning principle are: determining the dangerous source line; simplifying the traveling wave positioning network; performing the elimination of the invalid initial traveling wave arrival time; calculating the dangerous source Distance; set weights for all hazard source distances; calculate hazard source distances and output hazard source location results.

2.4   Realizing On-Line Monitoring of Hazard Source

After the hazard source is located, the hazard source is monitored online by the hazard source indicator. The hazard source indicator is composed of a detection circuit, an analysis algorithm circuit, a trigger circuit, a wireless transmission circuit module, a power supply circuit, etc., and mainly uses a hazard source indicator to pass the detected dangerous sources such as power-on, power-off, grounding, and short-circuit signals. The distance radio frequency module is transmitted to the signal transmission terminal. The three hazard source indicators are a set of detection terminals, which are responsible for detecting the data of the line hazard sources. Each group is equipped with a data communication device, which can transmit the collected data information to the indoor main station receiving switch. The hazard source indicator is mounted on the line and can be directly loaded and unloaded by the operating lever without power failure. It is installed at the following location: at the exit of the substation, to determine whether the source of danger is inside or outside the station; The entrance is used to determine whether the hazard is on the main line or on the branch line; at the junction of the cable and the overhead line, it is used to determine whether the hazard is in the cable segment or on the overhead; in the plain or open area to reduce the hunt Work pressure.

The alarm display function of the hazard source indicator: when the line is positioned to the hazard source, all hazard source indicators of the hazard source line from the substation exit to the hazard source point are activated or flashed, and the hazard source indicator after the hazard source point Then it does not work. In this way, the lineman can quickly determine the section of the source of danger by means of the alarm display of the indicator and can find out where the source of the danger occurred. At the same time, the hazard source indicator can also detect the running status of the line and the point of occurrence of the hazard source in real time, such as short circuit, power outage, power transmission, grounding, overcurrent and other dangerous sources. When the running status of the line changes, the on-duty personnel and the operation management personnel can be quickly notified. They can quickly make a treatment plan, which can greatly improve the reliability of power supply, ensure the stability of power supply, and improve user satisfaction. Firstly, the hazard source monitoring of the distribution network is carried out. The hazard source monitoring of the distribution network refers to the data collected and transmitted to the host computer by the control center according to the hazard source indicator installed in the distribution network after the occurrence of the hazard source. The actual structure of the distribution network uses the distribution network information and hazard source information to automatically determine the location of the hazard source and reflect the source of the hazard in the network structure and topology map.

Then, based on the graph-based overheating search algorithm, the online monitoring of the hazard source of the power system network is carried out according to the hazard source model of the power system network. It regards the outlet switch of the substation and the switchyard, the segmentation switch of the distribution feeder and the tie switch as the apex, the feeder line segment is regarded as the arc, the load supplied by the feeder is regarded as the arc load, and the current flowing through the switch is regarded as The load at the apex, and the ratio of the load to the rated load, multiplied by 100 is the normalized load, then the hazard source segment is obviously those arcs whose normalized load is much greater than 100. Using the network topology formed by GIS, at the same time, the expert system can initially determine the possible dangerous source nodes based on the user's dangerous source telephone information record database and retrieval knowledge base. The inference engine performs dynamic search and backtracking reasoning based on the dangerous source nodes to form multiple dangerous source sequences, and then determines the dangerous source monitoring area according to the intersection of the key user information and the dangerous source sequence.

3 Experimental Research

In order to detect the on-line monitoring method of power system network hazard sources based on mobile Internet, a comparative experiment is designed.

The parameters of this experiment are shown in Table 1:

Table 1. Experimental parameters

Project	Data	Environmental science
Collecting data	Hazard source data	Software environment: data collection software, data processing software
Data base	On-line monitoring database of hazard sources in power system network	
Data sources	On-line monitoring and management system of hazard sources in power system network	
Operating platform	On-line monitoring and management platform for hazard sources in power system network	Hardware environment: mobile internet hardware system
Port	Bidirectional operating port	
Technical support	Mobile internet technology	
Contrast method	On-line monitoring method of network hazard sources based on feeder automation technology, on-line monitoring method of network hazard sources based on power load and on-line monitoring method of power system network hazard sources based on mobile Internet	Software environment + hardware environment
Evaluation criteria	Monitoring accuracy	
Data source path	Obtain actual parameters	
Experiment flow	On-line monitoring of network hazards in power system	
Operating system	Microsoft Windows XP	

Using the Microsoft Windows XP operating system, the online monitoring and management platform for the hazard source of the power system network is used to conduct online monitoring of the hazard source of the power system network, and the monitoring accuracy is compared. In order to ensure the validity of the experiment, the online hazard source online monitoring method based on feature recognition technology, the online hazard source online monitoring method based on communication message parsing and the online mobile hazard source online monitoring method based on mobile internet are proposed, observe the experimental results.

Using the on-line monitoring method of network hazard source based on feature recognition technology and the on-line monitoring method of network hazard source based on the analysis of communication message. The on-line monitoring method of network dangerous source of power system based on mobile Internet is used to monitor the network dangerous source of electric power system. The accuracy of monitoring is compared as shown in Fig. 3.

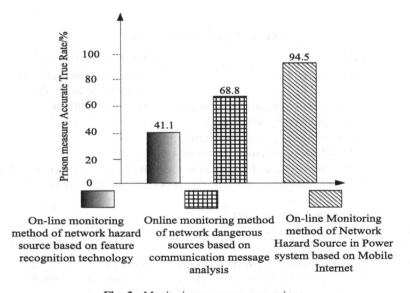

Fig. 3. Monitoring accuracy comparison

As can be seen from Fig. 3, the on-line monitoring accuracy of the on-line monitoring method based on feature recognition technology is 41.1%, that of on-line monitoring method based on communication message analysis is 68.8%, and that of on-line monitoring method based on the analysis of communication message is 68.8%. The accuracy of the on-line monitoring method based on mobile Internet is 94.5%. The comparison shows that the on-line monitoring method based on mobile Internet has the highest accuracy. The performance superiority of this method is proved.

4 Conclusion

The dangerous source of the power system network endangers the security of the power system. Its evolution or regeneration will lead to the large area interruption of the power supply network or the large area paralysis of the main business system. The on-line monitoring method of dangerous sources in power system network based on mobile Internet can realize the high-efficiency monitoring of dangerous sources, which has far-reaching significance for the smooth operation of power system networks. Due to space constraints, there is room for improvement in this study, especially in hazard monitoring, which will be my future research direction.

Acknowledgements. The research is funded by the Guangdong Power Grid Co., Ltd. Information Center Science and Technology Project "Research and Application of Key Technologies for Mobile Application Security Oriented to Three-dimensional Defense" (Project No: 037800K52180001).

References

1. Lin, J., Zhang, J., She, Z., et al.: On-line monitoring of dissolved gases in transformer oil based on BP network. Power Syst. Autom. **25**(28), 156–158 (2017)
2. Anonymous. Intelligent monitoring system for transmission line hazards based on in-depth learning. J. Nantong Univ. (Nat. Sci. Edn), **64**(71), 1418–1453 (2018)
3. Wang, X.: Software design scheme for on-line closure calculation of distribution network based on mobile Internet. Power Syst. Protect. Control **45**(54), 128–133 (2017)
4. Gu, K., Hu, W., Fu, C.: Design and implementation of handheld power data acquisition and analysis device based on mobile network. Power Syst. Protect. Control **46**(48), 110–116 (2018)
5. Zhu, Q., Dangjie, Chen, J., et al.: Power system transient stability assessment method based on deep confidence network. Chin. J. Electr. Eng. **38**(43), 123–123 (2018)
6. Huang, Y.: On-line monitoring method and implementation of EMS data quality based on multi-data source verification. Power Syst. Protect. Control **45**(57), 130–135 (2017)
7. Ma, J., River, W., Wang, X.: Current situation and improvement methods of network security of power monitoring system. Inf. Comput. (Theoret. Edn.) **21**(22), 187–188 (2017)
8. Hu, Z., Liu, J., Z, B., et al.: Environmental risk assessment of transmission line operation based on key risk characteristic quantities. Power Syst. Autom. **41**(48), 160–166 (2017)
9. Liu, S., Glowatz, M., Zappatore, M., Gao, H., Gao, B., Bucciero, A.: E-learning, e-education, and online training, pp. 1–374. Springer, USA
10. Fan, X., Yuan, Z., Guang, Z.: Research on the evaluation method of power communication transmission network operation state based on multi-strategy equilibrium. Power Inf. Commun. Technol. **10**(14), 109–113 (2017)
11. Zheng, P., Shuai, L., Arun, S., Khan, M.: Visual attention feature (VAF): a novel strategy for visual tracking based on cloud platform in intelligent surveillance systems. J. Parallel Distrib. Comput. **120**, 182–194 (2018). https://doi.org/10.1016/j.jpdc.2018.06.012
12. Liu, S., Li, Z., Zhang, Y., et al.: Introduction of key problems in long-distance learning and training. Mob. Netw. Appl. **24**(1), 1–4 (2019). https://doi.org/10.1007/s11036-018-1136-6

An Algorithm of Intelligent Classification For Rotating Mechanical Failure Based on Optimized Support Vector Machine

Yun-sheng Chen[✉]

Guangzhou Huali Science and Technology Vocational College,
Guangzhou 511325, China
pofjha@sina.com

Abstract. The classification algorithm of rotating machinery fault cannot effectively recognize the false components and true components in fault signal of rotating machinery. Therefore, an intelligent classification algorithm of rotating machinery fault based on optimized support vector machine was put forward. The K-L divergence was used to measure the nonlinear and symmetry of probability distribution of two processes in rotating machinery, and the error of information in the process of rotating machinery was measured to eliminate the false component of fault signal of rotating machinery. Meanwhile, the multi-value classification support vector machine algorithm based on decision directed acyclic graph was used to process the signal that only had a true component. Moreover, the value of each node in support vector machine decision function was calculated. Finally, based on calculation results, the fault categories were excluded. Thus, the intelligent classification of rotating machinery fault was completed. According to experimental results, the proposed algorithm can accurately eliminate false components in the rotating machinery fault signal. Meanwhile, the classification result is accurate.

Keywords: Support vector machine · Rotating machinery · Intelligent classification · Fault signal

1 Introduction

With the continuous growth of importance on innovation technology and the rapid development of modern science and technology, the scale of modern production is expanding increasingly and the degree of automation of industrial equipment is increasing. The structure of industrial equipment is more and more complex in the meantime [1]. Based on this background, the basis of modern industrial production is the safe operation of equipment. In the operation, a minor issue will cause casualties, economic losses, and baneful influence to the society and other consequences [2]. The improvement of the degree of automation and the integration level of equipment in modern industry makes the safe operation become the main effort way. With the large scale of modern industrial system and the complexity of industrial equipment, the problem of safety, reliability, maintainability and usability of industrial equipment and machinery has gradually emerged, which promotes the research on diagnosis technology

S. Liu and L. Xia (Eds.): ADHIP 2020, LNICST 348, pp. 146–153, 2021.
https://doi.org/10.1007/978-3-030-67874-6_14

of rotating machinery fault [3]. To classify the fault of rotating machinery is an important step in the intelligent diagnosis technology of rotating machinery fault. Therefore, it is necessary to analyze and study the fault classification algorithm of rotating machinery in depth [4].

Yang et al. have proposed a classification algorithm of rotating machinery fault based on FSVM. When classifying rotating machinery fault feature, this algorithm extracts characteristic index of rotating machinery fault signal and solves the fuzzy membership degree of sample. The classification of rotating machinery fault is completed through fuzzy support vector machine model. This algorithm cannot remove false component in fault signal of rotating machinery [5]. Wang et al. propose a kind of classification algorithm of rotating machinery fault based on KSLPP and RWKNN. This algorithm extracts information in rotating machinery fault feature set through KSSLPP method and uses category information to maximize the degree of separation between fault feature categories in the process of dimension reduction. Then, the feature set is identified by RWKNN to complete the classification of rotating machinery fault. The classification results obtained by the proposed algorithm has error [6]. Chen et al. propose a fault classification algorithm based on point-to-point principal component analysis. This algorithm uses amplitude-frequency characteristic technology to extract feature of rotating machinery fault. Then, the algorithm uses the point-to-point principal component analysis to construct three-layer neural network and classify the fault of rotating machinery. Thus, this algorithm cannot accurately classify the category of rotating mechanical fault [7]. Wang et al. propose a classification algorithm of rotating mechanical fault based on RSCNMF. This algorithm improves the decomposition algorithm of coefficient constraint non-negative matrix and classifies the rotating machinery fault through the fault model of robustness sparsity constraint. The proposed algorithm cannot accurately identify the real and false components existing in fault signal of rotating machinery and cannot accurately eliminate false components [8].

In conclusion, an algorithm of intelligent classification of rotating mechanical fault based on optimized support vector machine is proposed, which solves the problem existing in current algorithm. The specific steps are as follows:

(1) K-L divergence is used to eliminate the false components in the fault signal of rotating machinery and improve the classification accuracy.
(2) The multi-value classification support vector machine algorithm based on decision directed acyclic graph is used to complete the intelligent classification of rotating mechanical fault.
(3) Experimental results and analysis: the effectiveness of proposed algorithm is verified through the elimination of false components and the accuracy of classification results.
(4) Conclusion: summarize the full text and put forward the future direction of development.

2 Elimination of False Components

To eliminate the false components in the fault signal of rotating machinery can improve the accuracy of classification of rotating machinery fault [9]. The K-L divergence is used to measure the nonlinear and symmetry of probability distribution of two processes in rotating machinery. That is to say, similarity and information of two process are measured. Meanwhile, the threshold value is set to judge false components existing in the mechanical fault information, so as to eliminate the false ingredient in fault signal of rotating machinery.

Supposing that the probability density functions of the process $P = (p_1, p_2, \cdots, p_n)$ and the process $Q = (q_1, q_2, \cdots, q_n)$ are represented by $p(x)$ and $q(x)$, $\delta(P, Q)$ denotes the K-L distance from process P to process Q. The formula of $\delta(P, Q)$ is as follows:

$$\delta(P, Q) = \sum_{x \in n} p(x) \log \frac{p(x)}{q(x)} \tag{1}$$

According to the formula (1), the K-L distance $\delta(Q, P)$ from process Q to process P is obtained, and the formula of $\delta(Q, P)$ is as follows:

$$\delta(Q, P) = \sum_{x \in n} q(x) \log \frac{q(x)}{p(x)} \tag{2}$$

The K-L divergence between P and Q through formula (1) and formula (2) is obtained.

$$D = (P, Q) = \delta(P, Q) + \delta(Q, P) \tag{3}$$

The nonparametric estimation is used to calculate probability density, and the formula is as follows:

$$p(x) = \frac{1}{nh} \sum_{i=1}^{n} k \left[\frac{x_i - x}{h} \right] \tag{4}$$

The formula (4) is the density estimation of density function $p(x).x_i$ and x are the parameters in the density function $p(x)$. In the formula, $k(\cdot)$ denotes the kernel function, and h is known as the smooth parameter or the window width. Usually, the Gauss function is used to express the kernel function $k(\cdot)$. The formula is:

$$k(u) = \frac{1}{\sqrt{2\pi}} e^{-2} \tag{5}$$

In formula (5), $k(u)$ is Gauss function. the information difference in two processes of rotating machinery can be measured by K-L divergence. That is to say, the similarity between two processes is measured [10]. The bigger the K-L divergence between two processes, the greater the difference between two processes, and the smaller the similarity. The smaller the K-L divergence between two processes, the smaller the difference between two processes, and the greater the similarity. For the fault signal $x(t)$ of rotating machinery, the specific algorithm is:

The signal $x(t)$ is decomposed to get n components $c_i(t)$ of the signal $x(t)$. Where, $i = 1, 2, \ldots, n$. The probability density function $p_x(t)$ and $p_i(t)$ are calculated by the method of nonparametric estimation [11]. Through formula (1), formula (2) and formula (3), the K-L divergence value $D_i(x, c)$ between $x(t)$ and $c_i(t)$ is calculate. Then, $D_i(x, c)$ is normalized to get K-L divergence value λ_i. The formula of λ_i is:

$$\lambda_i = \sqrt{\frac{D_i^2}{\sum_{i=1}^{n} D_i^2}} \qquad (6)$$

The value of λ_i is $0 < \lambda_i < 1$. The smaller the value of λ_i is, the stronger the correlation between the original signal and component is. The bigger the value of λ_i is, the weaker the correlation between the original signal and component is, which may be a false component [12]. When judging false components, we set threshold value λ, and compare λ with λ_i. When $\lambda_i > \lambda$, it represents that component is the false components in the rotating machinery fault signal, and we need to remove it. When $\lambda_i < \lambda$, this shows that component is the true constituent of signal [13].

3 Support Vector Machine Classification Algorithm Based on Decision Directed Acyclic Graph

The rotating machinery fault after eliminating false components was classified. Then a k class of classification problem of rotating machinery fault was set. The training samples are $(x_1, y_1), (x_2, y_2), \cdots, (x_i, y_i), \cdots, (x_n, y_n)$. In training samples, $i = 1, 2, \cdots, n$, $x_i \in R^n, y_i \in [1, 2, \cdots, k]$. The support vector machine classifier based on decision directed acyclic graph is equivalent to a triple $< F, SVM, ST >$. Where, F is the leaf node of directed acyclic graph, and it is a set including k fault types of normal state. SVM denotes is the set of internal nodes of directed acyclic graph algorithm composed of $k(k - 1)/2$ support vector machines. The formula of SVM is:

$$SVM = (SVM_{12}, SVM_{13}, \cdots, SVM_{pq}, \cdots) \qquad (7)$$

In the formula, SVM_{pq} denotes a sub-classifier, which is used to differentiate the p-th category and q-th category, $p, q \in [1, k], p < q.ST$ denotes the training set, which includes k categories of rotating machinery failure mode. The formula of ST is:

$$ST = (ST_1, ST_2, \cdots, ST_i, \cdots, ST_K) \qquad (8)$$

In the formula, ST_i denotes is the i-th type of composition of sample. All the training samples are consist of $x_i, y_i, \sum n_i = n$.

The classifier SVM_{pq} is used to classify training sample S_{pq} by formula (9):

$$\begin{cases} S_{1k} = All\ the\ training\ samples \\ S_{pq}=\{ST_p, ST_{p+1}, \cdots, ST_q\}, p, q \in [1, k], p < q \end{cases} \quad (9)$$

According to formula (9), the number of training samples gradually decreases with the continuous increase of the number of layers during training.

The support vector machine classifier of directed acyclic graph corresponds to a table operation. All fault categories of rotating machinery are included in the table. After operating a node, we delete a class of rotating mechanical faults in the table. All of possible rotating machinery categories are arranged in a certain order under the initial state [14]. The first category and last category in the table are compared when they are classified. Then, the most impossible category is deleted to reduce the number of categories in the table. That is to say, for the inputting sample of rotating machinery fault category, we start from the root node, then calculate the value of each node by support vector machine decision function based on decision directed acyclic graph [15, 16]. When the value is one more than the original value, we enter the next node from the left. When the value is one less than the original value, we enter the next node from the right. And so on, after the $k - 1$-th calculation, the only category remaining in the table is the category of rotating machinery fault. Thus, the intelligent classification of rotating machinery fault is completed.

4　Experiment and Analysis

In order to prove the overall effectiveness of intelligent classification algorithm of rotating machinery fault based on optimized support vector machine, the intelligent classification algorithm of rotating machinery fault based on optimized support vector machine is tested. The operation system in this test is Windows 7.0. The instantaneous frequency of Chirp signal in different time is different. It is a typical non-stationary signal. The Chirp signal is decomposed into 6 components. As shown in Fig. 1, only the component 2 is the true component.

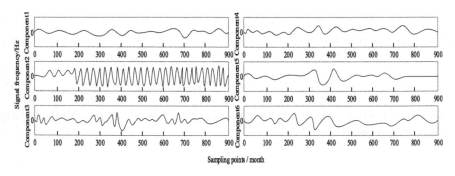

Fig. 1. Components of chirp signal

Many features can be extracted from vibration signals, such as various dimensionless indexes in time domain and amplitude spectrum, power spectrum and Cepstrum in frequency domain. The total amount of different frequency peaks in several frequency bands of vibration signal spectrum is the most commonly used characteristic quantity in rotating machinery fault diagnosis, which is widely used in rotating machinery system fault diagnosis. Extract the total amount of spectrum peaks of different frequencies in 9 frequency bands of vibration signal spectrum, including 0.01–0.39f, 0.4–0.49f, 0.5f, 0.51–0.99f, 1f, 2f, 3–5f, odd multiple F, high-power F, a group of 9 features, forming a multi-dimensional feature vector. The specific division of frequency band is shown in Table 1, where f represents the working frequency of rotor.

Table 1. Frequency band division

Frequency band	1	2	3	4	5
Frequency range	0.01–0.39 f	0.4–.049 f	0.5 f	0.51–0.99 f	1 f
Frequency band	6	7	8	9	
Frequency range	2 f	3–5 f	Odd multiple f	High fold f	

In the experiment, 80 groups of vibration signals of normal, rotor unbalance, rotor misalignment, rotor bending, oil whirl and oil film oscillation were collected respectively, and the data were processed and normalized as training data set.

The intelligent classification algorithm of rotating machinery fault based on optimized support vector machine (algorithm 1) and the classification algorithm of rotating machinery fault based on FSVM (algorithm 2) are used to test the false components in Chirp signal through K-L divergence value. The test results are shown in Table 2.

Table 2. Test results of two methods

K-L divergence value and threshold λ	Component1	Component2	Component3	Component4	Component5	Component6
Algorithm 1	$>\lambda$	$<\lambda$	$>\lambda$	$>\lambda$	$>\lambda$	$>\lambda$
Algorithm 2	$<\lambda$	$<\lambda$	$<\lambda$	$>\lambda$	$>\lambda$	$>\lambda$

From the analysis of the data in Table 2, we can see that using the intelligent classification algorithm of rotating machinery fault based on optimized support vector machine to test, only the K-L divergence of component 2 is less than the given threshold λ. However, by using the classification algorithm of rotating machinery fault based on FSVM, the K-L divergence values of component 1, component 2 and component 3 are less than the given threshold λ. It is known that when K-L divergence value is greater than the threshold value λ, the component is a false component, and when K-L divergence value is less than the threshold value λ, the component is a true component. The false components obtained by the intelligent classification algorithm of rotating machinery fault based on optimized support vector machine are consistent with the fact. To verify the intelligent classification algorithm of rotating machinery fault based on optimized support vector machine can remove the false components in fault signal of rotating machinery.

The intelligent classification algorithm of rotating machinery fault based on optimized support vector machine (algorithm 1) and the classification algorithm of rotating machinery fault based on FSVM (algorithm 2) are used to test. Then, classification results of two methods are compared. The test result is shown in Fig. 2. In the diagram, different shapes represent different types of rotating machinery faults.

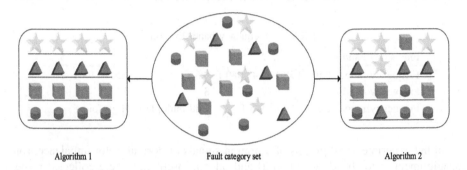

Algorithm 1   Fault category set   Algorithm 2

Fig. 2. Classification results of two algorithms

It can be seen from the analysis in Fig. 2 that different fault types can be distinguished when the rotating machinery fault intelligent classification algorithm based on optimal support vector machine is used for centralized fault classification of rotating machinery fault. So that the classification results are more accurate. The intelligent fault classification algorithm of rotating machinery based on FSVM can not accurately classify the fault types when classifying the concentrated fault types of rotating machinery. Therefore, the classification results are not accurate.

5   Conclusions

To accurately classify rotating machinery fault is the basis of the safe operation of rotating machinery. Using current algorithm to classify the fault categories of rotating machinery has a problem about inaccurate identification of false components and inaccuracy of classification results. An intelligent classification algorithm for rotating machinery fault based on optimal support vector machineis proposed. The first step is to remove false components in rotating machinery fault signal according to K-L divergence. The second step is to use support vector machine classification algorithm based on decision directed acyclic graph for the intelligent classification of rotating machinery fault.

To classify the faults of rotating machinery based on the support vector machine is a new technology, which has attracted the extensive attention of experts and scholars.

The support vector machine has many advantages and some disadvantages. Therefore, it is necessary to research on the following two points in future work.

(1) The selection problem of the kernel function parameters and the kernel function.
(2) Research on the learning method of support vector machine.

References

1. Shom, P.D., Sudarsan, P.: A novel hybrid model using teaching–learning-based optimization and a support vector machine for commodity futures index forecasting. Int. J. Mach. Learn. Cybernet. **9**(1), 97–111 (2018)
2. Zhou, Y.H., Zhang, H.L., Li, F.F., et al.: Local focus support vector machine algorithm. J. Comput. Appl. **38**(4), 945–948 (2018)
3. Björn, J.P., van der Ster Frank, C.B., Tammo, D., et al.: Support vector machine based monitoring of cardio-cerebrovascular reserve during simulated hemorrhage]. Front. Physiol. **8**(8), 1057–106 (2018)
4. Lyndia, C.W., Calvin, K., Jesus, L., et al.: Detection of american football head impacts using biomechanical features and support vector machine classification. Sci. Reports **8**(1), 15–21 (2018)
5. Liu, S., Li, Z., Zhang, Y., et al.: Introduction of key problems in long-distance learning and training. Mobile Networks Appl. **24**(1), 1–4 (2019)
6. Wang, X.D., Zhao, R.Z., Deng, L.F.: Rotating machinery fault diagnosis based on KSLPP and RWKNN. J. Vibration Shock **35**(8), 219–223 (2016)
7. Shuai, L., Weiling, B., Nianyin, Z., et al.: A fast fractal based compression for MRI images. IEEE Access **7**, 62412–62420 (2019)
8. Wang, S.L., Niu, Y.G., Han, P., et al.: RSCNMF algorithm and its application in fault detection of industrial process. Comput. Simul. **34**(8), 386–390 (2017)
9. Zhao, M., Lin, J.: Health assessment of rotating machinery using a rotary encoder. IEEE Trans. Ind. Electron. **12**(99), 1–7 (2017)
10. Sasaki, H., Igarashi, H.: Topology optimization using basis functions for improvement of rotating machine performances. IEEE Trans. Magn. **14**(99), 1–4 (2017)
11. Patel, D., Chothani, N.G., Mistry, K.D., et al.: Design and development of fault classification algorithm based on relevance vector machine for power transformer. IET Electric Power Appl. **12**(4), 557–565 (2018)
12. Fu, W., Liu, S., Srivastava, G.: Optimization of big data scheduling in social networks. Entropy **21**(9), 902 (2019)
13. Han, T., Liu, C., Yang, W., et al.: A novel adversarial learning framework in deep convolutional neural network for intelligent diagnosis of mechanical faults. **165**(1), 474–487 (2019)
14. Ertuncay, D., Costa, G.: An alternative pulse classification algorithm based on multiple wavelet analysis. J. Seismol. **23**(4), 929–942 (2019)
15. Brankovic, A., Falsone, A., Prandini, M., et al.: A feature selection and classification algorithm based on randomized extraction of model populations. IEEE Trans. Cybernet. **48**(4), 1151–1162 (2018)
16. Zheng, P., Shuai, L., Arun, S., Khan, M.: Visual attention feature (VAF): A novel strategy for visual tracking based on cloud platform in intelligent surveillance systems. J. Parallel Distr. Comput. **120**, 182–194 (2018)

Research on Anti-point Source Jamming Method of Airborne Radar Based on Artificial Intelligence

Zong-ang Liu[1,2,3](✉), Jia-guo Lu[4], Zhen Dong[3], and Yu-han Jie[5]

[1] No. 38 Research Institute of CETC, Hefei 230088, China
xingxingekl@163.com
[2] Unit 91550, Dalian 116023, China
[3] National University of Defense Technology, Changsha 410073, China
[4] No. 43 Research Institute of CETC, Hefei 230088, China
[5] Institute of Technology, East China Jiao Tong University,
Nanchang 330045, China

Abstract. Due to the coexistence of multiple electromagnetic interference, the operational performance of radar equipment will be seriously affected. Therefore, it is necessary to study the anti-jamming problem of airborne radar. In view of the problem that airborne radar is easily affected by point source signal interference under the traditional method, an airborne radar anti-jamming method based on artificial intelligence is proposed. The anti-jamming method is designed. Firstly, the airborne radar is detected by frequency shift, and the detected information is analyzed to judge the jamming environment and identify the point source target intelligently. Then the suppression jamming filter is generated based on the analysis of the point source jamming information, and then the suppression jamming signal is output. Finally, the anti-jamming method of airborne radar is obtained. The performance results of the airborne radar anti-point source jamming method are analyzed by simulation experiments. Compared with traditional method, the proposed anti-jamming method can effectively suppress the point source jamming information, the radar signal is clearer and the anti-jamming effect is better. The results verify the effectiveness of the proposed method.

Keywords: Artificial intelligence · Airborne radar · Anti-point source interference · Anti-jamming method

1 Introduction

Airborne radar is the general name of all kinds of radar mounted on aircraft. It is mainly used to control and guide weapons, implement air alert and reconnaissance, and ensure accurate navigation and flight safety. The basic principle and composition of airborne radar are the same as those of other military radars [1]. Its characteristics are as follows: generally, there are antenna platform stabilization system or data stabilization device; generally, the band less than 3 cm is used; it is small in size and light in weight; and it has good seismic performance [2, 3]. This radar device provides target data interception radar for air-to-air missiles, rockets and aerial guns, bombing radar for aiming at bombing

S. Liu and L. Xia (Eds.): ADHIP 2020, LNICST 348, pp. 154–164, 2021.
https://doi.org/10.1007/978-3-030-67874-6_15

surface targets, guiding air-to-surface missiles and providing target information for pilots, air reconnaissance and terrain mapping radar for providing position and topographic data of surface targets, and observation of meteorological conditions, air targets and ground targets. The shape and features ensure accuracy and safety [4]. With the development of electronic technology and information technology, electronic information equipment has played a more and more important role in modern warfare. The mode of modern warfare has changed from simple fire countermeasure to complex electromagnetic countermeasure. Many kinds of electromagnetic interference coexist in the battlefield, which seriously affects the operational performance of radar equipment [5].

At present, many scholars have done a lot of research on anti-jamming of airborne radar. In reference [6], a performance evaluation method of PCL radar based on frequency modulation is proposed. This method studies a PCL radar system based on FM, and attempts to quantify its performance under different jamming waveforms, that is, wideband noise and single tone jamming on carrier. The results show that the effective jamming can be achieved under relatively low jamming power, but the suppression effect is not good in the face of point source interference information. Reference [7] based on the geometric model of synthetic aperture radar (SAR), a fast algorithm for deception jamming in large scenes for different SAR systems is proposed. Firstly, the template deception image is transformed into time domain signal by inverse imaging algorithm. Then the transformed signal is convoluted with the SAR signal received by the enemy to deal with the electronic countermeasure (ECCM) technology. Finally, the jamming signal is transmitted to the enemy's SAR system to achieve the purpose of deception. The experimental results show that the deceptive jammer has the ability to deceive SAR system, but the radar signal obtained is not clear enough.

In view of the above problems, in order to meet the needs of modern warfare and improve the operational performance of weapon system in complex electromagnetic environment, airborne radar adopts advanced active phased array system. With the help of some principles and performances of AI, AI produces an intelligent machine that can respond in a similar way to human intelligence. The research in this field includes robots, language recognition, image recognition, natural language processing and expert systems. Since the birth of artificial intelligence, theory and technology have become increasingly mature, and the field of application has been expanding. Airborne radar is often disturbed by point source signals when it works, which leads to erroneous judgment of radar system and affects decision-making. Therefore, applying the working principle and technical characteristics of AI system to airborne radar system can effectively counter various electromagnetic interference modes and complete detection and tracking of incoming targets in complex electromagnetic interference environment. The normal operation of fire control system can effectively improve the survivability of weapon equipment on the battlefield.

2 Design of Anti-point Source Jamming Method for Airborne Radar

It is an inevitable trend for radar to develop towards cognitive and intelligent. Airborne radar can be regarded as the rudiment of intelligent radar. This kind of radar can work at the weakest frequency of the enemy's jamming power, or force jammers to implement

156   Z. Liu et al.

broadband jamming and reduce the jamming power density, so as to realize the function of anti-jamming. The intelligent anti-point source jamming method of airborne radar should have such a continuous cycle of recognition, determination, processing, re-recognition, re-determination and re-processing. This requires that radar should have several characteristics: Firstly, the comprehensive perception characteristics of jamming environment. Radar anti-jamming system can respond to the change of jamming environment in a specific way without external direct interference and guidance, and update the model base, characteristic parameter base and knowledge base of radar jamming continuously according to its internal state and perceived jamming environment information. Secondly, intelligent interference recognition and classification based on comprehensive features, with intelligent anti-jamming measures, intelligent anti-jamming will have more complex criteria and cognitive channels, can deal with more kinds of interference. Through strategy optimization deduction, anti-jamming strategies that can be applied to many complex scenarios are found, and various factors are parameterized. Computer quantitative analysis is used to solve complex coping strategies. By analyzing and extracting the jamming features in radar channel, the classification of jamming is completed, and corresponding jamming countermeasures are invoked for different types of jamming. The core of the Intelligent Airborne Radar anti-point source jamming method is to automatically identify the jamming type and take anti-jamming measures to complete the jamming countermeasure. Its main system structure is shown in Fig. 1.

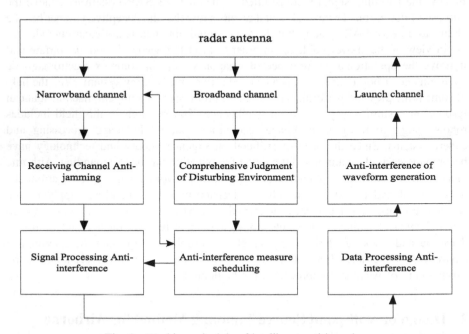

Fig. 1. Working principle of intelligent anti-jamming

From the working principle in the figure, it can be seen that the signals received by radar antennas between airborne radar systems are input into broadband channel and narrowband channel. After a series of analysis and processing of jamming environment, signal processing and anti-jamming measures scheduling, the anti-jamming waveform is finally generated, and returned to the airborne radar antenna through transmission channel, and finally the airborne radar antenna is realized. The ultimate purpose of radar system is to against the point source interference.

2.1   Frequency Shift Detection of Airborne Radar

Because the frequency modulation slope and pulse width of airborne radar transmitting signals at different periods are invariable, but the initial frequency is different, the position of target echo signal and jamming signal in frequency domain is different, and the frequency shift value is different, and the overlap in frequency domain is also different. According to the different overlap degree of target echo signal, jamming signal and matched filter reference function (impulse response function) in frequency domain, the purpose of jamming suppression can be achieved by distinguishing target echo signal, false target and deceptive jamming signal. The radar signal transmitted in the n pulse period of radar can be expressed as:

$$S_n(t) = rect(\frac{t}{T}) \exp\{j[2\pi f_0 t + 2\pi f_n t + \pi \mu t^2 + \varphi_0]\} \tag{1}$$

In the formula, μ is the signal frequency modulation slope, f_0 is the signal bandwidth, φ_0 is the signal carrier frequency, f_n is the initial phase, 6 is the frequency modulation starting frequency of the n pulse cycle signal. In different pulse periods, the initial frequency of FM signal is a set of random sequences known to radar. Since the radar first downconverts the received signal and then processes the signal, it can be assumed that $f_0 = 0$, $\varphi_0 = 0$. The pulse compression of pulse compression radar is generally large, so the spectrum function of the reference signal of matched filter can be approximately expressed as follows:

$$H(f) = \frac{1}{\sqrt{\mu}} rect\left(\frac{f - f_n}{B}\right) \bullet \exp\left[j\left(\frac{\pi(f - f_n)^2}{\mu} - \frac{\pi}{4}\right)\right] \tag{2}$$

In the n pulse period, the jamming equipment generates jamming signal by using the radar signal of the previous m pulse period, and the radar signal of the n pulse period enters the radar receiver at the same time. The received signal can be expressed as:

$$x_n(t) = s_n(t - \tau_n) + j_n(t - \tau_n) \tag{3}$$

After the radar signal is reflected by the target, the spectrum of the target echo signal received by the receiver is $S_n(f)$, and the spectrum of the jamming signal component is derived from the spectrum $J_n(f)$. Considering $f_m < f_n$, the spectrum function of the output pulse pressure signal of the echo signal is the product of the spectrum of the echo signal and the spectrum of the reference function of the matched filter. The interference signal enters the spectrum function of the output signal of the matched filter, and carries on the inverse Fourier transform to obtain the output pulse pressure signal of the interference signal as follows:

$$j_0(t) = \int_{-\infty}^{+\infty} J_0(f) \exp(j2\pi ft) df$$
$$= \frac{B+f_m-f_n}{\mu} \sin c \left[(t - \tau_{jn} - \frac{f_m-f_n}{\mu})(B+f_m-f) \right]$$
$$\cdot \exp\left[j\pi(t - \tau_{jn})(f_m+f) \right] \tag{4}$$

Because the target echo signal coincides with the reference function in frequency domain, the output signal bandwidth is B. Because the interference signal only coincides with the reference function in frequency domain, the output signal bandwidth is only $B+f_m-f_n$. When $f_n - f_m < B$, the output value of the interference signal can be obtained by pulse compression. When the radar received signal is processed by frequency forward shift, the coincidence of target echo signal and reference function in frequency domain will be reduced, while the coincidence of interference signal and reference function in frequency domain will be increased first and then decreased.

2.2 Intelligent Point Source Target Recognition

After the airborne radar frequency shift detection and pulse doppler processing, the target detection is carried out, and the range doppler information of the target is depicted on the doppler plane. For example, there are three point sources jamming in distributed networked airborne radar. Because the doppler of false target is modulated by jammer, the doppler offset received by radar at each station should be equal. In the first step of the algorithm, the doppler information of the target is compared. In order to avoid the tedious steps caused by the combination, only the doppler information is used to sort the target. The doppler information of the false target will be relatively concentrated in a small window and far away from the doppler information of the real target. The second step is to further recognize the jamming by using the distance dimension characteristics of the jamming, and finally determine the jamming target, and then suppress it. It should be pointed out that in the process of these two steps, the target distance and doppler information still need to be retained in the data. In order to facilitate the analysis without losing generality, considering that there is only one real target, and the jammer modulation produces a false target. The target detection algorithm is used to process each radar pulse and doppler. After detection, the doppler and range information of the target can be obtained. Because of the existence of jamming signals, radar can not distinguish between true and false targets before jamming detection, so the radar detects two targets at this time. Then all the doppler information is sorted. Because the doppler shift of the false targets is equal, and affected by noise and doppler resolution, the doppler shift of the processed false targets is approximately equal, then they will be clustered together and have a small interval between each other after sorting [8, 9].

The sorted data is processed by sliding window. The window length is ε and a threshold value is preset. It is related to doppler resolution. The sketch shows that when and only when the initial position of the window is located in the first false target doppler, the number of targets in the window will be N, and in other cases the number of targets in the window is less than N. When the number of targets in the window is N, consider these targets as false targets, and then find out the range doppler coordinates corresponding to this point, and mark them as interference. Due to the existence of multiple radars in distributed radars, with the decrease of signal-to-noise ratio, the doppler information may be deviate from the real value, which may lead to the absence of the number of targets in the window [10–12]. In order to be more consistent with the actual situation, broaden the number of false targets in the window to $n(n < N)$, then n targets falling in the window will be identified as false targets.

After the preliminary judgment of the jamming, in order to increase the accurate distance and speed deception jamming information, the judgment of the distance dimension feature is introduced to judge whether the real target delay is approximately equal, if equal, using the results of the first step, the final judgment is false target. Based on the range-velocity joint deception point source interference suppression method of airborne arrival, target detection and interference suppression are combined. Because of the characteristics of deception jamming, the jamming signal can also obtain processing gain at the receiving end, so that the energy can be accumulated. In the first step, do not distinguish the jamming target, but process the echo signal in two-dimensional range and velocity. After obtaining the range doppler information of the target and the jamming, the jamming is identified preliminarily by using the characteristics of the jamming in doppler, and the number of targets in the discriminant window is adjusted according to the possible errors. The second step, combined with the characteristics of the interference in the distance dimension, further identifies the interference [13–15]. Finally, the target is judged by combining the results of two steps. Using the result of judgment, the radar information of the sub-node which is identified accurately is utilized, the interference is eliminated, the target is retained, and the correct detection of the target is realized.

2.3 Generating Suppression Interference Filtering

The adaptive suppressed jamming beamforming based on the linear constrained minimum variance criterion is to sum the received signals of each airborne radar element by weighting, and minimize the output power of the array under the constrained condition that the signal gain in one direction is constant. The detailed steps of the adaptive suppression jamming beamforming method based on linear constrained minimum variance criterion are as follows: Knowing that the received carrier radar signal sequence $X(t) = \{x_1(t), x_2(t), \ldots x_k(t)\}$ is a N × M dimension matrix (where N is the number of antenna arrays and M is the number of snapshots), the search space of OTH radar is divided into P azimuths. The constraints are set and the covariance matrix of the array is obtained as follows:

$$R = E\{X(t)X^H(t)\} \tag{5}$$

In the formula, E is the unit matrix, and H represents the matrix constraints of airborne radar. In practical calculation, the covariance matrix of the array is estimated to be R by

the finite number of snapshots $X(t)$, and then the adaptive weight vector for suppressing interference is W. SVD decomposition of R is carried out. The first P-1
eigenvalue λ_1, λ_2,...λ_{P-1} and the eigenvector corresponding to the eigenvalue
v_1, v_2,...v_{P-1} are obtained. The weight vector W is projected into the interference
subspace of v_1, v_2,...v_{P-1} signals, and the weight coefficient W_e is obtained. Thus,
when the number of interference sources and sources is known, the output value of
beamforming can be obtained by formula 6 without knowing the direction of interference and sources.

$$Y(t) = W_e^H x(t) \tag{6}$$

2.4   Output Suppression of Interference Signal to Realize Anti-jamming Method

The received signal of airborne radar enters the matched filter. Its essence is to receive
the signal and convolute it with the reference function of the filter. The conversion to
the frequency domain is the multiplication of two frequency functions. Figure 2 is a
frequency domain schematic diagram of matched filter reference signal, target echo
signal and deception jamming signal.

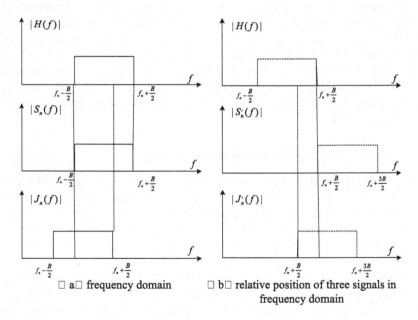

□ a□ frequency domain   □ b□ relative position of three signals in
                                        frequency domain

Fig. 2. Signal frequency domain diagram

Because the interference signal has the same bandwidth as the reference function,
but the starting frequency is different, the overlap part decreases when the two signals
multiply in the frequency domain, which makes the amplitude of the output signal of

the interference signal decrease, while the target echo signal coincides with the reference function in the frequency domain completely, so it has a good energy aggregation characteristic. If the radar received signal is frequency shifted, the frequency of the target echo signal will not coincide with the frequency domain of the reference signal. With the increase of frequency shift, the frequency domain of the two signals will not coincide at all. When the frequency shift is signal bandwidth B, the relative position of the signal in frequency domain is shown in Fig. 2 (b). At this time, the target echo signal and the reference function will not overlap in frequency domain, and the output signal will be very small, while the interference signal is different from the target signal's starting frequency. When the frequency shift is made, it will still overlap with the reference signal in frequency domain. Therefore, the radar received signal can be frequency shifted and the frequency shift value can be obtained $\Delta f = B$. According to the peak position t_j° of the output signal, the peak position $t_j = t_j^\circ - \Delta t$ of the output interference signal of pulse pressure before frequency shift is determined, so that the interference can be suppressed by setting a certain width of time domain. The interference suppression methods are summarized as follows:

Step 1: The radar received signal $x(t)$ is frequency shifted, the frequency shift value is $f_M = B$, and the frequency shift signal $x^\circ(t)$ is obtained.

Step 2: Identifies the interference environment by passing the frequency-shifted signal through a matched filter, determines the point source information to obtain the pulse compression output signal, and determines the time delay t_j° corresponding to the peak value.

Step 3: Pulse compression is applied to the received signal. When the time delay of the output signal is $t_j = t_j^\circ - \Delta t$, the initial frequency of the time domain will be greater than the initial frequency of the target echo signal.

Step 4: Frequency shift of the received signal $x(t)$ of airborne radar is carried out. The value of frequency shift is used to get the signal after frequency shift. Step 2 and Step 3 are repeated to realize the whole airborne radar anti-point source jamming method.

3 Experimental Analysis

The performance of airborne radar anti-point source jamming method is tested and analyzed. Firstly, the simulated complex jamming experimental environment is constructed, which consists of shielding darkroom, turntable, radiation array, complex electromagnetic environment signal simulation system, radio frequency simulation laboratory control system, data recorder, demonstration and verification computer evaluation system, radar display control system and physical radar.

Target simulation and clutter simulation equipment in complex electromagnetic environment signal simulation system receives radar synchronization signal, and simulates target signal and clutter signal interfering with radar point source; electronic support equipment in complex electromagnetic environment signal simulation system detects radar signal by arrays of horns, and transmits it to jamming simulation equipment as samples to simulate various deceptive jamming required. With suppressing interference; In the complex electromagnetic environment signal simulation system, the radar emitter simulator can simulate various radar signals independently to help simulate the realistic electromagnetic environment. At the same time, it can also evaluate the performance of the radar against the co-frequency asynchronous jamming and the performance of the reconnaissance signal when the airborne point source phased array radar starts the EsM function. The three-axis turntable is used to simulate the three-axis motion of the airplane. The received signal passes through the radar receiving channel and enters the radar processor for signal processing. The information of radar searching and tracking target is displayed on the avionics screen. At the same time, the digital data received by the radar is recorded by the recorder for post-analysis and evaluation. The traditional anti-jamming method is used as the experimental contrast group. The experimental environment of the experimental group and the control group is point source jamming environment, and the airborne radar equipment used is the same. The purpose is to ensure the uniqueness of the experimental variables, so that the experimental data obtained is more accurate, and the final experimental analysis conclusion is more valuable. Start up the complex point source jamming environment, using different anti-jamming methods, the radar signal results are shown in Fig. 3.

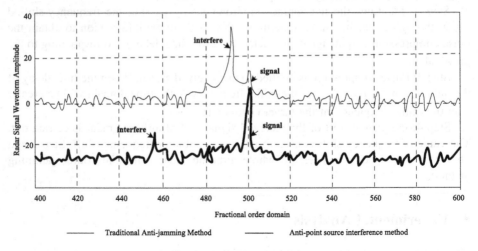

Fig. 3. Experimental results

From the radar waveform of the experimental results, it can be concluded that the waveform of interference points in traditional anti-jamming methods is more obvious than that of radar signals, and the distance between the two waveforms is relatively close. It is easy to mistake the jamming signals as radar signals for recording, resulting in errors in the information obtained. The designed anti-point source jamming method, from the experimental results, effectively suppresses the intensity of the jamming signal, makes the performance of the radar signal clearer, and pulls the jamming information and radar information apart for a certain distance, thus realizing the anti-jamming function more efficiently. This is because in the process of airborne radar anti-jamming design, firstly, the airborne radar is detected by frequency shift, and then the jamming environment is judged and the point source target is identified intelligently. According to the analysis of the point source interference information, the suppression jamming filter is generated, and then the suppression jamming signal is output, so as to effectively suppress the interference information.

4   Conclusion

Faced with the increasingly complex electromagnetic environment and interference, radar anti-jamming technology needs to be developed continuously. Although the radar intelligent anti-jamming technology is still in its infancy, the urgency of the radar anti-jamming situation objectively requires the radar to develop in the direction of intelligence. Radar intelligent anti-jamming technology, as the concrete technology realization of intelligent radar and cognitive radar, will surely get considerable development. In order to overcome the shortcomings of traditional methods, an artificial intelligence-based airborne radar anti-point source jamming method is proposed. The experimental results show that the proposed method has better anti-jamming performance and can effectively solve practical problems.

References

1. Shi, C., Wang, F., Sellathurai, M., et al.: Low probability of intercept-based distributed MIMO radar waveform design against barrage jamming in signal-dependent clutter and colored noise. IET Signal Process. **13**(4), 415–423 (2019)
2. Shen, W., Xu, F., Wu, G.X., et al.: A multi-station angle fusion anti-jamming method. Modern Radar, **22**(1), 47–50 (2018)
3. Tao, Z., Shaoqiang, C., Huayu, F., et al.: Design and processing of a novel chaos-based stepped frequency synthesized wideband radar signal. Sensors **18**(4), 985 (2018)
4. Lee, G.H., Jo, J., Park, C.H.: Jamming prediction for radar signals using machine learning methods. Secur. Commun. Netwk. **2020**(3), 1–9 (2020)
5. Wen, C., Peng, J., Zhou, Y., et al.: Enhanced three-dimensional joint domain localized stap for airborne fda-mimo radar under dense false-target jamming scenario. IEEE Sensors J. **18**(10), 4154–4166 (2018)
6. Paine, S., O'Hagan, D.W., Inggs, M., et al.: Evaluating the performance of fm-based pcl radar in the presence of jamming. IEEE Trans. Aerospace Electronic Syst. **55**(2), 631–643 (2019)
7. Saeedi, J.: A new hybrid method for synthetic aperture radar deceptive jamming. Int. J. Microwave Wireless Technol. **4**(1), 1–14 (2019)
8. Hanbali, S.B.S.: A review of radar signals in terms of Doppler tolerance, time-sidelobe level, and immunity against jamming. Int. J. Microwave Wireless Technol. **10**(10), 1–9 (2018)
9. Liu, S., Li, Z., Zhang, Y., et al.: Introduction of key problems in long-distance learning and training. Mobile Netwk. Appl. **24**(1), 1–4 (2019)
10. Liu, S., Glowatz, M., Zappatore, M., et al.: e-Learning, e-Education, and Online Training. Springer International Publishing, Berlin (2018)
11. Fu, W., Liu, S., Srivastava, G.: Optimization of big data scheduling in social networks. Entropy **21**(9), 902 (2019)

12. Shi, C.G., Wang, F., Salous, S. et al.: Adaptive jamming waveform design for distributed multiple-radar architectures based on low probability of intercept. Radio Sci. **54**(1–2), 72–90 (2019)
13. Enzheng, Z., Benyong, C., Hao, Z., et al.: Laser heterodyne interference signal processing method based on phase shift of reference signal. Opt. Express **26**(7), 8656 (2018)
14. Lu, M., Liu, S.: Nucleosome positioning based on generalized relative entropy. Soft. Comput. **23**(19), 9175–9188 (2018). https://doi.org/10.1007/s00500-018-3602-2
15. Joshi, H.D., Kaur, R., Singh, A.K., et al.: An improved method for deceptive jamming against synthetic aperture radar. Int. J. Microwave Wireless Technol. **10**(1), 115–121 (2018)

Statistical Analysis of Catalytic Removal of Soot Particles Based on Big Data

Xiu-hong Meng, Ping Yang, Hui-bo Qin, and Lin-hai Duan[✉]

Guangdong University of Petrochemical Technology, Maoming 525000, China
xin1104100@163.com

Abstract. Different temperature, power, flow rate and other factors have different effects on the removal of soot particles in the tail gas of simulated diesel vehicles, and the removal effect of each kind of soot particle catalytic removal method is also different. In order to further improve the effect of soot particle catalytic removal, a statistical analysis method of soot particle catalytic removal method based on big data is designed. Using large data technology to extract catalytic removal methods of soot particles, detailed analysis of each method was carried out, and the soot combustion performance of soot particles catalytic removal method was compared. The results showed that the removal of soot particles based on perovskite catalyst was more effective than that of soot particle removal method based on sol-gel preparation method, and that soot particles were catalyzed by low temperature plasma. The combustion performance of the removal method is better, and the catalytic removal performance is more superior.

Keywords: Big data · Soot particles · Catalytic removal · Sol-gel preparation · Low temperature · Plasma · Perovskite type catalyst

1 Introduction

With the rapid development of industrialization and urbanization, the situation of regional air pollution is increasingly serious. At present, a large number of emissions caused by the substantial increase of vehicle ownership, destroy the ozone layer of the atmosphere, cause the global greenhouse effect, make the whole earth temperature warm and climate abnormal [1]. With the deepening and recognition of the concept of "low carbon life", energy saving and emission reduction of motor vehicles has been put on the agenda. Therefore, driven by the policy of energy conservation and emission reduction, diesel engine has been more and more widely used for its advantages of large power, high efficiency, good fuel economy, strong adaptability and wide power range. At present, almost all heavy vehicles have used diesel engines, and light vehicles have gradually used diesel engines. As the diesel vehicle is not pollution-free and zero emission, the exhaust emissions of diesel vehicles have caused serious environmental problems. In order to control the rapid development of diesel vehicles and continue to cause serious air pollution, countries around the world have developed relevant vehicle exhaust emission regulations, so as to urge vehicle manufacturers to improve their

S. Liu and L. Xia (Eds.): ADHIP 2020, LNICST 348, pp. 165–179, 2021.
https://doi.org/10.1007/978-3-030-67874-6_16

products from the source to reduce the serious environmental pollution caused by the emissions of CO, HC, NOx and soot particles [2].

At present, there are many catalytic removal methods for soot particles, each method has different catalytic removal effect. In order to further improve the catalytic removal effect of soot particles, a method of catalytic removal of soot particles based on big data is proposed. The method of catalytic removal of soot particles is extracted using big data technology. Each method is analyzed in detail and the catalytic removal of soot particles is compared. The flue gas combustion performance of the method provides a basis for the catalytic removal of soot particles. Big data refers to the data collection that can not be grabbed, managed and processed by conventional software tools in a certain period of time. Big data has five characteristics, that is, large amount, high speed, diversity, low value density and authenticity. It has no statistical sampling method, but only observes and tracks what happened. The use of big data tends to predict analysis, user behavior analysis or the use of some other advanced data analysis methods, so it is applied to the catalytic removal of soot particles.

2   Statistical Analysis Framework of Catalytic Removal of Soot Particles Based on Big Data

Big data is a large-scale data collection, and data analysts cannot extract, process, analyze, and manage it with general software within a certain period of time.The key technologies for processing big data include parallel processing technology for large-scale data sets, distributed databases, distributed file storage and processing systems, data mining, and cloud computing. Big data processing tasks cannot be processed in a single-machine serial computing mode. You must use a distributed architecture for calculation. Read the original data from HDFS and send it to the Spark cluster for efficient distributed parallel calculation. After analysis, write the result set to persistent Into the HBase cluster. In spark cluster, it is also divided into Worker nodes that are used to directly participate in calculation, interact with input and output interfaces, and Master nodes that are used for scheduling, task allocation and management of Worker nodes. The statistical process of catalytic removal of soot particles based on big data is shown in Fig. 1:

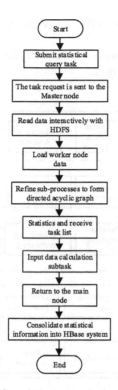

Fig. 1. Statistical flowchart of catalytic removal of soot particles based on big data

Step1: When the statistical query task of the soot particle catalytic removal method is submitted from the user client, the task request is sent to the Master node by the client, The Master node analyzes the size and distribution of the data collection. Because the data comes from HDFS, and HDFS has natural data segmentation (64 MB size per block area), the Master only needs to interact with the HDFS Master node to get the location of each block of data, then assign appropriate reading strategies according to the data location. These strategies include which data of each HDFS Slave node that each Worker node reads from;

Step2: In the process of loading the data of the Worker node, the Master node further refines the task into smaller sub-processes according to the query statement. These sub-processes have data dependencies, forming a directed acyclic graph, and The sub-processes are descriptions that the Worker node can directly execute within it;

Step3: After loading the statistics of the soot particle catalytic removal method and receiving the task list, the Worker node can start calculation based on the input data of the task and the task. After each subtask is calculated, it will be returned to the Master node A completed message is convenient for the Master node to track the progress of task processing in real time;

Step4: After each Worker node completes the calculation separately, according to the user's options when submitting the task, you can choose to schedule and merge the statistics of the soot particle catalytic removal method through the Master's scheduling, and then write it together into the HBase system, or Each Worker directly writes its own output data to HBase.

The statistical analysis framework for the catalytic removal method of soot particles based on big data is shown in Fig. 2:

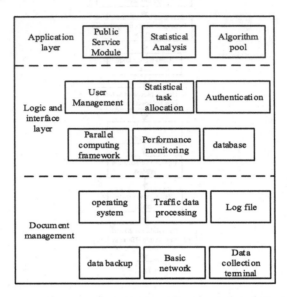

Fig. 2. Statistical framework of catalytic removal of soot particles based on big data

Through the above process, the relevant data about the catalytic removal method of soot particles are extracted, and then the data mining technology in big data technology is used to rank the relevant technologies, and the most commonly used removal methods are calculated. The expression is as follows:

$$f = \frac{g}{j} + \frac{x}{b} \tag{1}$$

In formula (1), g represents statistical data, j represents the definition of statistical indicators, b represents behavioral influencing factors, x represents the dimension of statistical indicators.

Through analysis, the most commonly used removal methods are the following three, the soot particle catalytic removal method based on the sol-gel preparation method, the soot particle catalytic removal method based on low temperature plasma, and the perovskite type catalyst Soot particle removal method.

3 Analysis of Catalytic Removal of Soot Particles

3.1 Catalytic Removal of Soot Particles Based on Sol-Gel Process

The soot particle catalytic removal method based on the sol-gel preparation method is mainly through alumina produced by Wenzhou Alumina Plant, zirconia produced by Sinopharm Group Chemical Reagent Co., Ltd., zirconia produced by Qingdao Ocean Chemical Group Company, Qingdao Ocean Chemical Silicon oxide produced by the group company, manganese acetate produced by Aladdin Reagent Co., Ltd., ammonia water produced by Sinopharm Group Chemical Reagent Co., Ltd., and ethanol produced by Sinopharm Group Chemical Reagent Co., Ltd. as the main experimental reagents [3]. Adopt gas chromatograph produced by Zhejiang Kexiao Instrument Co., Ltd., plasma power supply produced by Nanjing Suman Electronics Co., Ltd., dual dielectric discharge reaction tube, electronic balance produced by Shanghai Mingqiao Precision Scientific Instrument Co., Ltd. The temperature controllers produced by Beijing Huibolong Instrument Co., Ltd., mass flow meters produced by Shanghai Yifeng Electric Furnace Co., Ltd., muffle furnaces produced by Shanghai Yifeng Electric Furnace Co., Ltd. and Shangpu Compact Equipment are the main catalyst making instruments. Mainly used to prepare potassium titanate series catalysts with different K/Ti mass ratios, the precursors of the catalysts $Ti(OC_2H_5)_4$ and CH_3COOH, and the precipitant is H_2O [4]. Dissolve a certain amount of $Ti(OC_2H_5)_4$ in absolute ethanol, and add a certain amount of ethylene glycol to it as a plasticizer, and continue to stir at room temperature to obtain solution A; Then weigh a certain amount with an electronic balance Potassium acetate was placed on a magnetic stirrer and stirred well until potassium acetate was completely dissolved and dissolved in absolute ethanol. A certain amount of glacial acetic acid and a small amount of deionized water were added and stirred to obtain solution B. The solution at this time was a clear solution. Under constant stirring conditions, the solution B was added dropwise to the solution A, and the PH value of the solution was controlled to about 6 during the preparation process; stirring was continued for a while to obtain a sol-like substance, and the gel was obtained by standing. Then it is aged at room temperature for a period of time, placed in an electric blast drying oven at 100 °C for drying, then ground, and placed in a muffle furnace at 800 °C for roasting, and finally obtained $K_2Ti_2O_5$, $K_2Ti_4O_9$, $K_2Ti_6O_{13}$, $K_3Ti_8O_{16}$ and $K_4Ti_3O_8$ five potassium titanate catalysts [5]. Potassium titanate is a new type of high-performance material enhancer. It exists in the form of whiskers. This whisker is a new type of inorganic polymer compound with micro needle-shaped short fibers. Its chemical formula is $K_2O \cdot nTiO_2$ or $K_2Ti_2O_{2n+1}$. When n = 2,4,6,8, it is called potassium dititanate, potassium tetratitanate, potassium hexatitanate and potassium octatitanate respectively [6].

Potassium titanate belongs to monoclinic system and C2/m point group. Ti the crystal structure of potassium dititanate with Ti coordination number of 5 is a chain-like layered structure formed by connecting the triangular double cones of TiO_6 through a common vertex. The layers are separated by 6.5 angstroms. Parallel, and K^+ is located between the layers of the layer structure, so it has chemical activity. Similarly, in the crystal structure of potassium tetra titanate whose coordination number of Ti is 6, it is a lamellar structure formed by the connection of the TiO_6 octahedron through common edges and common apical angles. The layers are also parallel to the crystal axis, with a spacing of 8.5 angstrom. K^+ is located between layers, so it also has chemical activity. The K^+ ions in the middle layer of the structure of potassium dititanate and potassium tetra titanate crystals can be exchanged with other cations to replace the heavy metal ions in the wastewater, so they can be applied to the treatment of metal ions. In the crystal structure of potassium octatitanate different from the first two, the coordination number of Ti is 6, which is a tunnel structure formed by octahedron through common edges and common angles, rather than a layered structure. Among them, K^+ is in the middle of the tunnel, separated from the environment, and the K^+ in potassium octotitanate crystal has almost no chemical activity. Figure 3 is a schematic diagram of the cell structure of $K_2Ti_2O_5$ established by molecular simulation. The combination of the red and white spheres represents the triangularis of TiO_6, and the purple spheres represent potassium ions. It can be seen that they are connected by common vertices. TiO structure is uniformly distributed at different levels, with large space gaps and obvious regularity. However, potassium ions are distributed between the levels of TiO structure, which provides excellent conditions for the good chemical activity of potassium ions.

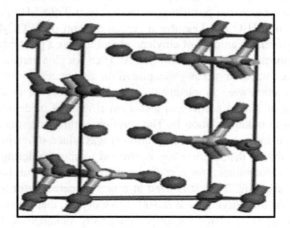

Fig. 3. Crystal cell structure of $K_2Ti_2O_5$

In addition, the following means of characterization were used for the removal effect of characterization:

First, XRD test was performed to determine the composition of the catalytic active agent by Rigaku D-max-rA Xray diffractometer. Experimental conditions: at room temperature, Cu target Kα beam, graphite monochromator, X-ray wavelength = 0.1548 nm, step length: 0.02, tube pressure: 40 kV, tube flow: 100 mA, scanning range: $10° \sim 90°$, scanning rate: $10°·min^{-1}$.

Second: BET test was conducted on Micromeritics ASAP2000. The sample was pretreated at 623 K and adsorbed in a liquid nitrogen environment. The cross-sectional area of nitrogen molecule was set at 0.162 m^2. The specific surface area was obtained by the BET equation and the nitrogen adsorption isotherm.

Thirdly, SEM test was performed to observe the morphology and size of potassium titanate catalyst particles prepared by scanning the electroscope of JSM-6700 from Japan Jeol co., LTD.

Fourth: thermogravimetric (TG) determination. The combustion temperature of the catalyst for the catalytic oxidation of carbon smoke particles is determined on the thermogravimetric analyzer. The tga is a type TAC/DX thermal analyzer manufactured by the United States company. First, a certain amount of catalyst samples (about 10 mg) were taken and pretreated at 300 °C for 60 min in N2 atmosphere. Then, after the temperature dropped to 100 °C, the gas passing through the catalyst sample was switched to a mixture of $20\%O_2/N_2$ and heated up to 800 °C at a heating rate of 10 °C/min.

Fifth, the morphology of the catalyst before and after the reaction was observed by transmission electron microscopy (TEM). The instrument was HT-7700 high-resolution electron microscopy (ht-7700) of Hitachi high-tech co., LTD. Before the test, the sample powder was dispersed evenly by ultrasound in anhydrous ethanol, and then a small amount of suspension was taken and transferred to the carbon film made of copper mesh. After being irradiated to complete drying by infrared lamp, the sample powder could be put into the instrument for testing. The soot combustion curve after catalytic removal of soot particles prepared based on sol-gel is shown in Fig. 4:

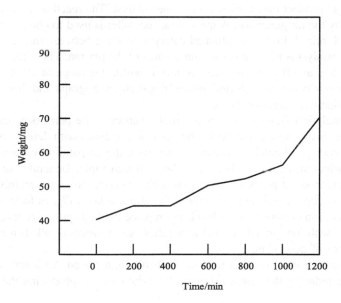

Fig. 4. Combustion curve of soot after catalytic removal of soot particles prepared by sol-gel process

The starting temperature and maximum oxidation rate temperature of different catalyst contents are shown in Table 1:

Table 1. Starting temperature and maximum oxidation rate temperature of catalyst

Catalyst content/%	The light-off temperature/°C	Maximum combustion temperature/°C
20	276	413
40	279	420
60	295	450
80	270	368
100	480	580

3.2 Catalytic Removal of Soot Particles Based on Low Temperature Plasma

The plasma-enhanced catalysis process is also known as the "two-stage process", in which the plasma and the catalyst are placed separately and the catalytic reactor is located behind the plasma reactor. In the reaction system, the reaction process is mainly divided into two steps: first, the reaction gas is activated in the plasma reactor, which can produce relatively more reactive intermediate products and or produce odor; Then these intermediates leave the plasma reactor and enter the catalytic reactor to produce the final target product under the action of the catalyst. The reaction atmosphere is the air purified by the air generator. A mass flow controller is used to control the gas flow rate of 30 mL.min^{-1}. In the coordinated catalytic reaction between the plasma and the catalyst, the catalyst is filled in the discharge area of the plasma. The filling mass of the catalyst is 0.5 g, and the reaction temperature is within the range of 20 °C $\sim$ 200 °C. The reaction products were detected online by gas chromatograph and then detected by thermal conductivity detector TCD.

In the catalytic oxidation of carbon smoke particles, the perovskite catalyst with ideal structure is not very active, while the perovskite catalyst with defect structure after distortion shows good catalytic activity. Therefore, the preparation of perovskite catalyst with defect structure has become a hot research topic. Generally speaking, the preparation methods of perovskite catalysts with defective structures mainly include A and B site ion doping and non-stoichiometric ratio synthesis [7], as follows:

a. Conduct ion doping at the a level and replace the high-priced A level ions (rare earth metals) with low-priced A level ions (alkali earth metals or alkali metals) doped into the lattice of perovskite.

b. Replace the low-priced A-bit ions with the high-priced A-bit ions at the a-bit ions, thereby reducing the valence state of the A-bit ions and producing the low-priced b-bit ions;

c. Carry out ion doping at the A-bit, and replace the high-priced A-bit ions with low-priced b-bit ions.

The following device was used to evaluate the performance of the catalyst. The first part is a catalytic reaction unit with a quartz tube reactor [8, 9] and a heating furnace. The second part is the exhaust gas detection device, namely gas chromatograph. The third part is the computer data output device.

The co-catalytic removal of soot particles by plasma was performed on the following plasma-driven catalytic devices.

The dielectric barrier plasma discharge was adopted, and the quartz tube with an internal diameter of 8.0 mm was used as the reactor and the barrier medium. A 3.0 mm diameter unembroidered steel rod is used as the high voltage electrode. At the same time, the stainless steel wire mesh wound on the surface of the quartz tube is used as the grounding electrode. The reaction device in the experiment is controlled by a programmed temperature rise controller in an open-type tubular resistance furnace. The high voltage power supply device is used to control the discharge voltage of the plasma.

The low temperature plasma discharge is dielectric barrier discharge. The catalyst is placed within the low-temperature plasma discharge region, that is, the combination mode of plasma and catalyst is plasma-driven catalytic reaction mode. Low-temperature plasma discharge power supply is high frequency and high voltage power supply (CTP-2000 K, nanjing suman electronics co., LTD.), which can provide 0–30 kV,9–16 kHz sinusoidal ac voltage. The discharge power of dielectric barrier discharge plasma is an important parameter affecting plasma discharge. Due to the phase imbalance between the current and voltage of dielectric barrier discharge plasma, its power calculation and measurement are quite complicated. The power of dielectric barrier discharge can be determined accurately by analyzing the power of dielectric barrier discharge by using the pattern of discharge voltage and charge li sayuan in plasma.

The discharge voltage and current waveform and discharge frequency in the process of low-temperature plasma discharge were measured by oscilloscope [10, 11]. The discharge power of the plasma was measured by the lisaru method:

$$p(w) = \frac{s(c)}{j} \qquad (2)$$

In formula (2), w represents the total charge on the capacitor, which is connected to the grounding electrode; s Represents the voltage at both ends of the capacitor; c Represents the discharge voltage, j represents the discharge frequency.

After removal of soot particles by catalytic removal method based on low-temperature plasma, the soot combustion curve is shown in Fig. 5:

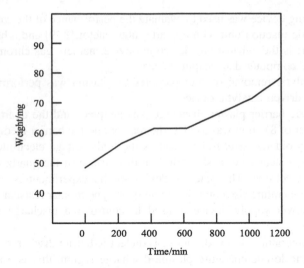

Fig. 5. Combustion curve of soot particles after catalytic removal based on low temperature plasma

The starting temperature and maximum oxidation rate temperature of different catalyst contents are shown in Table 2:

Table 2. Starting temperature and maximum oxidation rate temperature of catalyst

Catalyst content/%	The light-off temperature/°C	Maximum combustion temperature/°C
20	125	300
40	145	320
60	295	200
80	100	250
100	120	320

3.3   Removal of Soot Particles Based on Perovskite Catalyst

The catalytic activity of the catalyst was studied in a TPO unit, and the reaction product was analyzed online by gas chromatograph. The reaction device mainly consists of three parts: the catalytic reactor of the programmed temperature control device, the gas chromatograph (SP-3400, Beijing beifenruili co., LTD.) and the computer for recording data [12, 13]. The schematic diagram of the reaction device is shown in Fig. 6:

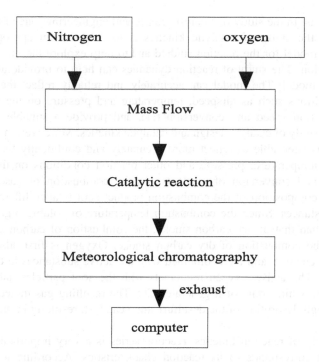

Fig. 6. Schematic diagram of temperature programmed reaction device

The performance of perovskite catalyst was tested by the simulated tail gas evaluation method. The specific process is as follows:

First of all, the contact mode between the soot and the catalyst is an important factor affecting the catalytic performance, and the catalyst always exhibits better activity under tight contact conditions than under loose contact conditions. But in practice, the contact between the carbon smoke and the catalyst in the tail gas device is loose. In order to comply with the actual conditions, the carbon smoke and the catalyst were directly mixed and ground in this experiment, so that the carbon smoke could be attached to the catalyst and fully contacted.

Second, in the process of the catalytic reaction, at a rate of about 1 L/min to device zhongtong into the air to ensure there is plenty of oxygen, the heating rate of 5 °C/min^{-1}, using the mass flowmeter control O_2 volume fraction (10%) and N_2 (gas) as the balance of flow rate of 200 ml/min, using the mixed gas to simulate the rich oxygen condition of diesel exhaust, the purpose is to test whether potassium titanate catalysts system has low temperature catalytic oxidation of carbon smoke particles combustion of diesel exhaust good activity. After the reaction, the gas was detected online by SP-3400 gas chromatograph, and the catalytic activity of the catalyst was evaluated according to the data.

Chemical kinetics is a branch of science that studies the reaction rate and reaction mechanism of the transformation of one chemical substance into another. However, the study of catalytic kinetics has become one of the most important components in the field of catalysis. In the study of catalyst science and engineering, one of the important objectives of the study of catalytic kinetics is to establish an appropriate kinetic mathematical model for the reaction studied and to help explore the mechanism of its catalytic reaction. The study of reaction dynamics can help to provide an appropriate mathematical model. The model can accurately and reliably reflect the influence of reaction conditions such as airspeed, temperature and pressure on the selectivity of reactants, reaction speed and conversion rate, and provide a suitable mathematical model for the study of catalyst design and catalytic kinetics. Moreover, in a large range, this model has been able to reflect more accurately and confidently the influence of space speed, temperature, pressure and other reaction conditions on the selectivity, reaction rate and conversion of reactants. As the combustion of gasoline is more complex, the composition of the combustion product soot is also different under different circumstances. Since the combustion temperature of soluble organic matter is much lower than that of dry carbon smoke, the combustion of carbon smoke can be regarded as the combustion of dry carbon smoke. Oxygen is first adsorbed on the surface of the catalyst, and then combined with the active substances in the catalyst to form [−MO]. The active oxygen reinteracts with the soot particles, and some of it generates CO_2, while some of it generates CO. The resulting gas molecules are then released through desorption of the desulfurization catalyst, resulting in the reduction of the catalyst.

In the study of reaction kinetics, reaction series is a very important factor. Each reaction level corresponds to its reaction characteristics. According to the reaction series, the reaction rate equation can be known. Based on the first order, second order and third order differential diagrams of carbon smoke oxidation, the reaction progression of carbon smoke combustion is determined in this paper.

Figure 7 shows the soot combustion curve after removal of carbonaceous smoke particles by the removal method based on perovskite catalyst:

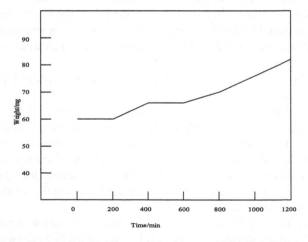

Fig. 7. Combustion curve of soot after removal of soot particles based on perovskite catalyst

The starting temperature and maximum oxidation rate temperature of different perovskite catalysts are shown in Table 3:

Table 3. Starting temperature and maximum oxidation rate temperature of catalyst

Catalyst content/%	The light-off temperature/°C	Maximum combustion temperature/°C
20	300	600
40	350	650
60	400	700
80	450	650
100	500	600

4 Discussion of Results

Using the programmed heating and oxidation device as the potassium titanate catalytic soot reaction device and the meteorological chromatograph, the post-reaction gas was detected online, and the experimental results were systematically analyzed and studied. Satisfactory results were obtained (Table 4):

Table 4. Experimental results

Different methods	Method for catalytic removal of soot particles based on sol-gel method	Low-temperature plasma-based catalytic removal method of soot particles	Method for removing soot particles based on perovskite catalyst
Burning time/min	1200	1200	1200
Burning weight/mg	70	78	82
Catalyst content/%	100	100	100
The light-off temperature/°C	480	120	500
Maximum combustion temperature/°C	580	320	600

First: when the burning time reaches 1200 min, based on the sol-gel preparation methods of carbon smoke particle catalytic removal method, based on the low temperature plasma carbon smoke particle catalytic removal method, based on perovskite catalyst of carbon smoke particle removal method after removing carbon burning smoke gradually rising trend curve. Among them, the combustion weight of the soot particle catalytic removal method based on the sol-gel method is 70 mg, the combustion weight of the soot particle catalytic removal method based on low-temperature plasma is 78 mg, and the soot particle catalytic removal based on the perovskite catalyst the burning weight of the method is 82 mg. After comparing, the perovskite catalyst based carbon smoke particle removal method of removal effect is best.

Second: when the catalyst content reaches 100%, the starting temperature of the method for catalytic removal of soot particles based on the sol-gel method is 480, and the maximum combustion temperature is 580, respectively. The start of the method for catalytic removal of soot particles based on low-temperature plasma the temperature is 120, the maximum combustion temperature is 320, the starting temperature of the perovskite catalyst-based soot particle removal method is 500, and the maximum combustion temperature is 600. It can be seen that, although the removal method of carbon smoke particles based on perovskite catalyst has the best removal effect, it requires higher starting temperature and maximum combustion temperature. The method of catalytic removal of carbon smoke particles based on low temperature plasma has the lowest requirements for the starting temperature and the maximum combustion temperature.

Thirdly, it is found that the potassium titanate catalyst is not easy to absorb water and has good stability and anti-aging performance. Moreover, the catalytic performance of carbon smoke is still good under loose contact. It is also found that a certain amount of NO_x catalysis is beneficial to the oxidation of carbon smoke.

Fourth, potassium titanate catalyst can still effectively catalyze the combustion of carbon smoke particles at low temperature even when it is in loose contact with carbon smoke, and has good catalytic activity, which fully proves that potassium titanate catalyst is an excellent catalyst for the oxidation of carbon smoke.

Fifthly, the mass ratio between catalyst and soot particles did not significantly change the catalytic oxidation of soot by potassium titanate through the experiment, which fully verified the good catalytic performance of the removal method of soot particles based on perovskite catalyst.

5 Conclusion

This paper presents a statistical analysis method based on big data for the catalytic removal of soot particles. Using big data technology, extract the catalytic removal method of soot particles, and compare and analyze the flue gas combustion performance of catalytic removal of soot particles according to different methods. The results show that the perovskite-type catalyst is more effective in removing soot particles than the sol-gel preparation method, and the low-temperature plasma catalyzed soot particle catalytic removal method has better combustion performance and better catalytic removal performance.

References

1. Li, Z., Qiu, L., Cheng, X., et al.: The evolution of soot morphology and nanostructure in laminar diffusion flame of surrogate fuels for diesel. Fuel **211**(1), 517–528 (2018)
2. Zhao, F., Yang, W., Yu, W., et al.: Numerical study of soot particles from low temperature combustion of engine fueled with diesel fuel and unsaturation biodiesel fuels. Appl. Energy **211**(1), 187–193 (2018)
3. Liu, S., Bai, W., Liu, G., et al.: Parallel fractal compression method for big video data. Complexity **2018**, 2016976 (2018)
4. Joo, P.H., Gigone, B., Griffin, E.A., et al.: Soot primary particle size dependence on combustion pressure in laminar ethylene diffusion flames. Fuel **220**(15), 464–470 (2018)
5. Abian, M., Martin, C., Nogueras, P., et al.: Interaction of diesel engine soot with NO2 and O-2 at diesel exhaust conditions. Effect of fuel and engine operation model. Fuel **212**(15), 455–461 (2018)
6. Shen, J., Feng, X., Liu, R., et al.: Tuning SnO 2 surface with CuO for soot particulate combustion: The effect of monolayer dispersion capacity on reaction performance[J]. Chinese J. Catalysis **40**(6), 905–916 (2019)
7. Alcan, G., Yilmaz, E., Unel, M., et al.: Estimating soot emission in diesel engines using gated recurrent unit networks. IFAC Proc. Vol. **52**(3), 544–549 (2019)
8. Liu, S., Li, Z., Zhang, Y., et al.: Introduction of key problems in long-distance learning and training. Mobile Netwk. Appl. **24**(1), 1–4 (2019)
9. Jerez, A., Cruz Villanueva, J.J., Figueira, D.S.L.F., et al.: Measurements and modeling of PAH soot precursors in coflow ethylene/air laminar diffusion flames. Fuel **236**(15), 452–460 (2019)
10. Duvvuri, P.P., Sukumaran, S., Shrivastava, R.K., et al.: Modeling soot particle size distribution in diesel engines. Fuel **243**(1), 70–78 (2019)
11. Shuai, L., Weiling, B., Nianyin, Z., et al.: A fast fractal based compression for MRI images. IEEE Access **7**, 62412–62420 (2019)
12. Zhang, R., Pham, P.X., Kook, S., et al.: Influence of biodiesel carbon chain length on in-cylinder soot processes in a small bore optical diesel engine. Fuel **235**(1), 1184–1194 (2019)
13. Liu, S., Glowatz, M., Zappatore, M., et al.: e-Learning, e-Education, and Online Training. Springer International Publishing, Berlin (2018)

Research on Electric Drive Control Method Based on Parallel Computing

Lin-ze Gao[(⊠)]

Guilin University of Electronic Technology, Guilin 536000, China
gaolz20@126.com

Abstract. In order to solve the problems of low control accuracy and poor stability of traditional electric drive control methods, a parallel computing based electric drive control method is proposed. According to the selected motor parameters, the transmission point with the maximum power or the strongest mechanical rigidity is selected as the main node, so that the parameters are equal to the load rate of the load distribution motor, and the load distribution is completed by parallel calculation; the air gap magnetic field is generated inside the motor, and the electromagnetic thrust is generated under the interaction with the excitation magnetic field generated by the permanent magnet by using the concepts of coordinate transformation and space vector Force is used to push the motor to move synchronously and linearly at the same speed to complete the motor vector control and the electrical drive control based on parallel computing. The experimental results show that the control accuracy of the proposed method is high and the operation is stable. It can effectively reduce the position tracking error and improve the control accuracy.

Keywords: Parallel computing · Electric drive · Control method · Motor parameters · Load rate · Air gap magnetic field

1 Introduction

Parallel computing is also known as parallel computing, as opposed to serial computing. It is an algorithm that can execute multiple instructions at a time. Its purpose is to improve the computing speed, and to solve large and complex computing problems by expanding the problem solving scale [1]. Parallel computing refers to the process of using multiple computing resources to solve computing problems at the same time, and is an effective means to improve the computing speed and processing power of computer systems. Its basic idea is to solve the same problem with multiple processors, the problem to be solved is decomposed into several parts, each part is calculated by an independent processor in parallel [2–4]. A parallel computing system can be a specially designed supercomputer containing multiple processors, or it can be a cluster of several independent computers interconnected in a certain way. Complete the data processing through the parallel computing cluster, and then return the processing results to the user.

Parallel computing can be divided into time parallel and space parallel. Time parallel refers to assembly line technology. For example, the steps of food production in the factory are divided into: cleaning: washing the food. Disinfection: disinfect the

S. Liu and L. Xia (Eds.): ADHIP 2020, LNICST 348, pp. 180–191, 2021.
https://doi.org/10.1007/978-3-030-67874-6_17

food. Cutting: Cut food into small pieces. Packaging: pack the food in the packaging bag. If the assembly line is not used, after a food completes the above four steps, the next food is processed, which takes a long time and affects efficiency. However, using pipeline technology, multiple foods can be processed simultaneously. In parallel algorithm, time is parallel, and two or more operations are started at the same time to improve the computing performance. Spatial parallelism refers to the concurrent execution of computation by multiple processors, that is, connecting more than two processors through the network to simultaneously compute different parts of the same task, or a large-scale problem that cannot be solved by a single processor.

In parallel computing, computing problems usually show the following characteristics:

1. Separate work into discrete parts, which helps to solve at the same time;
2. Execute multiple program instructions at any time and in time;
3. Solve under multiple computing resources.

The problem takes less time than a single computing resource. Parallel computing science mainly studies spatial parallel problems. From the perspective of programmers and algorithm designers, parallel computing can be divided into data parallel and task parallel. Generally speaking, data parallelism is mainly to resolve a large task into the same sub tasks, which is easier to handle than task parallelism. Spatial parallelism has led to the creation of two types of parallel machines. According to flynn, it is divided into: single instruction stream multiple data streams and multiple instruction streams multiple data streams. Commonly used serial machines are also called single instruction streams and single data streams. MIMD machines can be divided into the following five common categories: parallel vector processors, symmetric multiprocessors, large-scale parallel processors, workstation clusters, and distributed shared storage processors. Parallel computers connect each processor or processor by network. Generally, there are several ways as follows: a kind of network with fixed connection between processing units. During program execution, the point-to-point connection remains unchanged; typical static networks include one-dimensional linear array, two-dimensional mesh, tree connection, hypercube network, cubic ring, shuffle switching network, butterfly network, etc.

In recent years, with the development of motor manufacturing technology and power electronics technology, and the increasingly mature motor control technology, its comprehensive performance has been greatly improved [5]. In foreign literature, the electric load simulation research for testing motor and electric power rotating load capacity is proposed. At present, many foreign scholars study a load motor control method by means of the mechanical load simulation experimental platform of the drive motor load motor to realize the simulation of the dynamic mechanical load of the experimental platform, and finally realize the verification and test of various advanced control algorithms; it can also test the performance of the motor, electric drive system and driver in the experimental platform. Since the 1960s, there has been research work in this area. Until today, electric drive control has been greatly developed, but there is still some room for improvement. The following research will use parallel computing in the original electric drive control method. It is optimized on the basis to improve the effect of electric drive control.

2 Electric Drive Control Method Based on Parallel Computing

2.1 Load Distribution Based on Parallel Computing

The simple speed chain control can not meet the requirements of transmission control, so it needs to use different load distribution methods to control. The basic principle of load distribution is that the load rate of each transmission motor is the same, and each motor should distribute the overall load according to the load rate.

The load factor α of the motor has the following relationship:

$$\alpha = \frac{Q_a}{Q_{ab}} \qquad (1)$$

In the formula, Q_a represents the actual power of the a motor, and Q_{ab} represents the rated power of the a motor.

In formula (1), the motor power represents the actual load, which can also represent the parameters of the actual load as well as the current and torque. The torque is selected as the parameter, and the load distribution of torque is as follows:

$$Y_a = \frac{Y_{ab} \times Y}{\sum\limits_{a=1}^{n} Y_{ab}} \qquad (2)$$

In the formula, Y_a is the expected torque of the a th motor; Y_{ab} is the rated torque of the a th motor; Y is the total torque of the load; n is the number of motors participating in the load distribution.

In summary, the load distribution is based on the selected parameters, so that each parameter is equal to the motor load rate of the load distribution. There are many different distribution methods for load distribution, which can be summarized as two types of control methods: load distribution based on speed control and load distribution based on torque control. Select one of the inverters as the main node for load distribution. Generally, the transmission point with the largest power or the strongest mechanical rigidity is selected as the main node. Turn on the speed control mode and hang it on the speed chain to form the speed chain with other inverters [6–8]. Then load distribution is carried out according to the actual distribution situation.

The first case is the remaining point of felt ring, the main node is set as grid driven roller, and the torque is set to 1, the torque of the remaining nodes is given, and then given according to the actual power of the motor. For example, the short former is 0.7 and the vacuum roll is 0.5, then the relationship between the torque setting of the short former and the mesh drive roller and the mesh drive roller is 0.7: 0.5: 1, that is, all the loads are distributed to each point. The second case is the two-point contact load distribution, such as pressing the upper and lower rollers. Taking the press lower roll as the main point, the moment value of the press lower roll multiplied by a coefficient is directly taken as the given moment of the press upper roll. This coefficient can be calculated according to the actual situation. The flow chart is as follows:

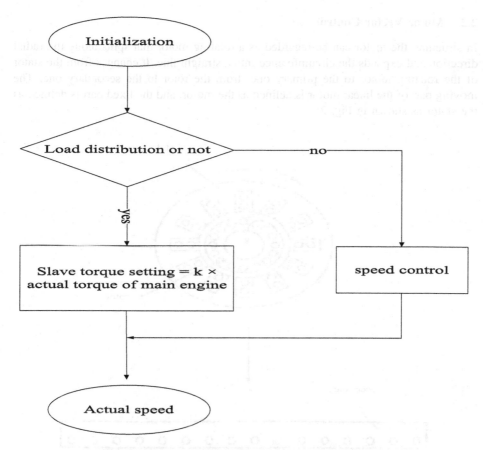

Fig. 1. Flow chart of load distribution program structure

In the Fig. 1, *K* is a constant. When designing the software for load distribution, you need to pay attention to the following issues: First, it cannot be disconnected from the main speed chain. The load distribution can only be hung on the main speed chain as a subsystem. The load distribution process cannot affect the other motors. Speed and status [9–11]. Second, the load distribution should consider the very short case, such as the runaway caused by the loose contact surface, and the load should be limited. Third, according to the actual production, some points in the paper introduction process are open, so it is necessary to install conversion device. For example, pressure sensor is used for conversion, speed chain control is used when opening, and load distribution is changed when the pressure reaches.

2.2   Motor Vector Control

In structure, the motor can be regarded as a rotating motor that splits along the radial direction and expands the circumference into a straight line. It changes from the stator of the rotating motor to the primary one, from the rotor to the secondary one. The moving part of the linear motor is defined as the motor, and the fixed part is defined as the stator as shown in Fig. 2:

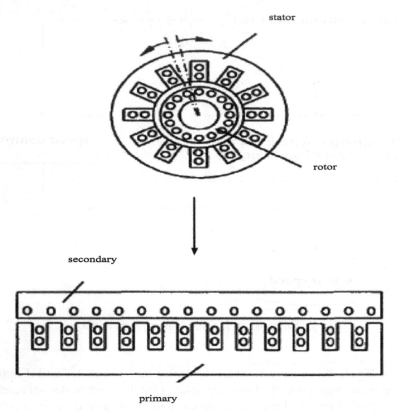

Fig. 2. Schematic diagram of rotating motor evolving into linear motor

From the structural point of view, the linear motors are mainly disc type, flat type, arc type and cylindrical type. For flat linear motors, if the primary is installed only on the secondary side, it is unilateral linear Motor, it will have a relatively large normal force; If the primary is installed on both sides of the secondary (armature winding part is installed on both sides of the pole) or the secondary is installed on both sides of the primary (armature winding part is installed on both sides) In the middle of the magnetic pole), it is a bilateral linear motor. There are short primary and short secondary structures for unilateral and bilateral linear motors. Among them, the operation cost and manufacturing cost of the short primary linear motor are relatively low, so in most cases, the short primary linear motor is used [12, 13]. The following research focuses on the electric drive control of the short primary bilateral flat AC permanent magnet synchronous linear motor.

For the AC permanent magnet synchronous linear motor, a three-phase symmetrical and time-varying sinusoidal current is input into the winding, and an air gap magnetic field is generated inside the motor, as shown in Fig. 3:

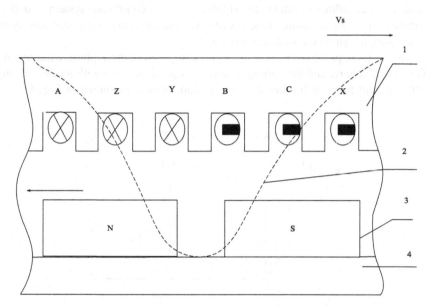

Fig. 3. Working principle of permanent magnet synchronous linear motor

In Fig. 3, 1 represents the mover and primary; 2 represents the traveling wave magnetic field; 3 represents the permanent magnet; 4 represents the stator and secondary. If the longitudinal side effect is ignored, the air gap magnetic field is also sinusoidal and moves along a straight line according to the phase sequence a, B and C, which is called traveling wave magnetic field. Under its interaction with the excitation magnetic field generated by the permanent magnet, an electromagnetic thrust is generated, which pushes the motor mover to perform synchronous linear motion at the same speed as it. The speed of the traveling wave magnetic field is:

$$B_\beta = 2f\beta \tag{3}$$

In the formula, B_β is the traveling wave magnetic field speed; f is the three-phase alternating current frequency; β is the magnetic pole center distance of the permanent magnet synchronous motor.

Under certain assumptions, the concepts of coordinate transformation and space vector can be used to simplify and calculate the nonlinear system with multi input and multi output, time-varying and multi variable coupling. It is assumed that: the magnetic circuit is linear, ignoring hysteresis, eddy current, remanence and saturation effects; the winding magnetic potential and air gap magnetic density are sinusoidal distribution, and ignoring spatial harmonics.

From the point of view of motor structure, linear motor can be considered as a kind of radial evolution of rotating motor. The following assumes that the permanent magnet synchronous linear motor is a rotating motor, and imagines its magnetic field as a rotating space magnetic field, so as to complete the Vector control. Three reference coordinate systems are commonly used in the analysis and establishment of their mathematical models, which are $q - w - e$ coordinate system, $\partial - \wp$ coordinate system and $x - y$ coordinate system, of which $q - w - e$ coordinate system and $\partial - \wp$ coordinate system are armature static coordinate systems, and $x - y$ coordinate system is space rotation dynamic coordinate system.

$q - w - e$ three phase coordinate system is composed of three-phase windings A, B and C of the armature, and the corresponding axes are q, w and e with an electric angle of 120° different from each other, forming a plane vector, as shown in Fig. 4:

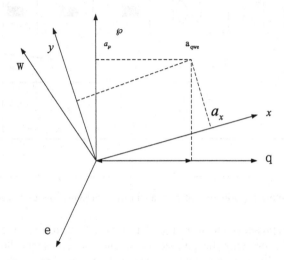

Fig. 4. Coordinate systems of AC permanent magnet synchronous motor

From the mathematical point of view, the plane vector can be described by using a rectangular coordinate system, so in the motor coordinate system, set $\partial - \wp$ rectangular coordinate system to simplify the calculation; where ∂ and q axis coincide, $\wp$ is 90° ahead of e axis.

The space current vector in $\partial - \wp$ coordinate system is:

$$\vec{a}_{qwe} = a_\partial + a_\wp b = \frac{2(a_q + a_w b^{120?} + a_e b^{240?})}{3} \qquad (4)$$

In the formula, the current components in the $\partial - \wp$ rectangular coordinate system where a_∂ and $a_\wp$ are stationary; a_q, a_w, and a_e are the current values input in the $q - w - e$ three-phase coordinate system. With the movement of the motor, the space magnetic field rotates correspondingly with $\vec{a}_{qwe}$. Imagine a rotating two-phase rectangular coordinate system $x - y$, which rotates at the same angular velocity as $\vec{a}_{qwe}$, so the current can be divided into thrust current a_y and excitation current a_x in the $x - y$ coordinate system. If a_y and a_x are controlled separately, it can be completed Current decoupling control. The position angle μ of the $x - y$ coordinate system relative to the $\partial - \wp$ coordinate system is:

$$\mu = \mathrm{mod}\left(\frac{u}{2\beta}\right) * (2\pi) + \mu_0 \tag{5}$$

In the formula, u is the displacement; μ_0 is the initial position angle; mod is the modulus operator.

The changing process of $q - w - e$ to $\partial - \wp$ coordinate system is:

$$Z_{qwe-\partial\wp} = \frac{2}{3}\begin{bmatrix} 1 & -\frac{1}{2} & -\frac{1}{2} \\ 0 & \frac{\sqrt{3}}{2} & \frac{\sqrt{3}}{2} \end{bmatrix} \tag{6}$$

or

$$Z_{qwe-\partial\wp} = \frac{2}{3}\begin{bmatrix} 1 & 0 & \frac{1}{2} \\ -\frac{1}{2} & \frac{\sqrt{3}}{2} & -\frac{\sqrt{3}}{2} \\ -\frac{1}{2} & \frac{1}{2} & \frac{1}{2} \end{bmatrix} \tag{7}$$

Changes from $\partial - \wp$ to $q - w - e$ coordinate system:

$$Z_{\partial\wp-qwe} = \frac{2}{3}\begin{bmatrix} 1 & 0 \\ -\frac{1}{2} & \frac{\sqrt{3}}{2} \\ -\frac{1}{2} & -\frac{\sqrt{3}}{2} \end{bmatrix} \tag{8}$$

or

$$Z_{\partial\wp-qwe} = \frac{2}{3}\begin{bmatrix} 1 & 0 & \frac{1}{2} \\ -\frac{1}{2} & \frac{\sqrt{3}}{2} & \frac{1}{2} \\ -\frac{1}{2} & -\frac{\sqrt{3}}{2} & \frac{1}{2} \end{bmatrix} \tag{9}$$

Through the above process, the current is converted from the three-phase $q - w - e$ coordinate system to the right-angle $x - y$ coordinate system to realize the decoupling control process, and the study of the electric drive control method based on parallel calculation is completed.

3   Experimental Analysis

3.1   Experimental Environment

In order to verify the effectiveness of the proposed method, simulation experiments are carried out. The experiment was carried out on MATLAB platform, using Windows 10 system, running memory of 8 GB. In this experimental environment, a comparative experiment is designed to verify the effectiveness of the proposed method. The composition of the experimental platform is shown in Table 1:

Table 1. Main components of the experimental platform

Name	Model	Quantity	Performance parameter
Motion controller + adapter board	Turbo PMAC2 Clipper + DTC-8B	1 + 1	DSP56303 4 axis board
Linear encoder	RGH22Y	2	Resolution is 0.1 μm Normal reading Allowable speed is 4 m/s
Driver	D1 MD-36-S	2	Frequency Range 47～63 Hz
X axis guide rail + slider	QHH15H1026Z-1082 + QH15	2 + 4	Self-lubricating
Y axis guide rail + slider	QHH20H1026Z-5068 + QH20	2 + 4	Self-lubricating
X axis permanent magnet synchronous linear motor	LMCB6	1	Stroke 750 mm
Y-axis permanent magnet synchronous linear motor	LMCC8	1	Stroke 1000 mm
Limit switch	EE-SX674	6	–

3.2   Experimental Parameters

When the load of the CNC platform is set to 17 kg and the external environment is relatively constant, the parameters of the position ring of x-axis linear motor and y-axis linear motor are as follows:

Table 2. Parameter settings

Motor shaft	X axis	Y axis
load	17 kg	48 kg
Kp	122	103
Ki	52	31
Kd	185	147
Kvff	1650	1950
Kaff	0	0

3.3   Experimental Scheme

In order to verify the rationality of the proposed method, the KP parameters of the proposed method and the traditional method are compared, and the stability of the proposed method and the traditional method in electrical drive control is analyzed. In order to ensure the credibility of the experiment, the experimental results are obtained through many experimental iterations and analysis.

3.4   Analysis of Experimental Results

3.4.1   Control Accuracy Analysis by Different Methods

In order to verify the comprehensive effectiveness of the proposed method, experimental analysis of the proposed method and traditional methods to adjust the Kp parameter, the experimental results are shown in Fig. 5:

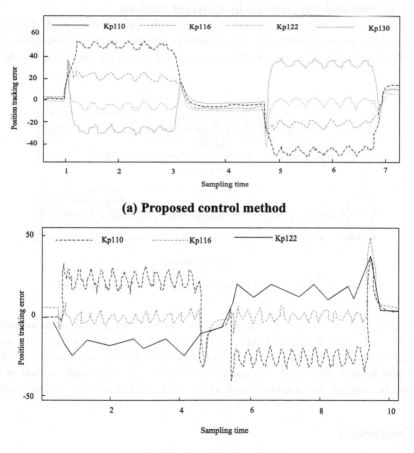

(a) Proposed control method

(b) Original control method

Fig. 5. Comparison of the effects of adjusting Kp parameters by different methods

Analysis of Fig. 5 shows that the proposed method uses the X-axis parameters in Table 2 under a load of 17 kg, and makes Kp equal to the errors obtained by 110, 116, 122, and 130, respectively, as shown in Figure (a). With the increase of Kp, the trend of the error changes firstly decreases in the positive direction and then increases in the negative direction. Under this control method, the optimal Kp = 122, and the position tracking error at this time is (−8, 9) μm.

Adopt the original control method, adopt the x-axis parameters in Table 2, and make KP be the error of 110, 116 and 122 respectively, as shown in figure (b). The trend of error change is also with the increase of KP, which first decreases in the positive direction and then increases in the negative direction. Under the traditional control method, the optimal KP = 116, at this time, the position tracking error is (−7,10) μ m, while when KP = 122, the position tracking error is (−24,22) μ M. In contrast, the proposed method has smaller tracking error and higher accuracy.

3.4.2   Control Stability Analysis by Different Methods

In order to further verify the scientific validity of the proposed method, the stability analysis of the proposed method and the traditional method in the electric drive control is analyzed in the experiment, and the experimental results are shown in Fig. 6:

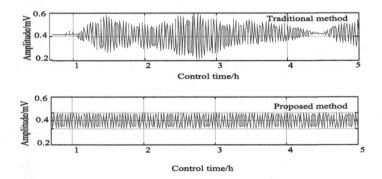

Fig. 6. Comparison of control stability with different methods

Analysis of Fig. 6 shows that under the same experimental environment, the control stability of the two methods is somewhat different. Among them, the control stability of the proposed method is always between 0.35 mV and 0.45 mV, which is better and more stable; while the control stability of the traditional method is more volatile. In contrast, the proposed method has good control stability and is feasible.

4   Conclusion

The research of the electric drive control method based on parallel computing is put forward. Through the selection of motor parameters and effective load distribution, the effective control of electric drive is realized.

Experimental results show that the proposed method can effectively improve the control accuracy and has good stability. However, due to the short research time, there are still some imperfections that need to be optimized in subsequent studies.

References

1. Shaoyan, H.E., Chunhui, W.U., Jianjun, T.I.A.N.: Review of the current transducer techniques. Electric Drive **48**(01), 65–75 (2018)
2. Le, H.A.N.: The application of plc in the integration of electric power drive. Telecom Power Technol. **36**(04), 89–90 (2019)
3. Xingtian, F., Huihui, H., Rong, C.: Teaching reform of engineering quality for automatic control system of electric drive. J. Electrical Electronic Educ. **40**(03), 47–49 (2018)
4. Yan, L.I.: Research and application of single chip microcomputer in electric drive control system. Telecom Power Technol. **35**(02), 169–170 (2018)
5. Fuzhong, W., Yuanyuan, L., Sumin, H., et al.: Status assessment of mine hoist electric drive system based on fuzzy synthetic evaluation. Power System Protect. Control **47**(09), 166–172 (2019)
6. Liu, S., Lu, M., Li, H., et al.: Prediction of gene expression patterns with generalized linear regression model. Front. Genetics **10**, 120 (2019)
7. Yue, P.: Selection of Fan and pump motor and electric drive system. Moder Archit. Electric **9**(04), 13–16 (2018)
8. Yuening, J., Hongguang, J., Ming, L.: CFD numerical simulation of unmanned aerial vehicle based on multi-core parallel computation. Comput. Eng. Appl. 54 (07), 221–225 (2018)
9. Liu, S., Liu, G., Zhou, H.: A robust parallel object tracking method for illumination variations. Mobile Netwk. Appl. **24**(1), 5–17 (2019)
10. Yaning, Y.: Unmanned aerial vehicle trajectory control simulation based on gpu parallel algorithm. Comput. Simul. **36**(03), 69–72 (2019)
11. Ying, C.: The construction and development of electric drive control training room in higher vocational colleges. Liaoning Higher Vocational Techn. Inst. J. **20**(03), 78–80 (2018)
12. Liu, S., Liu, D., Srivastava, G., et al.: Overview and methods of correlation filter algorithms in object tracking. Complex Intell. Syst. (2020) https://doi.org/10.1007/s40747-020-00161-4
13. Lu, M., Liu, S.: Nucleosome positioning based on generalized relative entropy. Soft. Comput. **23**(19), 9175–9188 (2018). https://doi.org/10.1007/s00500-018-3602-2

Community Discovery Algorithm Based on Parallel Recommendation in Cloud Computing

Jian-li Zhai$^{(\boxtimes)}$ and Fang Meng

Huali College Guangdong University of Technology, Guangzhou 511325, China
zhaijianli2033@163.com

Abstract. In the cloud computing environment, traditional social network community discovery algorithms have low accuracy in social network community discovery, leading to information waste, community overlap and low scalability, and unable to achieve ideal computing results. Therefore, a social network based on parallel recommendation is proposed. Network community discovery algorithm. By mining the candidate trusted user set, the number and composition of the community are obtained, and the communication units are divided into overlapping communities and non-overlapping communities according to the different numbers of communities belonging to the nodes in the network. Combining the mining of candidate trusted user sets and community division, social networking is realized Network community discovers and calculates. Experiments show that the algorithm improves the accuracy and stability of social network community discovery, and has good application value.

Keywords: Social networks · Recommendation system · Overlapping communities

1 Introduction

Cloud computing is an addition, use, and interaction model of the Internet-based services, often involving the use of the Internet to provide dynamic, easily scalable and often virtualized resources. Cloud is the network, a metaphor for the Internet. In the past, clouds were often used to represent telecom networks [1], and later to represent the abstraction of the Internet and underlying infrastructure. Cloud computing consists of a range of resources that can be dynamically upgraded and virtualized, shared by all cloud computing users and easily accessed over the network, without the need to master cloud computing technology. It only need to rent cloud computing resources according to the needs of individuals or groups. With the rapid development of Internet technology and the gradual popularization of mobile terminals and mobile Internet, more and more people have become a member of social networks. The continuous development of social networks has led to explosive growth in the size of social networks.

With the development of social networks, various forms of social media [2], such as facebook, twitter, Sina Weibo, WeChat, and so on, have realized peer-to-peer

© ICST Institute for Computer Sciences, Social Informatics and Telecommunications Engineering 2021
Published by Springer Nature Switzerland AG 2021. All Rights Reserved
S. Liu and L. Xia (Eds.): ADHIP 2020, LNICST 348, pp. 192–201, 2021.
https://doi.org/10.1007/978-3-030-67874-6_18

information interaction among network users, and users can record everything in their lives anytime and anywhere. Including what you see and hear in your work and what you think about hot topics. Users are not only recipients of information, but also publishers and communicators of information, so as to gain more opportunities to be recognized by others. The change of the information transmission mode greatly reduces the communication cost among the network individuals, it is also possible to spontaneously form different communities by participating in line-on-line activities in a specific area with those who own common attributes. Real social network is a multidimensional network, users have emotional and preference attributes, there are various types of contacts between users. If the multiple information can be used to detect the community structure in the social network correctly, then the other members can be inferred from the known members.

In recent years, social network community discovery algorithms based on parallel recommendation have been put forward [3]. However, most of these methods only focus on the information of single dimension in the network. The parallel recommendation algorithm is faced with a large amount of data and high complexity. Since most of the traditional social network community discovery algorithms can only be applied to small-scale networks or experimentally generated networks, when the number of users in the network is large. Using the traditional parallel recommendation algorithm is limited by the complexity of the hardware and the algorithm itself. Therefore, it is very difficult to deal with such a large amount of data efficiently, which seriously restricts the social network community discovery algorithm for parallel recommendation in large-scale social networks. Based on this, the design of social network community discovery algorithm based on parallel recommendation in cloud computing is proposed. Because of the large amount of data in cloud computing environment, the stability is low, the information is wasted, the overlapping community is not high and the scalability is not high when traditional classification is used. To this end, a social network community discovery algorithm based on parallel recommendation is proposed and designed. The validity of the method is verified in the simulation platform [4]. The results show that LER algorithm can improve the computational accuracy of social network community discovery, reduce the waste of resources, and improve scalability.

2 Design of Social Network Community Discovery Algorithm

Early community discovery algorithms are implemented by hierarchical clustering, but traditional hierarchical clustering methods tend to ignore the members who are less connected to the community. In order to avoid the disadvantages of the traditional method, the LER algorithm uses a new way to detect the community.

2.1 Mining Candidate Trusted User Sets

At present, trust relationship is the most widely used socialized relationship in social network community discovery algorithm. In real life, people ask friends for advice, users rely on the items recommended by their friends, and target users add explicit trust relationships on their own on social platforms. The target user's evaluation of the item depends to some extent on the trust user's rating of the item. This feature is used to mine candidate user sets. Usually, when users add trust relationships on social platforms [5], they are influenced by the user's communication circle and the user's own character and habit. Trust information that can be collected is often sparse. Therefore, only directly using explicit trust relationships is a very limited set of families to improve the accuracy of recommendations and alleviate the problem of sparse trust information.

Figure 1 shows that trust relationship is represented by graph, there are many connections between users, there will be a lot of common ground, there will be different trust relations, with strong expansibility [6].

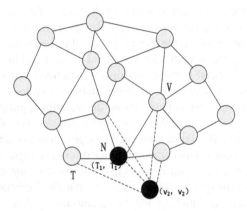

Fig. 1. Trust relationship Association Graph

For this trust relationship characteristics, cluster technology is used to calculate. The social friends in the same trust community as the target user are extended to the candidate trust user set of the target user. For a user set of n users $U=\{u_1, u_2, \ldots, u_n\}$. The item set for m projects is $I=\{i_1, i_2, \ldots, i_m\}$, and the project category set of l categories $C=\{c_1, c_2, \ldots, c_l\}$. We use indicator variables $T_{u,v} = 1$ or 0 indicates whether there is an explicit trust relationship between user u and user v. Use symbol N_v^{ci} represents the number of scores for items of user v in item classification ci. when the user u explicitly trusts the user v, and when user v has scored the item under a project category, it can be thought that user u initial trust user v under item classification ci. In this way, the initial trust relationship under each category is obtained. In the next step, the initial trust relationship is used to mine trust communities through community discovery methods under each project category. The initial trust conditions are as follows:

$$T_{u,v} = 1 且 N_v^{ci} > 0 \tag{1}$$

After calculating the initial trust condition, when user v_1 initially trusts v_2, there is a directed edge from v_1 to v_2. The algorithms are as follows:

First step: Specify a unique tag value for all user nodes, such as the user tag initially for the user id.

$$\forall_u \in V : I_u = U_{id} \tag{2}$$

Of which, $\forall_u$ represents the result of tag value; I representing the simplified coefficient of the network model j, this calculation does not do the directional analysis; k_j represents the training features representing the j convolution layer; U_{id} represents the commonality of the i trusted user.

Step two: to adjust the label values of all user nodes, each user needs to traverse and count the current tab values of all his neighbors, and take the label value with the largest number of occurrences as the new label for the user. When multiple label values appear in the largest number of times, a label value is randomly selected as the user's new label.

$$\forall_u \in V, 1_u = angmsc_k \left| N^K \right| \tag{3}$$

Of which, k represents the label, N^K represents a collection of u-trusted users with a tag value of k, 1_u represents the value of the user tag, this calculation does not do directional analysis.

The number of iterations required to divide a community to a stable state is generally five, when more than 90% of the nodes in the network have been divided into the correct communities. After mining the trust user set, there are some overlapping communities, and there are some differences in the community division [7]. Therefore, after mining the candidate trust user set, the community division is carried out.

2.2   Community Division

Through the community discovery, we can dig out the number and composition of the community from the network. It can help us to discover the deep laws in the network, analyze the realistic significance of the community division representation in the network, and thus help solve the practical problems. According to the different number of communities belonging to the nodes in the network, the communities are roughly divided into overlapping communities and non-overlapping communities [8]. In overlapping communities, each node can belong to multiple communities. Non-overlapping communities only allow each node to belong to one community, and there are no nodes that exist across multiple communities. This paper defines the community based on the idea of edge partition, and puts forward the HCL algorithm. In the HCL algorithm, the overlapping community merge and partition error correction are incorporated into the two steps. In each iteration, the similarity between edges is calculated, and the similarity between edges is calculated in each iteration. All the edges are divided into specific communities, and the combination is applied to community discovery of large-scale networks. The edge similarity algorithm is formulated as follows:

$$S(eik, ejk) = \frac{|n+(i) \cup n+(j)|}{|n+(i) \cup n+(j)|} =$$

$$\frac{|n+(i) \cap n+(j)|}{|n+(i)| + |n+(j)| - |n+(i) \cap n+(j)|} \qquad (4)$$

Of which, $n+(i)$ represents a collection of neighbor nodes for node S. And ejk and jk are S sides, this calculation does not do directional analysis.

The basic process of the HCL algorithm is: through the hierarchical clustering algorithm, the edges of the network are clustered into a community, because each node may have multiple connected edges, and these edges are divided into different communities, and each node may be divided into multiple different communities. The overlapping community structure is obtained. HCL algorithm can calculate the similarity between edges [9], and the larger the similarity is, the better the edge can be merged into a community.

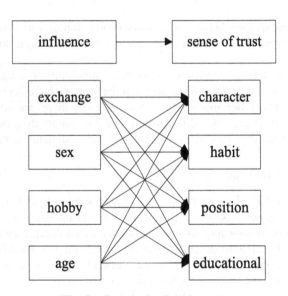

Fig. 2. Community Partition proc

Figure 2 mainly demonstrates the process of community division by analyzing the age distribution, sex, education level, interests and interests of community members, this paper studies the interaction between individuals in the community and the evolution of community structure. Through the analysis of the community structure, the flawlessness, robustness and stability of the entire community network are studied, the impact of key nodes in the community on information dissemination is analyzed, and the stability of the community structure is analyzed through the analysis of the community. The structure completes the design of the social network community discovery algorithm.

The HCL algorithm improves the accuracy and processing speed of community partition to a certain extent, which is of great significance to the design of social network community discovery algorithm.

2.3   Implement Social Network Community Discovery Computin

Combined with mining candidate trust user set and community partition, the implementation of social network community discovery calculation is realized. Next, the implementation process of each step will be described in detail in this paper.

According to the influence degree of attribute features on the classification results, the features are classified. In classification, a large weight is given to certain features of nodes with greater impact on community discovery, and a smaller weight is given to certain types of features of nodes with less impact on community discovery. Based on this, the LER algorithm is proposed for calculation and generation. The formulas are as follows:

$$X_i = \text{minc} \left(\sum_{m=1}^{n} mxc_j \right) + k\mu_j \tag{5}$$

Of which, n represents the number of nodes in the simulated network; m represents the number of edges in the simulated network; k represents the average node degree of each node; μ_j represents the proportion of the edges of each node connected to a node that is not in the same community as the sum of all the edges of the node, the larger the value, the less obvious the community structure of the simulated network is; minc represents the number of nodes owned by the smallest community in the generated simulation network; mxc_j represents the maximum number of nodes owned by the community in the generated simulated network.

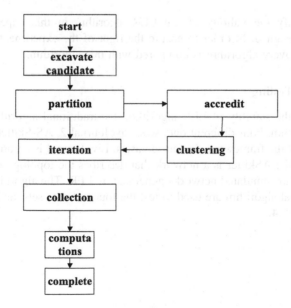

Fig. 3. LER Algorithm flow-process diagram

Figure 3 shows the LER algorithm to discover the social network computing process for the community. This calculation method can reduce the degree of community ambiguity, calculate in the case of large community size, and improve the precision of community calculation. This is because the LER algorithm uses multiple iterations to calculate the spanning tree to distinguish the community structure in the network, and the cohesion within the community is expressed by the number of spanning trees. This makes the real community size of the LER algorithm smaller in the network [10–12] and facilitates the generation of more spanning trees that cover the nodes within the entire community. Therefore, it is easier to detect the community structure in the network effectively. It is shown that the LER algorithm can improve the accuracy of the calculation and can carry out the effective calculation under the condition of big data. It can effectively highlight the topological structure of the network, thereby effectively mining the community structure in the network, and realize community discovery calculations based on parallel recommendations in cloud computing [13–15].

The LER algorithm also has a huge storage effect. Generally, the data is a graph with more edges than points, so the graph data is stored by the way of point segmentation. It can avoid too much redundancy in the stored procedure of edges, and the interaction between nodes and their neighbors only needs to satisfy the exchange law and union law. This method can effectively reduce the network transmission and storage overhead. The underlying implementation process is to store the edges in each node, and when data interaction occurs, it can be transmitted by broadcasting the nodes between each machine. The cost is that multiple redundant backups are needed for each node's attributes, and there is data synchronization overhead when the node update operation is needed.

3 Experimental Demonstration and Analysis

In order to verify the validity of the LER algorithm in this paper, a simulation experiment is designed. In order to ensure the rigor of the experiment, the traditional community discovery algorithm is compared with this algorithm.

3.1 Stability Testing

In order to test the stability of LER algorithm, the traditional algorithm is compared with LER algorithm. Four different data sets, LiveJournal 2, AS-Skitter 3, D1 and D2, were used to test the framework. In four datasets, LiveJournal is an online dating blog network of friends; ASkitter is a network that describes the topology of Internet; data sets D1 and D2 are simulated networks generated by LFR. The algorithm in this paper and the traditional algorithm are used to test the four test data sets, and the test results are shown in Fig. 4.

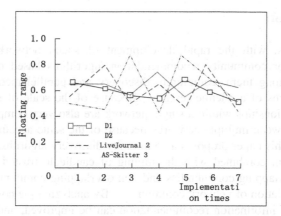

(a) Floating range of traditional algorithm

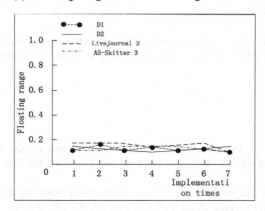

(b) The floating range of the algorithm in this paper

Fig. 4. Stability comparison

According to the comparative analysis of Fig. 4, when the computing time of the LER algorithm is less affected by the change of the number, the stability is high and the number of tasks is large, the running time of the algorithm is relatively small, and the proportion of the total running time of the calculation is smaller when the number of tasks is more than the number of tasks. The amplitude of the fluctuation produced by the algorithm in this paper is small and basically fluctuates about 0 amplitude. But in the traditional image classification process, the fluctuation is large, the highest is about ±5 dbs. When the number of algorithms increases and the total running time of the algorithm decreases, the traditional algorithm floats greatly and its stability is low.

Through the above analysis, we can basically determine the validity of this algorithm. When designing community discovery algorithms in cloud computing environment, the data can be classified and calculated accurately and stably, which is convenient for further management and analysis, and the stability is very high.

4 Conclusion

In recent years, with the rapid development of social networks, there are new requirements for community recommendation algorithms based on parallel recommendation. Among them, community discovery and parallel recommendation have become the focus of academic research. Because of the scale of today's social networks, the relationships within a single network are also very complex, and there is a relationship between multiple networks because of the same account number.

Therefore, this paper proposes a community discovery algorithm based on parallel recommendation, combined with the mining of candidate trusted user sets and the division of overlapping communities and non-overlapping communities, to realize the discovery calculation of network communities. By analyzing the community structure, the accuracy of information recommendation can be improved, and it is of great significance to study the parallel recommendation algorithm recommended by social network communities.

Acknowledgments. T his work is supported by Second batch of school-level scientific research projects of Huali College of Guangdong University of Technology (project number: HLKY-2018-ZK-07)

References

1. Xu, W., Lin, B., Lin, S., et al.: Research on a method of social network community discovery based on user interaction behavior and similarity. Inf. Network Secur. **33**(7), 77–83 (2015)
2. Lu, Y., Cao, B., Rego, C., et al.: A Tabu search based clustering algorithm and its parallel implementation on Spark. Appl. Soft Comput. **63**, 97–109 (2018)
3. Longyuan, Zhang, J.: Parallelization of community discovery algorithms based on local expansion. Mod. Comput. **21**(11), 41–45 (2016)
4. Wang, S., Xu, G., Shi, S.: Label propagation overlapping community discovery algorithm based on large groups. Electron. Measur. Technol. **23**(2), 45–49 (2016)
5. Liu S, Liu D, Srivastava G, et al. Overview and methods of correlation filter algorithms in object tracking. Complex Intell. Syst. 1–23 (2020). doi:10.1007/s40747-020-00161-4
6. Wang, S., Xu, G., Shi, S.: Maxima-based label propagation overlapping community discovery algorithm. Electronic Measur. Technol. (2): 45–49 (2018)
7. Wahab, O.A., Bentahar, J., Otrok, H., et al.: Towards trustworthy multi-cloud services communities: a trust-based hedonic coalitional game. IEEE Trans. Serv. Comput. **PP**(99), 184–201 (2018)
8. Liu, S., Lu, M., Li, H., et al.: Prediction of gene expression patterns with generalized linear regression model. Front. Genet. **10**, 120 (2019)
9. Zhao, R., Wu, Y., Chen, X.: Community discovery and visualization algorithm for large-scale social networks. J. Comput.-Aided Des. Graphics **29**(2), 328–336 (2017)
10. Wen, R.: Research and implementation of personalized recommendation system based on social network. Zhejiang University of Technology **9**(3), 28–36 (2015)
11. Wang, J.: Research on social network community discovery method based on local expansion. Yanshan University **9**(3), 28–36 (2015)

12. Zhao Runqian, WuYu., Xin, C.: Large-scale social network community discovery and visualization algorithm. J. Comput. Aided Des. Graphics **29**(2), 328–336 (2017)
13. Liu, S., Wang, S.: Trajectory community discovery and recommendation by multi-source diffusion modeling. IEEE Trans. Knowl. Data Eng. **29**(4), 898–911 (2017)
14. Liu, S., Liu, G., Zhou, H.: A robust parallel object tracking method for illumination variations. Mob. Networks Appl. **24**(1), 5–17 (2019)
15. Shuai, L., Gelan, Y.: Advanced Hybrid Information Processing Springer International Publishing, USA, 1-594

Deployment Optimization of Perception Layer Nodes in the Internet of Things Based on NB-IoT Technology

Rui Liu[1], Jie-ran Shen[1], Feng Jiao[1], and Ming-hao Ding[2(✉)]

[1] State Grid Liaoning Electric Power, Slwply Co. Ltd., Shenyang 110006, China
huangdongping8962@163.com
[2] Department of Computer and Software Technology, Tianjin Electronic Information College, Tianjin 300350, China

Abstract. The traditional deployment optimization method of perception layer nodes in the Internet of Things has the drawbacks of poor optimization performance. Therefore, this paper proposes a research on deployment optimization of perception layer nodes in the Internet of Things based on NB-loT technology. The genetic algorithm is used to code the nodes in the perception layer of the Internet of Things, and the initial population is determined. Based on the coding of the nodes in the perception layer and the initial population, the fitness function is designed, and the NB-loT technology is used to optimize the deployment of the nodes in the perception layer of the Internet of Things. Experiments show that the average coverage of the proposed method is 24% higher than that of the traditional method, which shows that the proposed method has better optimization performance.

Keywords: Internet of things · Perception layer · Node · Deployment · Optimization

1 Introduction

The Internet of Things (IOT) is a kind of network that uses modern communication technology to connect sensors, people and objects in a new way to form information and intelligence. It can be divided into four layers: perception layer, network layer, middle layer and application layer [1]. The perception layer is a bridge connecting the physical world and the information world, and is also the bottom layer of the entire Internet of Things. It connects sensors on various devices through wired or wireless ways to form a network of information collection and control, and transmits the collected information to the upper layer of the Internet of Things. In the application process of the Internet of Things, it is necessary to monitor and reconnaissance the ecological environment and battlefield environment. It is required that all the targets moving along any path in the monitoring area can be found by the perception layer nodes. In order to fulfill the above requirements, multiple nodes must work together. This requires that the perception layer nodes deploy with high coverage and good node communication. At the same time, during the work process, the layout of network

S. Liu and L. Xia (Eds.): ADHIP 2020, LNICST 348, pp. 202–210, 2021.
https://doi.org/10.1007/978-3-030-67874-6_19

nodes should be automatically adjusted according to the location and importance of the goods to optimize the role of the Internet of Things. Therefore, it is the most important problem for the Internet of Things to design an effective optimization method for deployment of sensor layer nodes, to achieve reasonable and dynamic deployment of sensor layer nodes according to the monitoring area, to improve coverage and target detection probability of wireless sensor networks, and to reduce network energy consumption and maximize network lifetime.

NB-loT technology, known as narrowband Internet of Things, is an important branch of the Internet of Things. NB-IoT is built in cellular network and consumes only about 180 kHz bandwidth. It can be directly deployed in GSM network, UMTS network or LTE network to reduce deployment cost and achieve smooth upgrade. NB-IoT is a new technology in the field of IoT. It supports low power devices'cellular data connection in WAN. It is also called low power wide area network (LPWAN). NB-IoT supports efficient connection of devices with long standby time and high network connection requirements. It is said that the battery life of NB-IoT devices can be increased by at least 10 years, while providing a very comprehensive coverage of indoor cellular data connections [2]. The traditional method of node deployment optimization at the perception layer in the Internet of Things has the disadvantage of poor optimization performance and cannot meet today's increasing demand. Therefore, this paper proposes a research on deployment optimization of perception layer nodes in the Internet of Things based on NB-loT technology.

2 Design of Optimal Deployment Method for Perception Layer Nodes in the Internet of Things

The Internet of Things (IOT) has the characteristics of large scale, complex environment, limited network resources, random deployment and self-organization. These characteristics determine the importance of node deployment and conflict prevention technology in the implementation of the Internet of Things. Node deployment is a key low-energy technology, which can not only ensure the quality of Internet of Things, but also improve the reliability of the entire Internet of Things and ensure the safety of goods in the transport process [3]. In the Internet of Things, first of all, we should solve the deployment problem of nodes, that is, how to deploy the perception layer nodes of the Internet of Things, achieve the maximum coverage of the perception layer nodes, and optimize the performance of the Internet of Things.

In practical applications, the deployment of sensor nodes needs to consider many factors such as application environment, label distribution density, interference and so on. The distribution of nodes not only ensures the maximum coverage of the target area, but also satisfies the requirement of reading rate. It also reduces the cost of equipment and additional economic costs. A reasonable node topology will greatly reduce the cost of network construction. Therefore, the design of node deployment optimization in the Internet of Things is a multi-objective combinatorial optimization problem. According to the characteristics of the Internet of Things, genetic algorithms are used to optimize the deployment of sensing layer nodes. The specific process is as follows.

2.1   Perception Layer Node Coding

The genetic algorithm is used to code the nodes in the perception layer of the Internet of Things to prepare for the optimization of node deployment.

The genetic algorithm is proposed by Professor Yu of the University of Michigan. With the in-depth study of other scholars, genetic algorithm has become more perfect. At present, genetic algorithm is widely used in machine learning, industrial optimization control, biology, pattern recognition, image processing, software technology, neural network, genetics and so on. Generally speaking, genetic algorithm constructs a fitness function according to the objective function to solve the problem, evaluates and operates on a population composed of multiple individual solutions. After many iterations, the individuals with the best fitness function value are taken as the output of the optimal solution [4].

Genetic algorithm is an important branch of artificial intelligence. It combines "genetic" and "algorithm" well. It is an adaptive, global optimization and probabilistic search algorithm, which simulates the natural selection and genetic mechanism of organisms by using Darwin's evolutionism for reference. It is based on the natural evolutionary rules of "survival of the fittest, survival of the fittest", searching in the solution space and solving the problem.

The genetic algorithm introduces the principle of biological evolution in nature into the solving process of practical complex problems. Firstly, the candidate solutions of the problem to be optimized are coded to form a population containing several individuals. According to the fitness function corresponding to the optimization objective, the individuals with higher fitness function values are screened. Then the new individuals are crossed and mutated by genetic operation. The fitness function of the individual in the population will be improved continuously until the termination condition of the algorithm is satisfied by the continuous iteration operation. The individual with the highest value of the final output fitness function is regarded as the optimal solution of the problem to be optimized [5]. The specific coding process of the perception layer nodes is shown below.

Coding is the expression of solution space. According to the workflow of genetic algorithm to solve the problem, the candidate set of the problem should be coded first. Encoding is a mechanism that converts the parameters of practical problems into direct operation with genetic operators. The appropriateness of coding directly affects the speed and quality of problem solving. Binary encoding, Gray encoding and real number encoding are currently the three main encoding methods.

The problem of deployment optimization of perception layer nodes in the Internet of Things studied in this paper is how to deploy perception layer nodes reasonably. For the problem of deployment of perception layer nodes, this paper uses real coding method [6, 11]. Therefore, the possible solution $X_n (n = 1, 2, \cdots, n)$ of the problem can be set to a random array in the working area, that is:

$$X_n = \{X_1, X_2, \cdots, X_N\}, X_N \in (0, 30) \tag{1}$$

In the formula, individual X_n is randomly generated, and N represents the number of populations. $X_{n1} = \left\{X_1, X_2, \cdots, X_{\frac{m}{2}}\right\}$ represents the abscissa of the n th node, $X_{n2} =$

$\left\{ X_{\frac{m}{2}}, X_{\frac{m+1}{2}}, \cdots, X_m \right\}$ represents the ordinate of the n th node, and $\frac{m}{2}$ represents the number of nodes in the perception layer of the Internet of Things.

Since genetic algorithm operates mainly on a group of individuals, it is necessary to prepare an initial population consisting of multiple candidate solutions. Initial population is the starting point of genetic algorithm optimization, and its construction is related to the execution efficiency of genetic algorithm. Initial population generation is generally divided into random method and priori-based method. This paper optimizes the deployment of the perception layer nodes in the Internet of Things. The initial deployment of the nodes is based on the random spraying method, so the initial population is randomly generated [7, 12].

Through the coding of the above perceptual layer nodes and the determination of the initial population, data support is provided for the following fitness function design.

2.2 Design of Fitness Function

The fitness function is designed based on the coding of the perception layer nodes and the initial population. The specific process is shown below.

Fitness function is mainly used to simulate the evolutionary selection process in nature. In order to implement the principle of "natural selection, survival of the fittest", genetic algorithm uses fitness function value to evaluate each individual in the population. Fitness function is a measure of the quality of each individual in the population. The selection of fitness function will directly affect the efficiency of the algorithm. Generally, individuals with higher fitness function have better survival ability, and the probability of inheritance to the next generation is higher; on the contrary, poor individuals have lower fitness function value, and their survival ability is weaker [8–10]. As fitness function is the only deterministic index to measure whether an individual in a population can inherit to the next generation, it directly determines the evolution of the population, so the selection of fitness function is very important in genetic algorithm.

The deployment optimization of perception layer nodes in the Internet of Things is a multi-objective optimization problem. It not only achieves the maximum coverage of goods, but also minimizes the level of interference between nodes. Therefore, the optimal objective function of node deployment is designed as follows:

$$f(\tau) = w_1 \cdot f_1(\tau) + w_2 \cdot f_2(\tau) \tag{2}$$

The objective function of node deployment optimization is the weighted sum of two sub-objective functions. Among them: $f_1(\tau) = Ind(\tau)$. $f_2(\tau) = Cov(\tau)$, w_1, w_2 is the weight of the two sub-objective functions in the whole optimization objective function. It mainly determines the size of these two weights according to the comprehensive requirements of the deployment of the perception layer nodes in the Internet of Things, and they satisfy $w_1 + w_2 = 1$. Larger w_1 guarantees the maximum coverage of goods, while larger w_2 can minimize the interference between intelligent nodes. In order to ensure the maximum coverage of cargo and minimize the interference between nodes, according to the empirical value, this paper sets $w_1 = 0.9, w_2 = 0.1$.

Through the above process, the fitness function is designed, and the goal of node optimization is determined, which provides the goal for the realization of the following node deployment optimization.

2.3   Implementation of Node Deployment Optimization

Based on the above fitness function, i.e. the optimization objective function of node deployment, NB-loT technology is used to optimize node deployment. The specific optimization process is shown below.

In the genetic algorithm, if the fitness function value of the individual X_i in the group is $f(X_i)$, the probability that the individual X_i is selected is:

$$P(X_i) = \frac{f(X_i)}{\sum_{i=1}^{q} f(X_i)} \tag{3}$$

Where q is the number of candidate solutions in each generation group. In order to ensure that the genetic algorithm can effectively converge to the optimal solution in the search space, certain protection measures can be taken when selecting all individuals in the population, that is, the individuals whose current fitness function value is the largest are not involved in the genetic operation. Instead, let it directly replace the individuals with low fitness function values in the next generation of populations, thus preserving the optimal individuals in the population.

The crossover probability P_c is defined as the ratio of the number of subalgebras produced by the crossover operation to the total number of individuals in the population in each generation of population. Obviously, a higher crossover probability can achieve a larger solution space, thereby reducing the probability of selecting a non-optimal solution individual; but if the crossover probability is too high, it will waste a lot of time because too many other solution spaces are searched. Reduce the search speed of the algorithm.

Let a pair of parent individuals be X_A and X_B respectively, then the pair of offspring individuals after the intersection are:

$$\begin{cases} X_A' = \alpha X_B + (1 - \alpha)X_A \\ X_B' = \alpha X_A + (1 - \alpha)X_B \end{cases} \tag{4}$$

Where α is a random number on [0, 1].

The mutation operation mainly simulates the mutation of a certain gene in an individual, thereby changing the characteristics and attributes of the individual. The probability of variation P_v is defined as the percentage of the number of individuals in the population as a percentage of the total number of individuals in the population. The probability of mutation controls the proportion of new individuals entering the population.

The optimization process of the IoT awareness layer node deployment is as follows:

Step 1: First, encode every possible point in the search space of the Internet of Things perception layer node deployment problem;

Step 2: Determine the size N of the initial population, the number of iterations DT, the fitness function $f(f > 0)$, the crossover probability P_c, and the mutation probability P_v;

Step 3: randomly generate N individuals in the search space to form the initial group $P(k)$, and let $k = 0$;

Step 4: Calculate the fitness function value of each individual in $P(k)$;

Step 5: Calculate whether the current population performance meets certain indicators. If yes, go to step 8; if not, go to step 6.

Step 6: According to the genetic strategy, the selection, crossover and mutation operations are applied to the current population to generate the next generation populations $P(k)$, $k = k + 1$;

Step 7: Determine whether k reaches the predetermined number of iterations, if yes, go to step 9; if not, go to step 4;

Step 8: Select the individual with the highest fitness function value in the population as the optimal solution output found by the algorithm;

Step 9: The algorithm ends and no optimal solution is found.

According to the above basic idea, a flowchart for optimizing the deployment of the Internet of Things sensing layer node as shown in Fig. 1 is obtained.

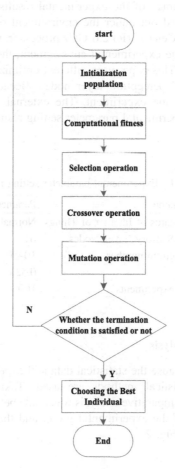

Fig. 1. Flow chart of optimization of IoT awareness layer node deployment

3 Performance Optimization Analysis of IoT Sensing Layer Node Deployment Optimization Method

In order to ensure the optimization performance of the IoT sensing layer node deployment optimization method proposed in this paper, the design experiment verifies it. In the experiment process, the Internet of Things sensing layer is taken as the experimental object, mainly to optimize its node deployment. In order to ensure the accuracy of the experimental process and the results, the proposed IoT perception layer node deployment optimization method is compared with the traditional IoT perception layer node deployment optimization method, and the experimental comparison results are observed. In the experiment process, the traditional IoT perception layer node deployment optimization method is called the control group, and the proposed IoT perception layer node deployment optimization method is called the experimental group.

3.1 Experimental Parameter Number Preparation

In order to ensure the accuracy of the experimental results as much as possible, the simulation analysis is carried out under the environment of Microsoft Windows XP operating system, Intel (R) Celeron (R) 2.6 GHz processor, matlab simulation tool and 24.0 gb memory. And in the experimental environment, the parameters in the experimental process are set up. This paper uses different optimization methods to optimize the deployment of the IoT perception layer nodes. Because the methods used are different, therefore, during the experiment. The external environmental parameters must be consistent. The experimental parameter setting results in this paper are shown in Table 1.

Table 1. Experimental parameter setting results

Parameter name	Parameter setting
Operation Status of Internet of Things	Normal operation
Number of Sensory Layer Nodes	6T
Population group number	10–50
α	0.523
Number of experiments	100 s

3.2 Node Coverage Analysis

During the experiment, because the statistical data and experimental data are recorded in different methods, statistical methods are used. Taking node coverage as an experimental indicator, a comparative analysis was made between the node coverage of the IoT perception layer of the experimental group and the control group. The comparison result is shown in Fig. 2.

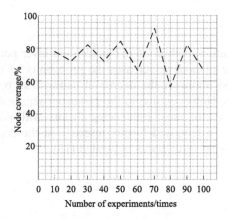

(a) Node coverage of experimental group

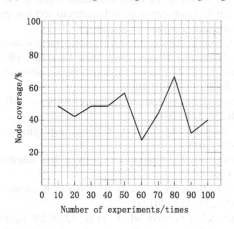

(b) Node coverage of control group

Fig. 2. Comparison of the results of the node coverage experiment

According to the data in Fig. 2, the experimental group's IoT perception layer node coverage is between 59% and 92%, and the maximum coverage rate is as high as 92%, while the control group's IoT perception layer node coverage rate is 66%. The node coverage of the IoT perception layer of the experimental group is higher than that of the control group, indicating that the proposed IoT sensor node deployment optimization method has better optimization performance.

4 Conclusions

The proposed optimization method of IoT perception layer node deployment greatly improves the coverage of nodes, increases the probability of cargo detection of the Internet of Things, and improves the degree of deployment optimization of the IoT perception layer nodes. However, due to the parameter setting during the experiment. Ignoring the interference of most influencing factors will affect the results of the experiment to a certain extent, and further research on the deployment optimization method of the IoT sensing layer nodes is needed [13].

References

1. Zhang, Q.: Technical performance and application of the honeycomb-based Narrow Band Internet of Things (NB-loT). Sci. Technol. Commun. **9**(20), 12–15 (2017)
2. Zhu, W., Yao, Y.: Analysis on coverage enhancement technology of NB-loT Internet of Things. Wirel. Interconnect. Technol. **16**(8), 28–29 (2017)
3. Wu, J., Cheng, W., Liang, Y.: Discussion on the deployment strategy of NB-IoT and eMTC for operator cellular Internet of Things. China New Commun. **18**(23), 64–65 (2016)
4. Ye, Y., Jiang, S., Xing, L., et al.: End-to-end deployment strategy of NB-loT [J]. Guangdong Commun. Technol. **26**(2), 51–53 (2018)
5. Wang, J.: Discussion on coverage enhancement technology of NB-LoT Internet of Things. Commun. World **31**(23), 3–4 (2017)
6. Xing, Y., Hu, Y.: Narrow-band Internet of things deployment strategy. Inf. Commun. Technol. **62**(1), 33–39 (2017)
7. Fu, W., Liu, S., Srivastava, G.: Optimization of big data scheduling in social networks. Entropy **21**(9), 902 (2019)
8. Mingjun, L., Shiping, J.: Business Application of Narrow Band Internet of Things (NB-loT). Inf. Commun. **50**(10), 254–255 (2017)
9. Negash, B., Rahmani, A.M., Westerlund, T., et al.: LISA 20: lightweight internet of things service bus architecture using node centric networking. J. Ambient Intell. Humaniz. Comput. **7**(3), 305–319 (2016)
10. Liu, S., Bai, W., Srivastava, G., Machado, J.A.T.: Property of self-similarity between baseband and modulated signals. Mob. Networks Appl. **25**(4), 1537–1547 (2019). https://doi.org/10.1007/s11036-019-01358-9
11. Lu, M., Liu, S.: Nucleosome positioning based on generalized relative entropy. Soft. Comput. **23**(19), 9175–9188 (2018). https://doi.org/10.1007/s00500-018-3602-2
12. Yongbin, W., Zhongping, Z.: Low power, Dalian Wide Area Internet of Things access technology and deployment strategy. Inf. Commun. Technol. **52**(1), 27–32 (2017)
13. Shuai, L., Weiling, B., Nianyin, Z., et al.: A fast fractal based compression for MRI images. IEEE Access **7**, 62412–62420 (2019)

Analysis of Energy Saving Method for Multiple Relay Nodes in Wireless Volume Domain Network

Tian-bo Diao, Hong-e Wu, and Shuo-yu Zeng$^{(\boxtimes)}$

School of Mathematics and Statistics, Nanyang Institute of Technology,
Nanyang 473000, China
zengshuoyu321@163.com

Abstract. In order to solve the problem that the transmission link of wireless volume domain network is likely to be interrupted and consume unnecessary energy, this paper introduces probability statistics and proposes a research on energy saving of wireless volume domain network multi-relay nodes based on probability statistics. The energy consumption of network is analyzed and the formula of total energy consumption per bit network is derived. The simulation results show that compared with the traditional multi-path multi-relay node forwarding method, this method can greatly reduce the overall energy consumption of the network. Consumption also plays a role. This method can reduce the overall energy consumption of the network and prolong the life cycle of the network. When the optimal relay node is used for transmission, the transmission power is greatly increased.

Keywords: Probability statistics · Wireless body area network · Multi-relay nodes · Energy saving

1 Introduction

Wireless body area network is a kind of wireless network based on RF technology, which is composed of micro nodes with sensor or actuator functions distributed on the surface of human body or implanted in human body. It is an application of wireless sensor network in biomedical field [1]. Wireless body area network can be used to monitor, collect and manage different vital signs (such as body temperature, blood pressure, heart rate, blood sample concentration, etc.) of human body. It has been widely used in disease monitoring, telemedicine diagnosis, home care and other aspects, and has gradually become a research hotspot. Aiming at the problem of signal attenuation in human body area transmission, a wireless body area network channel model based on path loss and power delay is proposed [2]; A UWB channel model of wireless body area network is proposed to solve the path loss problem in normal distributed small-scale fading environment; In this paper, a cooperative mechanism of network coding is proposed to explore the energy-saving way in lossy channel, and to give enlightenment to the energy-saving way in wireless body area network; A new network coding multicast routing algorithm based on link sharing degree is proposed to

S. Liu and L. Xia (Eds.): ADHIP 2020, LNICST 348, pp. 211–222, 2021.
https://doi.org/10.1007/978-3-030-67874-6_20

improve the performance of multicast transmission, which has certain guiding significance for multi-path and multi relay multicast transmission; The outage probability of direct transmission, single relay cooperation and multi relay cooperation, as well as the energy consumption and power consumption based on the outage probability are compared and analyzed in wireless body area network; The concept of cooperative communication between nodes is proposed, and the transmission link assisted by double relay is analyzed by using spatial diversity gain technology, but the energy efficiency of wireless body area network is not solved. Network coding is introduced to study the throughput of wireless body area network, but the energy consumption of network after network coding is not considered [3–5].

The traditional multi-path and multi relay forwarding method consumes too much network energy considering the transmission link interruption probability in wireless body area network. By introducing probability statistics into wireless body area network, this paper proposes a research on multi relay energy saving in wireless body area network based on probability statistics. Compared with multi-path multi relay forwarding method, this method can greatly reduce network energy consumption and improve the overall performance of the network.

2 Energy Saving Method for Multiple Relay Nodes in Wireless Volume Domain Network

2.1 Wireless Body Area Network

According to the various application forms of wireless sensor networks, researchers have designed different network structure forms. The most basic structure includes the following parts:

(1) Sensor network

The network is the core part of wireless sensor network. In the sensing area, a large number of sensor nodes monitor and perceive information, process the information and send it to the sink node; at the same time, receive the operation command from the sink node and execute [6].

(2) Sink node

The sink node has enough energy and transmitting power to process the information sent by the received sensor node and forward it to the transmission medium, or to transmit the user operation instructions to the sensor node [7].

(3) Transmission mode

The transmission mode includes satellite communication and network transmission, which realizes the information interaction between wireless sensor network and users, and is the communication medium [8].

(4) Network users

Network users are responsible for collecting the required data from the network, analyzing and processing the data, and monitoring the wireless sensor network [9].

The wireless body area network node consists of four modules. The network node module is mainly responsible for information collection and data conversion of the sensing object, such as measuring the physical properties of the surrounding light, electromagnetic, acoustic and so on, so as to obtain the corresponding information; The processor module is mainly responsible for storing and processing the data collected by itself and the data sent by other nodes; the wireless communication module is mainly responsible for wireless communication with other sensor nodes, exchanging control information and receiving and transmitting the collected data; The energy module is mainly responsible for providing energy for network nodes [10–12].

The wireless body area network protocol framework mainly defines and describes the functions that the network and its components should complete. The wireless volume LAN protocol architecture is shown in Fig. 1.

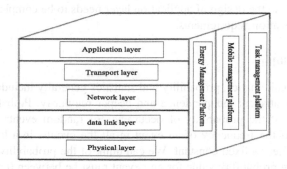

Fig. 1. Wireless body area network protocol architecture

The architecture of the protocol includes physical layer, data link layer, network layer, transmission layer and application layer, corresponding to the five layers of TCP/IP protocol. In addition, the protocol architecture also includes energy management platform, mobile management platform and task management platform [13].

(1) Physical layer

The main function of the physical layer is to evaluate the channel, select the better channel, detect the wireless signal, complete the transmission and reception of the signal and so on. The design goal is to obtain the larger link capacity and reduce the energy consumption as much as possible [14]. At present, the main problems in the physical layer of wireless sensor network are: how to design an integrated, digital and general circuit while reducing the cost of hardware; how to design a modulation algorithm with high data rate and low symbol rate while reducing the energy consumption.

(2) Data link layer

The main functions of data link layer are data framing, frame detection, media access and error control. This layer can be subdivided into media access control sublayer and logical link control sublayer. The main task of media access control sublayer is to share channel resources with users, and the main task of logical link control sublayer is to provide a standard and unified interface to the network [15].

(3) Network layer

The network layer is mainly responsible for route generation and route selection; its main functions include packet routing, network interconnection, congestion control, etc. The main purpose of routing protocol is to establish the route between sensor node and sink node, and to transmit data reliably and safely.

(4) Transport layer

The main function of transmission layer is to complete the transmission control of data flow in wireless sensor network. The aggregation node collects data, and uses network, satellite and other ways to communicate with the external network. The operation of the transmission layer is an important part to ensure the quality of service.

(5) Application layer

The main function of the application layer is to acquire and process the transmitted data. The design of application layer is closely related to the actual application situation and environment, so the design of application layer needs to be completed according to the specific application requirements.

2.2   Model Building

As a branch of mathematics, probability and statistics generally includes probability of random events, statistical independence and deeper regularity. Probability is a quantitative indicator of the probability of occurrence of random events. In independent random events, if the frequency of an event in all the events, in a larger range, it is obviously stable near a fixed constant. We can think of the probability of this event as this constant. The probability value for any event must be between 0 and 1. The multi relay cooperation model of wireless body area network based on probability statistics is shown in Fig. 2.

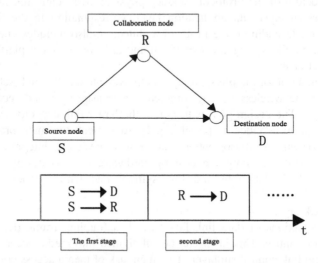

Fig. 2. Multi relay cooperation model of wireless body area network based on probability and statistics

The research on cooperative transmission technology is based on this model. The model includes a source node s, a destination node D, and a collaboration section R. In the process of data transmission, the source node s will forward the data relay to the destination node d with the assistance of the collaboration node R. The whole data transmission is divided into two stages. In the first stage, the source node s sends data signals in the form of broadcast. If the link sad exists, the cooperative node R and the destination node D can receive the data signals, otherwise only the cooperative node r receives the data signals; In the second stage, the source node s suspends the transmission of data signals, and the cooperative node r processes the received data information according to a certain cooperative transmission protocol and forwards it to the destination node D. At this time, the destination node D receives two identical data signals from different paths. It can decode the backup data comprehensively through certain criteria to obtain the spatial diversity gain.

In order to increase the reliability of network transmission, all relay nodes use a cluster based cooperative forwarding method to transmit data. At the same time, the relay node introduces random network coding to transmit data, which improves data throughput and reduces the overall energy consumption of the network.

In the network model shown in Fig. 3, a multi relay cooperative energy-saving algorithm based on probability statistics is proposed. The description process is as follows:

First, M source node sends its original packet $R_1, R_2, R_3, \cdots, R_M$ to all relay nodes in this cluster, and each relay node will receive the original packet $R_1, R_2, R_3, \cdots, R_M$ of M different source nodes at the same time.

Then, N relay node randomly encodes M original packets. The encoded packets are as follows:

$$Z_j = \sum_{i=1}^{M} \lambda_{ij} R_i, j = 1, 2, 3, \ldots, N \tag{1}$$

The coding coefficient λ_{ij} is randomly selected from the finite field g_{ij}, R_1 is the original data packet of the i-th source node, and Z_j is the j-th relay node coding packet.

In consideration of transmission interruption, no matter how many original packets of different source nodes are received by the a relay node, all received packets are encoded. The number of times the relay node encodes depends on how many packets of different source are received in j-th certain period of time.

Finally, the sink node decodes all the received encoding packets. When the sink node receives at least N encoding packets and the received encoding coefficients are linearly independent, the original data can be decoded. If the decoding is successful, the sink node will feed back the confirmation information to all relay nodes, and all relay nodes will lose the rest of the coding packets. If the decoding is not successful, the sink node will retain the first round of transmitted coding packets and feed back the denial information to all relay nodes. All relay nodes will send the second round of coding packets to the sink node until the sink node can successfully decode or send the second round of coding packets to the N round of sending.

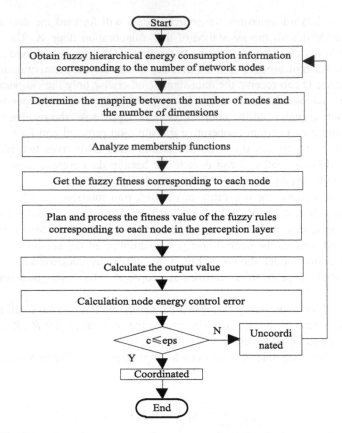

Fig. 3. Multi relay cooperative energy-saving process of wireless body area network based on probability and statistics

2.3 Study on Energy Saving of Relay Cooperation

According to the optimal coverage model, the scheduling algorithm is improved, and the minimum transmission period of all nodes is T; The time between the sending periods of two adjacent adjusted nodes is, which represents the minimum common multiple of the sending time of all periodic information. In each adjustment period, the smaller the minimum common multiple of information transmission time is, the faster the adjustment speed of each communication data transmission period is.

In the research process of multi relay cooperative energy-saving in wireless body area network, taking the obtained transmission information of relay nodes as the input data of probability statistics, the optimal energy-saving solution of multi relay writing nodes is calculated, and the multi relay cooperative energy-saving flow based on probability statistics is designed, as shown in Fig. 3.

The specific collaborative energy saving process is as follows:

When there is at least one frame of information in the distributed network carrier communication network and no data is successfully transmitted in the longest transmission time, the communication network can be considered to be in a non working state. At this time, the communication network idle time t_i shall meet the following requirements:

$$T < t_i \tag{2}$$

In formula (2): T represents the longest transmission time of a frame of information.

In this case, the minimum transmission period T can be expressed as:

$$T = T' - \left(t_i - t_j\right)/\alpha \tag{3}$$

In formula (3): T' represents the previous node regulation period; t_j represents the reference value of idle period.

When there are n nodes on the distributed network communication bus, the transmission cycle of each node is the same, and the change range of communication network utilization P is:

$$P \le \frac{(2n+1)T''}{(2n+1)T'' + T} \tag{4}$$

Where T'' is the average transmission time of information frame, and $T'' < T$ is:

$$P < \frac{(2n+1)}{(2n+2)} \tag{5}$$

When T in formula (4) is idle, the inequality is an equation, which can be expressed as follows:

$$P = \frac{(2n+1)T''}{(2n+1)T'' + T} \tag{6}$$

Therefore, when the average transmission time of information frame is the minimum, the communication network utilization P is the highest.

When n value is 1, the minimum sending period is T, which can be expressed as:

$$T \ge (2n+1)T'' + T_k = 3T'' + T \tag{7}$$

Therefore, when the minimum transmission period is less than $3T'' + T$, the distributed network carrier communication will no longer work. Therefore, this period is the minimum period allowed for transmission.

In the distributed network, the transmission frequency in the same period increases with the number of information nodes, and the network delay becomes very serious. In order to improve the problem, the hybrid scheduling algorithm is improved.

Send in T minimum period. Only one time of data sent is sent through the network. Set in a certain transmission period, select any time as the transmission time, then the data to be sent will be sent successfully, and will not be sent in other time. After such optimization, the impact of network delay on data transmission can be reduced, and the real-time performance of distributed network carrier communication can be improved.

3   Experimental Results and Analysis

Under the environment of MATLAB 7.0, the Internet of things experimental simulation platform is built, and the experimental research is carried out under glomosim simulator. The network simulator is used as information transmission simulator, and C++ language is used as network protocol to verify the effectiveness of the research of multi relay energy saving based on probability statistics.

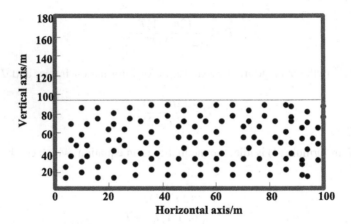

Fig. 4.   Node distribution diagram

3.1   Initial Environment Setup and Parameter Setup

In the research experiment of the energy consumption control method of the nodes in the perception layer of the Internet of things, the node distribution diagram is shown in Fig. 4.

From Fig. 4, we can see the node distribution. Use the energy consumption equalization protocol to set the experimental parameters, as shown in Table 1.

Table 1. Experimental parameter settings

Parametric	Value	Parametric	Value
Number of communication network nodes	100 nodes	Initial energy	10 J
Monitoring range	100 m × 100 m	Communication energy consumption	40 bit
Initial power	0.40 J	Data fusion energy consumption	10 bit
Receiving power	0.30 J	Multipath model magnification	0.010
Control layer protocol	MAC-SENSOR	Experimental time	1000 s
Data packet	550 B	Baotou size	25 B

According to the experimental environment and parameters, the experimental results are analyzed.

3.2 Comparison of Experimental Results

The traditional method is compared with the probabilistic method, and the results are as follows:

(1) Scheduling time

Scheduling time is an important indicator to judge the scheduling speed of the two methods. In this experiment, the number of communication tasks was taken as an independent variable. During the experiment, the set communication task was increased from 0 to 80. The more communication tasks, the more scheduling time is required. The simulators used in the process are ready to sleep and ready on duty. With the support of glomosim simulator, the scheduling time of the two methods is compared and analyzed. The results are shown in Fig. 5.

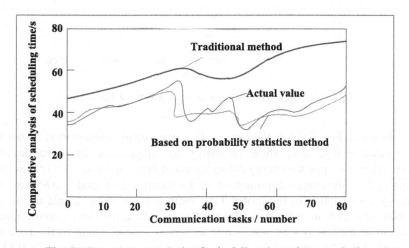

Fig. 5. Comparative analysis of scheduling time of two methods

It can be known from Fig. 5 that the curve based on probability statistics method is closer to the actual value curve, while the curve of the traditional method is far from the actual value curve. When the number of communication tasks is 35, the actual value and the scheduling time based on the probability statistics method both reach the maximum, which are 53 s and 49 s in order. The traditional method is that when the number of communication tasks is 80, the maximum scheduling time is 73 s. It can be seen that the scheduling time based on the probability statistics method is closer to the actual value, and compared with the traditional method, the scheduling time of the method is shorter.

The analysis of the traditional node cooperation method and the network survival time based on the probability statistics method, the results are shown below.

It can be known from Fig. 6 that with the increase of time, the number of surviving nodes of both methods decreases. When the time is 600 s, the number of surviving nodes under the traditional method is 0, and the number of nodes surviving based on the probability statistics method is 40. The experimental results show that the number of viable nodes in the proposed method decreases less and more slowly. Therefore, for a node with a research period of 400 to 600, under this condition, a comparative analysis of its energy consumption situation is shown below.

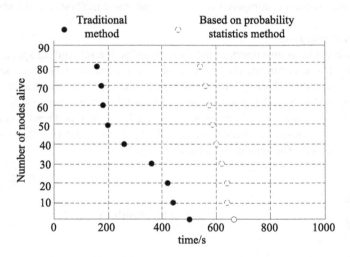

Fig. 6. Network survival time analysis results of two methods

As shown in Fig. 7:When the time is 10 s, the energy consumption of the traditional method is 0.92 mw, while the energy consumption of the probability based method is 0.66 mw, and the energy consumption of the actual node is 0.65 mw; When the time is 20 s, the energy consumption of the traditional method is 0.85 mw, while the energy consumption of the probabilistic method is the same as that of the actual node, both of which are 0.56 mw; When the time is 30 s, the energy consumption of the traditional method is 0.88 mw, while the energy consumption of the probability based method is 0.51 mw, and the energy consumption of the actual node is 0.53 mw;

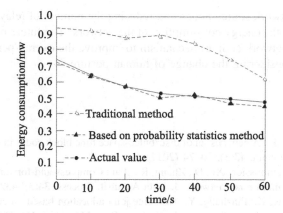

Fig. 7. Comparison of node energy consumption in two methods

When the time is 40 s, the energy consumption of the traditional method is 0.86 mw, while the energy consumption of the probabilistic method is the same as that of the actual node, both of which are 0.54 mw; When the time is 50 s, the energy consumption of traditional method is 0.76 mw, while that of probability based method is 0.52 mw, and the actual energy consumption of node is 0.49 mw; When the time is 60 s, the energy consumption of traditional method is 0.68 mw, while that of probability based method is 0.47 mw, and the actual energy consumption of nodes is 0.48 mw. Therefore, the method based on probability and statistics is basically consistent with the actual node energy consumption.

4   Summary

Wireless body area network is one of the focuses in the field of science and technology. It penetrates into every link of our life and every corner of the society. It is conducive to broaden human's understanding of the depth, breadth, accuracy and timeliness of the physical world, strengthen and close the relationship between human and the whole physical world, and to a large extent enhance human's remote monitoring of the physical world environment It has a wide application prospect. Energy limitation is one of the key problems in wireless body area network. How to control the energy consumption of multi relay cooperation in wireless body area network is very important for the application of wireless body area network.

The energy-saving scheme of multi relay cooperation in wireless body area network based on probability and statistics is studied. In the case of multi-source and multi relay, considering the influence of network environment fading, the transmission link is interrupted. The energy consumption formula under the corresponding network model is derived by using the link interruption probability. The experimental results show that, in the case of path loss, the energy-saving scheme of multi relay cooperation based on network coding can reduce the overall energy consumption of the network to a certain extent. At the same time, to meet the requirements of interrupt value, increasing the number of network source nodes and reducing the number of relay nodes also play a role in reducing the energy consumption of the network. The future research direction is how to apply network coding mechanism to improve the overall performance of the network when considering the change of human posture.

References

1. Chu, H., Meng, L.: A new design of distributed-space time block codes in relaying networks. Study Opt. Commun. **43**(3), 70–74 (2017)
2. Bao Military Commission, Xu, D., Zhang, R., et al.: Compress-and-forward system based on heterogeneous multi-relay network. J. Data Acquisit. Process. **34**(3), 432–441 (2019)
3. Yang, L., Huang, K., Pinzhang, Y.: Resource joint allocation based on energy efficiency in cognitive collaboration radio network. Comput. Eng. **44**(1), 134–138 (2018)
4. Wang, W., Zhao, S., Huang, G., et al.: The throughput analysis in cooperative communication based on relaying energy harvesting. Sci. Technol. Eng. **5**(35), 38–42 (2018)
5. Chu, W., Zhang, Q.: Power allocation optimization strategy based on full-duplex relay network. Comput. Eng. **44**(1), 149–153 (2018)
6. Yao, L., Jin, Y., Gu, J.: Research on private data retrieval algorithm in cloud for WBAN based on dynamic clustering and SS Tree. Comput. Dig. Eng. **47**(2), 360–366 (2019)
7. Chen, J., Zhou, Y., Wang, T., et al.: Body area network node sleep strategy based on sparse representation. Modern Electron. Tech. **40**(17), 15–19 (2017)
8. Tang, Z., Srivastava, G., Liu, S.: Swarm intelligence and ant colony optimization in accounting model choices. J. Intell. Fuzzy Syst. (Preprint):1–9
9. Lu, M., Liu, S.: Nucleosome positioning based on generalized relative entropy. Soft. Comput. **23**(19), 9175–9188 (2018). https://doi.org/10.1007/s00500-018-3602-2
10. Shuai, L., Weiling, B., Nianyin, Z., et al.: A fast fractal based compression for MRI images. IEEE Access **7**, 62412–62420 (2019)
11. Andreas, W.: Autonomous allocation of wireless body-sensor networks to frequency channels: modeling with repeated "balls-in-bins" experiments. Wirel. Networks **17**(5), 1–18 (2018)
12. Moravejosharieh, A.H., Lloret, J.: Mitigation of mutual interference in IEEE 802.15.4-based wireless body sensor networks deployed in e-health monitoring systems. Wirel. Networks, **26**(12), 258–266 (2020)
13. Chandrasekaran, B., Balakrishnan, R., Nogami, Y.: TF-CPABE: An efficient and secure data communication with policy updating in wireless body area networks[J]. ETRI J. **41**(10), 56–68 (2019)
14. Rubani, Q., Gupta, S.H., Pani, S., et al.: Design and analysis of a terahertz antenna for wireless body area networks[J]. Optik **179**(15), 684–690 (2019)
15. Liu, S., Liu, D., Srivastava, G., et al.: Overview and methods of correlation filter algorithms in object tracking. Complex Intell. Syst. (2020). https://doi.org/10.1007/s40747-020-00161-

Study on Probability Statistics of Unbalanced Cloud Load Scheduling

Shuo-yu Zeng$^{(\boxtimes)}$, Yu-jun Niu, and Hong-e Wu

School of Mathematics and Statistics, Nanyang Institute of Technology,
Nanyang 473000, China
wuhonghe321313@163.com

Abstract. Aiming at the problem of unstable equilibrium probability in modern load scheduling applications, a statistical method of unbalanced probability in cloud load scheduling is proposed. The weights and anti-saturation factors are calculated, the servers are grouped, the fuzzy cyclic iterative control of dynamic network resources is realized, and the network packet cloud load scheduling is designed. By comparing with the common methods, it is proved that the method designed in this paper can guarantee high equilibrium probability and good stability in a certain program.

Keywords: Cloud load · Scheduling imbalance · Probability statistics

1 Introduction

In recent years, the rapid growth of network applications, such as the rise of mobile Internet, cloud computing, big data, and other services, has a profound impact on people's daily life and brought great convenience to users. Since the 1970s, the demand for the network is only a simple end-to-end transmission. IP data packet mainly contains the network address of the source host and the destination host. Almost all the traffic on the Internet is based on the TCP/IP architecture. At that time, TCP/IP can also meet this demand well. Since the opening of the network, the network equipment is constantly updated, and the network traffic that can be handled is also growing. However, there is no breakthrough in the structure of the network. Today's network environment is complex, and various applications emerge in endlessly, such as the development of the cloud. Its computing, storage, service, and security are the most popular projects at present. Behind these rising commercial projects are various data centers and tens of thousands of server clusters. These switches and servers provide the most basic support for the upper layer services, while the data center is the bridge in the whole network operation [1]. With the growth of business and the continuous expansion of the data center, the computing power of the server is thousands of times higher than that of the original computer. In the process of data transmission between the server and the server, the link plays a role of connection. The whole data center, the network link is a complex topology. Today, the link utilization rate of the large-scale network center is only 30%–40%, which is just to deal with sudden large flow; The utilization of the whole link bandwidth is low. In addition to the extensive demand for

S. Liu and L. Xia (Eds.): ADHIP 2020, LNICST 348, pp. 223–231, 2021.
https://doi.org/10.1007/978-3-030-67874-6_21

the Internet, the traditional data center has not been able to fully support the bandwidth demand of all kinds of application processing, so the link transmission of the data center has become the bottleneck of cloud computing. The research of network link load balancing can greatly improve the network utilization, and at the same time, it can make the network accept more traffic requests and improve the network throughput, which will be more beneficial to the network data interaction and user experience.

The research of network load balancing has always been a hot spot in network research. It can guarantee the performance of the network and improve the user experience; The calculation of the optimal path is the basis of network interconnection [2]. At this time, the emergence of software-defined networking (SDN) can adapt to the current network requirements. Centralized control, separation of control and data, network programmability and monitoring capability of the whole network have made breakthroughs in network load balancing. Statistical method of unbalance probability in cloud load scheduling. The weights and anti saturation factors are calculated, the servers are grouped, the fuzzy cyclic iterative control of dynamic network resources is realized, and the network packet cloud load scheduling is designed.

2 Probability Statistics Method of Unbalanced Load Scheduling in Network Packet Cloud

2.1 Cloud Server Load Calculation

It overcomes the problem that static algorithm can not solve the load of cloud server, which leads to the imbalance of cluster. The controller should dynamically collect the load information of the server, including the utilization rate of the central processing unit (CPU), memory utilization rate, and network bandwidth utilization rate [3]. Represented by C_i, M_i, and B_i (i represents the i-th server), the load of the i-th server is:

$$L_i = \varepsilon_1 * C_i + \varepsilon_2 * M_i + \varepsilon_3 * B_i \tag{1}$$

Among them, $\sum_i \varepsilon_i = 1$; ε_i represents the impact factor. A larger value indicates that this performance parameter has a greater impact on this server. The size of ε_i can be statically set according to the business that the balanced service faces. To let the equalizer better understand the service capabilities of each service, the capacity of each server is collected here:

$$R_i = \mu_1 * F_i + \mu_2 * N_i * 1000 + \mu_3 * MT_i + \mu_4 * BT_i \tag{2}$$

Among them, $\sum_i \mu_i = 1$; μ_i larger value indicates that this parameter accounts for a larger proportion of the capacity of the computing server. The size of μ_i can be set according to the different services targeted by the balanced service. F_i is the CPU frequency; N_i is the number of CPU cores; MT is the total memory; BT_i is the total bandwidth.

According to the defined load L_i and capacity R_i, the weight of each server is obtained:

$$W_i = 1/\left((1 - L_i) * \sqrt{R_i}\right) \tag{3}$$

2.2   Anti-saturation Factor Calculation

When the number of sent requests reaches a certain value, the number of server response requests suddenly drops to zero and continues for some time. A measure taken by system software to prevent crashes, also known as "denial of service" or "fake death" [4]. When the server is "fake dead", the process is that the response time slowly increases, then jitters, and then increases sharply. Once it increases sharply, it is called to enter the saturation state. If an anti-saturation factor is added during the jitter, the server weight The value is falsely increased, and the threshold is used to prevent it from entering the saturation state, to effectively prevent the server from entering the saturation state and affecting the services provided.

$$WS_i = \begin{cases} \frac{1}{(1-L_i)*\sqrt{R_i}}, no - time - shake \\ \frac{1}{(1-L_i)*\sqrt{R_i}} + \varsigma_i, time - shake \end{cases} \tag{4}$$

Here, the resource utilization of the cloud host exceeds the threshold and the higher packet entered by the cloud host is determined as the response time jitter, plus the anti-saturation factor; otherwise it is determined that the response time is not jitter [5].

The anti-saturation factor ς_i is the result of multiplying the cloud host weight by the static scale. Because the configuration of each cloud host is different, the weights will be different and ς_i will also be different. When the load information of the cloud host is collected, first, the resource utilization of the cloud host will be judged whether it exceeds the threshold. If it exceeds, the weight will be set to 1, and the cloud host will enter the high load group. If it is not exceeded, it is determined whether the weight of the cloud host enters a group with a higher weight. If the weight of a cloud host is entered, the cloud host's weight will increase by ς_i. After the cloud host weight is increased by ς_i, the cloud host will make the average weight of the group to be higher, so that the probability that the group will be allocated more requests is smaller, thereby achieving the purpose of anti-saturation.

2.3   Cloud Server Grouping

The anti saturation factor can prevent the cloud server with large weight from entering the saturation state, so it will not significantly increase the service response time. However, for servers with low weight, if the equalizer does not update its load information in time, the equalizer will continue to distribute the load, resulting in a rapid increase of its load, which may lead to overload. Unbalance the entire cluster [6, 7]. To solve this problem, servers with similar weights are grouped into a group and a distribution request is issued in that group instead of being distributed to a server. According to the weight, the server is divided into five groups, as shown in Table 1.

Table 1. Grouping servers by weight

Group	Weights	Group	Weights
1	(0, 0.05]	4	(0.5, 0.75]
2	(0.05, 0.25]	5	(0.75, 1]
3	(0.25, 0.5]		

From the first group to the fifth group, the server will process more and more requests, and the processing time will be longer and longer [8]. Here, according to the different weights, the time for the equalizer in the controller to collect the servers in the packet will be different. The servers of the first group and the second group do not change much in weight in a certain period of time due to the small number of requests they process [9, 10]; In the same way, the fourth group and the fifth group need more time to process more requests, and the weight will not change greatly in a certain period of time, so the time interval for collecting the load information of these two groups can be larger; But for the third group of servers, which are in the transition stage between the second group and the fourth group, they are very sensitive to the weight information [11]. Once the weight changes, the group will not accept the request again, so the collection interval for the weight information of servers in this group will be shorter. Set five groups of weight information collection intervals as follows: $\tau_3 < \tau_2 = \tau_4 < \tau_1 = \tau_5$. In addition, when the virtual machine in the host pool enters the fourth and fifth groups due to too many requested resources, because of the existence of the anti-saturation factor, it can ensure that the virtual machine in the two groups is in the pseudo saturation state, and can still process the accepted requests normally. If there is a new request coming at this time, the request will not be sent to the back end, and the SDN controller will be triggered to inform the cloud platform to establish a new virtual machine instance to join the host pool. This will be reflected in the next research.

2.4   Fuzzy Cyclic Iterative Control of Dynamic Network Resources

In the cloud computing environment, firstly, the fuzzy cyclic iterative control algorithm is used to extract the dynamic network resource balance scheduling characteristic parameters [12, 13]. Compared with the recursive algorithm and the VC++ standard library function nth element, it shows that the algorithm is more efficient and reliable than the traditional recursive algorithm. Compared with the standard library function nth element, it has obvious advantages in time efficiency. Establish the dynamic network resource weight distribution mechanism [14, 15], calculate the orthogonal weighted constraint equilibrium ratio of network resources, and carry out the dynamic network resource fuzzy cyclic iterative control. The specific process is as follows:

Suppose that $\{x_n\}_{n=1}^{N}$ represents the collection of dynamic network resources in the cloud computing environment. According to the following vector control methods, a new resource mapping relationship x_n is constructed in the stack space. k represents the sampling point time series of dynamic network resource load in the cloud computing environment. τ represents the dynamic network time delay parameter.

Then formula (5) can be used to represent a time span of dynamic network resource operation:

$$h(\tau_i, t) = \sum_{i=1}^{N_m} a_i(t) e^{\phi} \varsigma(t - (\tau_i, t)) \tag{5}$$

Where, $a_i(t)$ represents the number of tasks currently running, e^{ϕ} represents that the physical machine is in saturated running state, (τ_i, t) represents that the server is not currently running, $\varsigma(t - (\tau_i, t))$ represents the fastest progress of resource task execution, and N_m represents the waiting time of virtual machine resources during task execution. Suppose that R represents the trust relationship function of network resources, establishes a dynamic network resource weight allocation strategy, calculates the orthogonal weighted constraint equilibrium ratio e_{ij} of network resources, and evaluates the attribute weight of network resources classification. The effective ratio function of dynamic network resource tasks in the cloud computing environment is given as follows:

$$E(i,j) = \begin{cases} (e_{ij} - e(i,j))/(e_{max} - e(i,j)), e(i,j) < e_{ij} \\ (e_{ij} - e(i,j))/(e(i,j) - e_{min}), e(i,j)) \geq e_{ij} \end{cases} \tag{6}$$

In the formula, $e(i,j)$ represents the attribute weight of multi-source heterogeneous resources, e_{max} represents the highest efficiency of the task, and e_{min} represents the lowest efficiency of the task. For a point on the dynamic network resource set a in the cloud computing environment, The equilibrium probability calculation parameter is:

$$R'' = \varpi_1 C_i + \varpi_2 D_i + \varpi_3 M_i + \varpi_4 N_i \tag{7}$$

Where, $N_i \in N_m$, $\varpi_\eta (\eta = 1, \ldots, 4)$ represents the weight coefficient, C_i represents the dynamic network resource weighted constraint equilibrium ratio, D_i represents the time parameter, and M_i represents the time sampling period in the cloud computing environment.

Suppose that $h(A, B)$ represents the space directed fuzzy distance between resource set A and B, and N_A represents the covariance vector of data information flow of dynamic network resource set A. The delay of information collection caused by the approximate calculation of data flow equilibrium scheduling is T, and the fuzzy cyclic iterative control is used to control the dynamic network resource scheduling process. Set a set of control parameters $G(V, E)$ and $\sin k \in V$, find a method of network resource scheduling set to minimize the dynamic network load parameter $TS(u)$. Define the dynamic network load balancing scheduling parameters, select the corresponding delay response parameters, and use Eq. (8) to calculate:

$$Y = \frac{h(A, B)}{N_A \times U_i} + D_i \frac{[\varsigma_{ik}(t) + \Theta(t)]}{k} \qquad (8)$$

Where, $\varsigma_{ik}(t)$ represents the information of the dynamic network control node, U_i represents the dynamic network resource classification attribute, k represents the network load intensity, and $\Theta(t)$ represents the trust value obtained by the dynamic network resource search at the latest time [10]. According to the construction of data information flow, the time scale characteristic parameter $f_i(t)$ of dynamic network load resource is extracted, and the request information of dynamic network load resource scheduling task with scale c is p_c. The calculation formula (9) of unbalanced probability of cloud load scheduling is:

$$S_{\varpi} = \sum_{a=1}^{c} p \frac{1}{T} \sum_{k=1}^{T} [U_i \cdot f_i(t)] \qquad (9)$$

In formula (9), a represents the number of network edge nodes, and $\frac{1}{T}$ represents the dynamic network node quota. So far, the unbalanced probability calculation of cloud load scheduling is completed.

3   Method Test Experiment

In order to verify the performance of this method, a comparative experiment is designed with the common probabilistic methods. Through comparison, the hypothesis of the experiment is verified.

3.1   Experimental Program

In order to make the designed scheme and algorithm have universality and scalability, the collection of load information and the processing of load by the load balancer are implemented in a Java modular manner. The collection of each kind of load information only needs to implement an interface and generate the corresponding class. The load balancer handles the load in the same way. Versatility is reflected in the fact that the classes currently implemented are the most basic load information collected, and factors that should be considered for load balancing; scalability is reflected in interfaces and classes. After the server receives the request, it consumes the server's CPU, memory, and bandwidth resources through the script. Each requested service will result in corresponding resource consumption. For example, the CPU will consume 10%–20% of the consumption. The specific CPU consumption can be changed Script tuning, and after a period of time, the load generated by each request will automatically exit after completion. The Linux performance monitoring tool dstat is used in the experiment to record each cloud host server at regular intervals, and the recorded results will be saved in a log. Each record will have time, which is convenient for later comparison when analyzing data The load of each cloud host. The network bandwidth test uses the iperf tool, and the CPU and memory tests are performed using scripts. First, use the resource utilization of each cloud host to compare and analyze various strategies. As time changes, observe the impact of each strategy on the cluster equilibrium and the impact of each strategy on the cluster equilibrium at the same time. Then use the standard deviation of resource utilization of the cloud host cluster to compare different strategies, and experiment with the effects of different strategies on the overall cluster balance. The main comparisons are three strategies: Random, RoundRobin, and ASGS. Experimental result analysis the experimental environment configuration is shown in Table 2. During the test of the three strategies, the same software and hardware were used, but configuration changes were made at the equalizer through the RESTful structure according to the different strategies tested.

Table 2. Experimental environment configuration

Physical machine & VM	Role	CPU core	Frequency/GB	Internal storage/GB	Drive
compute1	OpenStack compute1	4	3.2	64	e1000e
controller	OpenStack controller	4	3.2	64	e1000e
compute1_vm	Server 101, 103	1	3.2	16	virtio-pci
compute1_vm	Client 106	2	3.2	8	virtio-pci
compute1_vm	Server 107, 108	2	3.2	8	virtio-pci

3.2 Result Analysis

In order to compare the equilibrium effects of the three strategies, two indicators are used for measurement: the first is to compare the resource utilization of each cloud host in the cloud host cluster at each moment. The closer the resource utilization of each cloud host under the three strategies at each moment is, the higher the balance is.

As shown in Fig. 1, H1 to H4 represent four cloud hosts, which represent three strategies (Random (1), Round Robin (2), ASGS (3) strategy) The resource utilization of each cloud host at each moment. For example, at the first moment, under the Random strategy, the resource utilization of the fourth cloud host is much higher than the other three. This results in a severe imbalance in the cluster load at the first moment, and the balancing effect is not very good as the moment changes. In contrast, Round-Robin and ASGS do not produce this situation, and the balance of ASGS is significantly better than that of Random and Round Robin after 300 times. Therefore, the order of the equalization effect is getting better and better: ASGS. The second is to compare the standard deviation of resource utilization of each host at each time, so as to measure the balance moment of each virtual machine at this time. If the standard deviation is close to 0, the higher the balance. The experimental results in Fig. 1 show that the ASGS method also has a good balance. It shows that the proposed method can effectively improve the stability of cloud load scheduling.

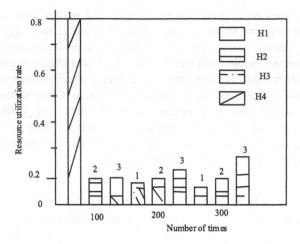

Fig. 1. Comparison of cloud host resource utilization in three strategies

4 Conclusion

Because of the unstable equilibrium probability of load scheduling in the modern application process, a statistical method for the unbalanced probability of cloud load scheduling is designed. Through the contrast experiment with the common methods, it is proved that the method designed in this paper can guarantee high equilibrium probability and good stability in a certain program.

References

1. Any fruit. Improving laser network load balance technology in cloud computing environment. Laser J. **38**(2), 137–141 (2017)
2. Ge, W., Ye, B.: Improved priority schedule scheduling algorithm for load balancing priority. J. Shenyang Univ. Technol. **39**(3), 241–247 (2017)
3. Fu, Y., Liu, B., Shu, Y.: Research on the multi-path load balancing algorithm in the data center of the schubby-tree data center based on SDN. Comput. Appl. Softw. **34**(9), 147–152 (2017)
4. Qu, H., Zhao, J., Fan, B., et al.: Application of ant colony optimization in software definition networks. J. Beijing Univ. Posts Telecommun. **40**(3), 51–55 (2017)
5. Fu, M., Wu, F., Huang, F., et al.: Research on performance of multipath routing algorithm based on software definition network. J. Fuzhou Univ. (Nat. Sci. Edn.) **45**(5), 628–634 (2017)
6. Parkhom, A.: virtual machine scheduling method based on load balancing in cloud computing environment. Inf. Comput. (Theoret. Edn.) **44**(21), 48–50 (2017)
7. Zhong, B., Jiang, L.: Distributed network load balancing strategy using IMKVS combined with NFV in software defined networks. Comput. Appl. Res. **36**(5), 1504–1509 (2019)
8. Xin, Z., Lei, T.: Simulation of dynamic network resource scheduling in cloud computing environment. Comput. Simul. **34**(12), 402–406 (2017)

9. Liu, S., Li, Z., Zhang, Y., et al.: Introduction of key problems in long-distance learning and training. Mob. Netw. Appl. **24**(1), 1–4 (2019)
10. Razzaghzadeh, S., Navin, A.H., Rahmani, A.M., et al.: Load balancing based on statistical model in expert cloud. Majlesi J. Electr. Eng. **13**(4), 61–71 (2019)
11. Fu, W., Liu, S., Srivastava, G.: Optimization of big data scheduling in social networks. Entropy **21**(9), 902 (2019)
12. Madni, S.H.H., Latiff, M.S.A., Ali, J.: Hybrid gradient descent cuckoo search (HGDCS) algorithm for resource scheduling in IaaS cloud computing environment. Cluster Comput. **22** (1), 301–334 (2019)
13. Pan, Z., Liu, S., Sangaiah, A.K., et al.: Visual attention feature (VAF): a novel strategy for visual tracking based on cloud platform in intelligent surveillance systems. J. Parallel Distrib. Comput. **120**, 182–194 (2018)
14. Chaudhary, D., Kumar, B.: Cloudy GSA for load scheduling in cloud computing. Appl. Soft Comput. **71**, 861–871 (2018)
15. Aruna, M., Bhanu, D., Karthik, S.: An improved load balanced metaheuristic scheduling in cloud. Cluster Comput. **22**(5), 10873–10881 (2019)

Intelligent Authentication Method for Trusted Access of Mobile Nodes in Internet of Things Driven by Cloud Trust

Shu Song[1] and Lixin Jia[2(✉)]

[1] Changjiang Polytechnic, Wuhan 430074, China
ss209809@126.com
[2] Jiangsu Fangtian Power Technology Co., Ltd., Nanjing 210096, China
lxj987600@126.com

Abstract. In order to solve the problem that traditional cloud trust-driven mobile nodes in the Internet of Things lack credible authentication, a cloud trust-driven intelligent authentication method for trusted access of mobile nodes in the Internet of Things is proposed. The mobile nodes in the Internet of Things are determined based on cloud trust-driven, relying on the processing of mobile nodes in the Internet of Things and the intelligent authentication of trusted access of mobile nodes in the Internet of Things. The cloud trust-driven Internet of Things migration is realized. Mobile node trusted access intelligent authentication. The experimental data show that the proposed intelligent authentication method can not only improve the credibility of the traditional authentication method, but also simplify and standardize the authentication process. It enhances the adaptability and flexibility of trusted access authentication of Internet of things driven by cloud.

Keywords: Cloud trust drive · Internet of Things · Mobile node · Intelligent authentication

1 Introduction

The Internet of Things (IOT) is a hybrid heterogeneous network composed of perceptual subnet, transmission subnet and application subnet. As an important part of the Internet of Things, wireless sensor has been widely used. Wireless sensor networks (WSNs) are composed of a large number of low-cost sensor nodes with weak computing and communication capabilities and limited power. After the sensor node collects the sensing data, the node sends the data to the background server of the base station in a mobile ad hoc manner, and the information acquisition, the processing and the analysis of a specific area at any time are realized [1]. In the future, there will be a large number of mobile nodes in the Internet of things, so it is necessary to deeply study the security of access authentication of mobile nodes in the Internet of things (Table 1).

The cloud platform of the Internet of things collects and uses data through the nodes of the Internet of things, performs data calculation and storage based on the cloud platform, improves the ability of the Internet of things to process data and the scope of data sharing, and enriches the content of cloud data. It promotes the

S. Liu and L. Xia (Eds.): ADHIP 2020, LNICST 348, pp. 232–241, 2021.
https://doi.org/10.1007/978-3-030-67874-6_22

penetration and integration of the Internet and the human world, and also brings new security issues. Due to the characteristics and limitations of IOT nodes, they are extremely vulnerable to attack. At present, scholars have carried out research on the node access of the Internet of things. Ben and others put forward the Internet of things chain is proposed to build trust in the Internet of things ecosystem. First, the system model is defined, including trusted Internet of things data server, authorization management server and semi trusted cloud re encryption proxy server. Secondly, describe the flow and algorithm of the system; finally, analyze and prove the security of pre-tuan. Based on proxy re encryption, pre-tuan will give full play to the computing power of the cloud. At the same time, ensure the security and reliability of Internet of things data sharing. In order to improve the efficiency of trusted proof of IOT nodes [2]. Gong and others proposed a threshold signature method for Internet of things based on credibility. When the sum of the credibility of the IOT nodes participating in the signature is greater than or equal to the threshold, the role of the nodes in the proof becomes greater, and vice versa. Security analysis and example analysis show that the scheme can resist the collusion attack of any member whose credibility sum is less than the threshold value, and can effectively reduce the burden of IOT nodes on the premise of ensuring the security of IOT [3].

This paper proposes an intelligent authentication method for trusted access of IOT mobile nodes driven by cloud trust. IOT mobile nodes are driven by cloud trust and determined by the processing of IOT mobile nodes and the intelligent authentication of trusted access of IOT mobile nodes. Realize the Internet of things migration driven by cloud trust. Intelligent authentication for trusted access of mobile nodes. Experimental data show that this method not only improves the reliability of traditional authentication methods, but also standardizes the authentication process.

2 Intelligent Authentication of Trusted Access of Mobile Nodes in Internet of Things Driven by Cloud

2.1 Mobile Node Determination of Internet of Things Based on Cloud Trust-Driven

In the perception subnet of the Internet of things, the roaming target domain of mobile nodes is usually random. In view of the future application trend of the Internet of things, the roaming phenomenon of mobile nodes is bound to exist in large quantities. After the mobile node joins the remote domain and passes the authentication, it can obtain all the network resources of the remote domain at will.In practical applications, the location of Internet of things mobile nodes based on cloud trust drive is generally unknown, so it is unreasonable to make the moving track of target nodes include all nodes, so it is necessary to determine the mobile nodes of Internet of things. Determining the location of each node is one of the fundamental problems in the field of wireless sensor networks. A two-step localization algorithm based on UKF filter and triangulation algorithm has been proposed. The algorithm uses a mobile node to traverse the whole network and periodically broadcasts information containing its current location. The self-localization process of the sensor node is realized by the target tracking method of UKF [4]. This algorithm proves that it can improve the special requirements of the moving trajectory of mobile nodes, and is more suitable for the actual situation, and obtain a better positioning accuracy. Assuming that the mobile node based on cloud trust-driven Internet of things has the capability of RSSI ranging, then the schematic diagram of the Euclidean location algorithm is shown in Fig. 1. In the figure, the known unknown node B and C are known to have known BC distance or can be obtained by RSSI measurement within the wireless range of the target node L, and node A is adjacent to B, C. Then, for all the edges of the quadrilateral ABCL, and a diagonal BC, the length of the AL (the distance between node A and L) can be calculated according to the properties of the triangle. Using this method, when the unknown node obtains the distance from three or more target nodes, the mobile node based on cloud trust-driven Internet of things can be determined.

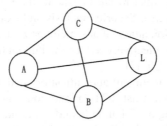

Fig. 1. Diagram of Internet of things mobile node determination based on cloud trust-driven

If a mobile node with the capability of RSSI ranging can move three times (or more) within the communication range of the sensor node and is not in a straight line, this is equivalent to meeting the requirements of the three target nodes in the Euclidean localization method [5].

The Euclidean distance-finding method uses mathematical calculation method to determine the mobile nodes of Internet of things driven by cloud trust. The mobile node determines the degree of deviation, assuming that X1, X2,. Xn is the mathematical expectation that the different mobile nodes, E (X) is a random variable, and the mobile node determination formula is as follows:

$$E\,(X) = \sum_{1=1}^{n} X_i \qquad (1)$$

$[X_i - E\,(X)]^2$ is the square deviation of mobile nodes, and the arithmetic mean of $[X_1 - E\,(X)]^2$, $[X_2 - E\,(X)]^2$, ... $[X_i - E\,(X)]^2$ is the average square deviation of this set of data., the expression formula is shown in formula 2:

$$\sigma = \frac{1}{n} \sum_{i=1}^{n} X_i [X_2 - E\,(X)]^2 \qquad (2)$$

σ is the standard deviation which measures the degree of dispersion between the measured value and the average value. The greater the standard deviation, the greater the degree of discretization of the random variables of the data, and the greater the degree of deviation in the determination of the moving nodes.

When the expected value is equal or close, the standard deviation can be used to compare the deviation degree directly [6]. If the expected value of the two groups of distributions is obviously different, the coefficient of variation should be used to compare it [7]. The coefficient of variation is the ratio of the standard deviation to the expected value and is expressed in formula 3:

$$V = \frac{\sigma}{E(X)} \tag{3}$$

In the whole positioning and ranging process, the trilateral localization algorithm can only measure the position of ordinary sensor nodes. When the sensor node has the ability of RSSI ranging, it cannot measure its position, and the Euclidean algorithm can improve this situation very well. This paper combines UKF filtering to eliminate the noise interference, that is, the Internet of things mobile node processing. As a result, more accurate results are obtained.

2.2   Filtering Process of Mobile Node in Internet of Things

The cloud trust-driven mechanism is defined as follows: in an open cloud environment, when the addressing service AS1 of the Internet of things across the domain strictly adheres to certain specific constraints, and acts according to the trust value of the addressing service AS2 through the sensor protocol management mechanism [8]. When the whole system can cooperate dynamically and reach the uniform state of the underlying addressing and positioning standard, it is called the trust-driven relationship between the addressing service AS1 and the addressing service AS2 [9].

For the $i(i = 1,\cdots,I)$ mobile node of the Internet of Things driven by cloud trust, the state equation in the $k(k = 1,\cdots,K)$ iteration cycle is:

$$Xi = (k-1) + wi(k) \tag{4}$$

In the formula, $wi(k)$ represents the noise of the Internet of things mobile node system driven by cloud trust. There are two kinds of factors affecting the processing of the Internet of things mobile node. The first is the selection of the initial state vector and the other is the selection of the measurement noise. The distance between the sensor node and the moving target node is unpredictable [10]. The filtering process must require an objective equation of state to predict the next moment, so the equation of state shown in formula (4) generally believes that the position of the next moment is basically the same as the current position, however, when the initial value is close to the real value, the state prediction equation is close to true. Therefore, in order to obtain accurate filtering results, the requirements for the selection of initial values need to be improved. For the measured noise Q, it will affect the filtering speed of each step of the iterative filtering estimation [11]. Generally, when the moving target node changes rapidly, a larger value should be taken, and a smaller value should be taken when the estimated value is close to the real value. For Q value, because the speed of moving node can be controlled, a moderate value can be set according to it. For the initial state setting, the Euclidean localization algorithm can usually be used.

2.3   Credit Value Evaluation and Processing Process of Internet of Things Mobile Node

The cloud trust value expresses the trust evaluation criterion of the underlying resource addressing service of the Internet of Things in the cloud environment, and the dynamic trust management model, the feedback-based trust driving mechanism, the trust benefit function (trust steepness function) heuristic algorithm can be adopted, The cloud-based trust evaluation algorithm is used to solve [12]. The credibility value depends on the dynamic information of the trust object which is difficult to capture, and it is difficult to verify the new time-effective weights of credit degree. To some extent, this affects the construction of trust relationship between cross-domain addressing services. In order to solve the above problems, Euclidean localization algorithm combined with cloud-based trust evaluation algorithm to improve the trust benefit function, and use the improved algorithm to solve the trust value.

According to the above discussion, the localization method can be divided into two steps: firstly, the initial position of sensor node is determined by Euclidean positioning and distance finding method, and then the UKF filtering method is used to locate the sensor node accurately. Since the work of each sensor is independent, for sensor I, the flow chart of each sensor is shown in Fig. 2 when the time threshold Tr meets certain conditions.

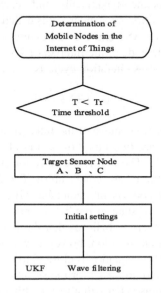

Fig. 2. Flow chart of Internet of things mobile node processing

This method reduces the requirements of moving trajectory of mobile nodes, and can obtain satisfactory positioning accuracy even if the nodes move at will. This method is improved at the expense of a certain amount of computation.

Based on the Euclidean positioning algorithm, the mobile node processing of the Internet of Things is finally realized by combining the cloud trust value. And lays the foundation for the trusted access intelligent authentication of the Internet of Things mobile node.

2.4   Intelligent Authentication of Trusted Access to Internet of Things Mobile Node

The intelligent authentication of trusted access of Internet of things is designed by determining the mobile node of Internet of things based on cloud trust drive and the processing of mobile node of Internet of things. On the basis of traditional authentication protocol of mobile node of Internet of things, intelligent authentication of trusted access of mobile node of Internet of things is carried out.

Traditional IoT mobile node authentication model mainly consists of Internet of things Management Center (CA-IoT), mobile aggregation terminal (Mobile Sink Node, MSN, base station (Base Station,BS), sensor node (Sensor) and Internet of things mobile node (Cluster Head,. CH) composition [13].

First, the traditional mobile node authentication protocol of the Internet of things is initialized. In the system initialization phase, when MSN registers with CA-IoT with its own real identity, CA-IoT provides a series of non-linked random pseudonym identity (PID),. PID = {pid1,pid2, …, pidn}. The local authentication server (Home Authentication Server,HAS) will be pre-assigned to each pseudonym identity Pidipid public key pkpidi and the corresponding private key skpidi and then CA-IoT will put all the tuples (pidi,pkpidi,skpidi) sent securely to MSN [14]. There has been a detailed and quantitative study of the storage space of the anonymous key and related certificate for long-term use of a pseudonym in advance. The pre-loaded pseudonym is applied and the storage overhead is within a reasonable range.

Modbus protocol is an important data to realize trusted access intelligent authentication of IoT mobile node, which determines the communication state of trusted access intelligent authentication of IoT mobile node [15]. If that communication protocol of the Modbus is not match with the trusted access intelligent authentication system of the Internet of Things mobile node, a separate remote communication can not be realized, and when the mobile node of the Internet of things is trusted to access the intelligent authentication response message, it is necessary to query in a broadcast mode, otherwise, no response message will be received. Since the Internet of things mobile node trusted access intelligent authentication response message is also composed of Modbus protocol, it is necessary to confirm the Internet of things mobile node trusted access intelligent authentication data and error detection domain. If the trusted access intelligent authentication of the Internet of things mobile node occurs in the process of receiving messages, then the trusted access intelligent authentication of the Internet of things mobile node will not execute its command, so it is necessary to establish the transmission process. Feedback the error message and send it out in a timely manner. The feedback circuit principle is shown in Fig. 3.

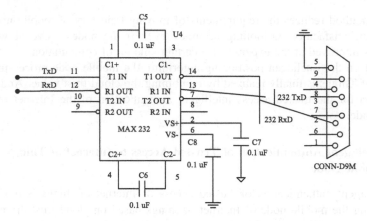

Fig. 3. Schematic diagram of error message feedback circuit

The application network makes the Modbus communication protocol match with the trusted access intelligent authentication of the Internet of things mobile node, so any action of the trusted access intelligent authentication of the Internet of things mobile node can be realized. In the process of trusted access intelligent authentication of Internet of things mobile nodes, the control system can be used not only as master system but also as slave system.

Improve the initial position accuracy of the filtering by using the Euclidean positioning method, thereby improving the processing effect of the Internet of Things mobile node, and simultaneously, on the basis of the traditional Internet of Things mobile node authentication protocol, conducting the intelligent authentication of the Internet of Things mobile node to the trusted access, And the trusted access intelligent authentication of the Internet of Things mobile node which is driven by the cloud trust is completed.

3   Results

In order to verify the validity of the proposed intelligent authentication method for trusted access of mobile nodes in Internet of things (IoT) driven by cloud trust, experiments are carried out to demonstrate and analyze the effectiveness of the proposed method. In order to ensure the accuracy of simulation test, the traditional iotchain: establishing trust in the Internet of things ecosystem using blockchain (reference [2] method) is used as the comparative experimental object, and the data generated by the two methods are given in the same data chart.

3.1   Experimental Environment and Test Data Setting

The specific experimental environment is as follows: Intel (R) core (TM) i5-6500 processor, 3.20 ghz CPU, 16 GB memory, windows 10 system version, 64 bit operating system.

In order to ensure the accuracy of the simulation test process, set the test parameters, first modify the address configuration mechanism of hierarchical mobile IP to make it use stateful address configuration, and access router 1 (AR1) in the home proxy (HA). AR2 and AR3 install dibbler software and package filtering software, and install RADIUS software on the server. It starts HMIPv6 program on MN, HMIPv6 program and DHCPv6 server program on HA, DHCPv6 server program and routing advertisement protocol on AR1, AR2 and AR3 respectively. In-service The AAA server-side program is started on the server. Where the user name option format, the password option format, and the authentication failure information are defined as follows:

Table 1. Lab parameter settings tabl

Identification	Character length
Username Option Identification (44)	Option Length (20 bytes)
User name (16 bytes)	
Password Option Identification (45)	Option Length (20 bytes)
Password ciphertext (16 bytes)	
Failure Information Option Identification (46)	Option Length (20 bytes)
Failure information (16 bytes)	

The experiment completes the access authentication of handoff process from home to AR1, AR2 in MAP1 and AR3 in MAP2, and the access authentication process from MAP1 to MAP2. If the user adopts the wrong user name or password, the IP address cannot be obtained during the handoff process, the access authentication fails, the handover process cannot be completed, and the MN is refused to join the phase. The network to which it is to be.

The simulated cloud trust-driven object-of-things mobile node, in the X–Y plane, the area[0, 1000 m] × [0, 1000 m], is randomly scattered by the airplane to the unknown sensor node and a movable, position-aware target node, and at the same time, it is assumed that the node can take off, then it will follow a predetermined trajectory, The sensor nodes on the air side to the ground periodically release their position information at a predetermined time interval. While the sensor node in the unknown position can measure its distance from the node. When the node is located with the GPS, it is assumed that the covariance mean value in the self-positioning of the moving target node is assumed in the simulation A positioning error of 50 m.

3.2   Analysis of Test Results

According to the setup of the experiment process, the experimental results of two kinds of authentication methods for mobile nodes of the Internet of things are obtained, and the experimental results are drawn into charts as shown in Fig. 4:

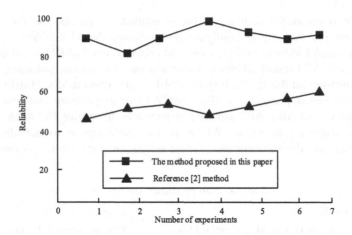

Fig. 4. Comparison table of experimental results

It is proved by experiments that when the server is secure and reliable, the configuration information of software and hardware of MSN platform will not be leaked to other legitimate users in the network, nor will it be leaked to CH, to effectively protect the privacy of the platform. This paper discusses the time needed to exchange information between the duplicate address detection process and the AAA protocol without increasing the additional delay and assigning a legal IP address to the MN as the online forwarding address when the authentication is successful.

The experiment and demonstration analysis show that the trusted access intelligent authentication method of the cloud trust-driven Internet-of-Things mobile node has high credibility, and at the same time, the self-adaptability and the flexibility of the trusted access authentication of the cloud trust-driven Internet-of-things mobile node are enhanced.

4 Conclusions

Because of the lack of trusted authentication in the traditional cloud trust driven mobile nodes in the Internet of things, the security and reliability of data processing in the Internet of things are improved. An intelligent authentication method for trusted access of mobile nodes in the Internet of Things driven by cloud trust is proposed. The method is based on the determination of mobile nodes in the Internet of Things driven by cloud trust, relying on the processing of mobile nodes in the Internet of Things and the intelligent authentication of trusted access of mobile nodes in the Internet of Things. Trusted Access Intelligent Authentication for Mobile Nodes of the Internet of Things Driven by Arbitrary. The experimental results show that the intelligent authentication method for trusted access of mobile nodes in Internet of things driven by cloud can effectively verify the validity of mobile nodes. While the authentication is successful, the mobile node (MN) is configured with a legal IP address, which can meet the needs of practical application and greatly reduce the hidden danger of information security.

Acknowledgment. "13th Five-Year Plan" for national social sciences education program-Education MinistryKey Subject in 2017. Research on the innovation ability of application-oriented talents training under the background of education supply-side reform: a case study of "The Internet of things".

References

1. Kim, D.Y., Kim, S., Park, J.H.: Remote software update in trusted connection of long range IoT networking integrated with mobile edge cloud. IEEE Access, PP(99), 1–1 (2017)
2. Bin, Y., Jarod, W., Surya, N., et al.: IoTChain: establishing trust in the Internet of Things ecosystem using blockchain. IEEE Cloud Comput. **5**(4), 12–23 (2018)
3. Gong, B., Wang, Y., Liu, X., et al.: A trusted attestation mechanism for the sensing nodes of Internet of Things based on dynamic trusted measurement. China Commun. **15**(2), 100–121 (2018)
4. Yuli, Y., Rui, L., Yongle, C., et al.: Normal cloud model-based algorithm for multi-attribute trusted cloud service selection. IEEE Access, 1–1 (2018)
5. Xu, D., Fu, C., Li, G., et al.: Virtualization of the encryption card for trust access in cloud computing. IEEE Access, PP(99), 1–1 (2017)
6. Lu, M., Liu, S.: Nucleosome positioning based on generalized relative entropy. Soft. Comput. **23**(19), 9175–9188 (2018). https://doi.org/10.1007/s00500-018-3602-2
7. Pawlick, J., Chen, J., Zhu, Q.: ISTRICT: an interdependent strategic trust mechanism for the cloud-enabled internet of controlled things. IEEE Trans. Inf. Forensics Secur. **14**(6), 1654–1669 (2018)
8. Tsai, J.L., Lo, N.W.: A privacy-aware authentication scheme for distributed mobile cloud computing services. IEEE Syst. J. **9**(3), 805–815 (2015)
9. He, D., Kumar, N., Khan, M.K., et al.: Efficient privacy-aware authentication scheme for mobile cloud computing services. IEEE Syst. J. 1–11 (2017)
10. Liu, S., Liu, D., Srivastava, G., et al.: Overview and methods of correlation filter algorithms in object tracking. Complex Intell. Syst. (2020). doi:10.1007/s40747-020-00161-4
11. Zhu, C., Rodrigues, J.J.P.C., Leung, V.C.M., et al.: Trust-based communication for the industrial Internet of Things. IEEE Commun. Mag. **56**(2), 16–22 (2018)
12. Yu, B., Wright, J., Nepal, S., et al.: Iotchain: establishing trust in the internet of things ecosystem using blockchain. IEEE Cloud Comput. **5**(4), 12–23 (2018)
13. Chen, J., Tian, Z., Cui, X., et al.: Trust architecture and reputation evaluation for Internet of Things. J. Ambient Intell. Humanized Comput. **10**(8), 3099–3107 (2019)
14. Fu, W., Liu, S., Srivastava, G.: Optimization of big data scheduling in social networks. Entropy **21**(9), 902 (2019)
15. Porambage, P., Okwuibe, J., Liyanage, M., et al.: Survey on multi-access edge computing for internet of things realization. IEEE Commun. Surv. Tutorials **20**(4), 2961–2991 (2018)

Visual Information Processing

Research on Dynamic Integration of Multi-objective Data in UI Color Interface

Ling-wei Zhu and Feng Zhai[✉]

Xi'an Eurasia University, Xi'an 710065, China
{zhulingwei541,zhaifeng5431}@sina.com

Abstract. On the traditional method of dynamic integration of multi-objective data in UI color interface, because of the single integration algorithm, it is easy to lose the target data when there is too much target data. Therefore, based on the use characteristics of UI color interface, a new integration method of multi-objective data is proposed. This method obtains the sampling target through deep web data, detects and tracks the target image, optimizes according to the multi-objective integration, realizes the optimal path multi-objective equilibrium integration. Experimental results: the proposed detection method is fully in place in data integration, the occupancy rate of arm is 0%, the load line of DSP is 20%, the system maintains reliable real-time, and achieves the ideal state of UI color interface operation. However, the traditional data integration method of SLR is not in place; it can be seen that the traditional integration method is not suitable for the requirements of UI color interface with large target data.

Keywords: Artificial intelligence · Network public opinion · Data abnormal behavior · Detection method

1 Introduction

With the development and evolution of multimedia technology in the digital era, UI interface design, as a means and way to enhance brand image, product value and human-computer interaction experience, gradually makes people feel the important role of interface design. In recent years, UI interface design has become a design research direction that both design field and computer field pay attention to [1]. In UI design, communication and communication are the essence of user interface, while in visual communication, 83% of human's reception of external stimuli is the function of visual media. UI interface is all the information sources that users touch when they use the software. Among these visual factors, color can form the effect of clarity, contrast and reflection, which has an impact on human visual system, can make people obtain a kind of impact in the first time, can also emphasize and convey information, as well as express the feelings and emotions of things. How to apply and organize these data has become the key of UI color interface processing.

This paper takes it as the research topic to analyze the dynamic integration of multi-objective data. By finding the candidate page of the data source, the query interface is obtained in the page; the USB camera of the ZC0301 microchip is used to collect images, and the moving targets and new targets newly appearing in the scene are

S. Liu and L. Xia (Eds.): ADHIP 2020, LNICST 348, pp. 245–254, 2021.
https://doi.org/10.1007/978-3-030-67874-6_23

detected and extracted from the video image sequence. Detect and extract; and use Mean-Shift algorithm to track the target and conduct experimental analysis.

2 Dynamic Integration Method of Multi-objective Data in UI Color Interface

2.1 Deep Web Data Acquisition Sampling

Data source discovery refers to the discovery of accessible databases in resources, which is generally divided into two steps: finding candidate data source pages and finding their query interfaces from the obtained pages [2]. The solution to the first step is to transform the technology that has focused on crawling. The goal of focused crawling is to identify those hyperlinks that are more likely to reach the target web page in the process of crawling, so that only part of the web pages that are closely related to the search topic can be crawled and the resource yield can be improved. A common focused crawler is usually composed of three parts: Web collector, web classifier and crawling queue. The key problems in focused crawling are the measurement of the correlation between the obtained web page and the target topic, and the ranking of crawling priority queues. Aiming at the former, this paper mainly studies the topic classification algorithm of web pages. For the latter, there are many different crawling scheduling strategies. One kind of method calculates the link value according to the link relationship between pages, such as algorithm, class algorithm, etc [3]. Another method, such as an algorithm, evaluates links according to the correlation between link information and subject area. The feedback information is used to train the classifier online and incrementally, so as to adjust the priority of the links in the candidate queue.

The second step is to find the query interface from the obtained home page. The current method is to lock the query interface based on the source code of the rule analysis home page. Through a large number of observations, starting from the home page of the website, crawling page links based on the principle of width first, the page where the query interface is located will not exceed the layer, and the query interface will not exceed the layer [4]. Based on the characteristics of query interface, three rules are put forward to judge whether there is a query interface in a page. Firstly, there should be a label in the page, secondly, there must be an input control in the label, thirdly, there should be at least one keyword in a group, such as "query", "search", etc. This method can achieve at least accuracy in its experiments.

2.2 Detection and Tracking of Target Image

Target image input this topic uses the USB camera of zc0301 microchip to collect image.

In this paper, the task of moving target tracking is to detect and extract such two types of targets in the video image sequence: moving targets and new targets appearing in the scene. Therefore, this task can be completed in two steps, the first is target detection, the second is target extraction [5]. The so-called target detection is to detect whether the monitored scene image in the video sequence image changes. If the image changes, it means that there are new targets, otherwise it means that there are no new targets. Target extraction is to segment and extract the target from the video sequence when the target detection algorithm detects the presence of the target, so as to provide data for the next target tracking. In the process of target tracking, one of the key technologies to track the moving target in detection [6]. The accuracy of target detection will have an important impact on the follow-up steps. In the experiment process, if the target detection is successful, the target detection module will know whether there is a moving target from the image sequence. If there is a moving target detection, the position and size of the moving target in the video image will be given through the algorithm, and then submitted to the target tracking module. The target tracking module will establish the tracking mode and the moving target template according to the given target position and size After extraction, the next step is target tracking. The first information needed to track the target is the target position information. The purpose of target tracking is to analyze the image sequence obtained by the camera, calculate the two-dimensional position coordinates of the target on each frame of the image, and associate the same moving target in different frames of the image sequence according to different characteristic values to get the complete motion track of each moving target, That is to establish the corresponding relationship of moving objects in the continuous video sequence. In short, it is to find the exact location of the target in the next image. In moving target tracking, the main work is to select good target features and use practical search algorithm.

The target tracking system designed in this paper adopts mean shift algorithm.

Mean shift algorithm is a local optimal search algorithm. By calculating the probability density index of the similarity between the candidate target and the target module directly, and then using the direction of probability density gradient decline to obtain the best path for matching search, accelerate the positioning of moving target and reduce the search time [7].

Mean shift algorithm starts from kernel density estimation (also known as Parzen window estimation), which is d relatively popular density estimation method at present. Given that the kernel density estimation of multidimensional variables of n sample data $x_i, i = 1, 2, \ldots, n$ in A-dimensional space R^d can be written as follows: Mean-Shift:

$$\widehat{f_{h,k}}(x) = \frac{c_{k,d}}{nh^d} \sum_{i=1}^{n} |H|^{-1/2} k \left[\left\| \frac{x - x_i}{H} \right\|^2 \right] \tag{1}$$

Where $c_{k,d}$ is the normalized constant, $k(x)$ is the kernel function or profile function, and H is the bandwidth matrix of $d \times d$ dimension. A complete representation of H parameters will increase the complexity of the algorithm. In practice, the diagonal matrix $H = diag \left[h_1^2, h_2^2, \ldots, h_n^2, \right]$ or $h^2 I$, I is usually the $d \times d$ - unit matrix. For simplicity, using the latter representation, the kernel density estimate can be written as:

$$\widehat{f_{h,k}}(x) = \frac{c_{k,d}}{nh^d} \sum_{i=1}^{n} k \left[\left\| \frac{x - x_i}{h} \right\|^2 \right] \tag{2}$$

Among them, kernel function $k(x)$ must satisfy the following conditions: 1) $k(x)$ is nonnegative; 2) $k(x)$ is monotonically decreasing from the center to the outside, if $0 \le a \le b$, then $k(a) \ge k(b)$; 3) $k(x)$ is bounded. When $k(x)$ is differentiable, the gradient can be obtained from Eq. (5–11):

$$\nabla \widehat{f_{h,k}}(x) = \frac{2c_{k,d}}{nh^{d+2}} \sum_{i=1}^{n} (x - x_i) k' \left[\left\| \frac{x - x_i}{h} \right\|^2 \right] \tag{3}$$

Definition $g(x) = -k'(X)$, obtained from the above formula:

$$\begin{aligned}
\nabla \widehat{f_{h,k}}(x) &= \frac{2c_{k,d}}{nh^{d+2}} \sum_{i=1}^{n} (x - x_i) k' \left[\left\| \frac{x - x_i}{h} \right\|^2 \right] \\
&= \frac{2c_{k,d}}{nh^{d+2}} \left[\sum_{i=1}^{n} g \left(\left\| \frac{x - x_i}{h} \right\|^2 \right) \right] \left[\frac{\sum_{i=1}^{n} x_i g \left(\left\| \frac{x - x_i}{h} \right\|^2 \right)}{\sum_{i=1}^{n} g \left(\left\| \frac{x - x_i}{h} \right\|^2 \right)} - x \right]
\end{aligned} \tag{4}$$

The second half of the formula is the mean shift algorithm:

$$\nabla m_{k,g}(x) = \frac{\sum_{i=1}^{n} x_i g \left(\left\| \frac{x - x_i}{h} \right\|^2 \right)}{\sum_{i=1}^{n} g \left(\left\| \frac{x - x_i}{h} \right\|^2 \right)} - x, \widehat{f_{h,g}}(x) \frac{c_{k,d}}{nh^d} \left[\sum_{i=1}^{n} g \left(\left\| \frac{x - x_i}{h} \right\|^2 \right) \right] \tag{5}$$

Where $c_{g,d}$ is the normalization constant, the above formula can be written as:

$$\nabla \widehat{f_{h,k}}(x) = \widehat{f_{h,g}}(x) \frac{2c_{k,g}}{h^2 c_{g,d}} \nabla m_{h,g}(x) \tag{6}$$

There are:

$$\nabla m_{h,g}(x) = \frac{1}{2} h^2 c \frac{\nabla \widehat{f_{h,k}}(x)}{\widehat{f_{h,g}}(x)} \tag{7}$$

Among them, $c = c_{g,d} l c_{k,d}$. The above formula shows that the local mean value moves towards the dense area of nearby data samples, so there is an iterative formula:

$$y_{t+1} = y_t + \nabla m_{h,g}(y_t) \tag{8}$$

Among them, y_t represents the sample data of step t, y_{t+1} represents the sample data of step $t+1$. After replacement and simplification, the iteration formula of mean shift algorithm is obtained:

$$y_{t+1} = \frac{\sum\limits_{i=1}^{n} x_i g\left(\left\|\frac{x-x_i}{h}\right\|^2\right)}{\sum\limits_{i=1}^{n} g\left(\left\|\frac{x-x_i}{h}\right\|^2\right)} \tag{9}$$

There are:

$$y_{t+1} = y_t + \lambda_t \bullet d_t \tag{10}$$

Among them, $\lambda_t = h^2 c / 2\widehat{f_{h,g}}(y_t > 0); d_t = \nabla \widehat{f_{h,k}}(y_t)$ The above formula shows that the mean shift algorithm iterates along the ladder direction, so that each point to be processed "floats" to the local maximum point of the distribution density function, and its step size λ_t Changes adaptively with the iteration process, that is, when the current data is relatively low density, the iteration step size is larger; near the local maximum, the iteration step size is smaller.

2.3   Multi Objective Integration

At present, many multi-objective data integration algorithms only consider the current scheduling, without considering the use of data information before task execution, which is easy to cause too many tasks to be allocated to some advantage information, making the advantage information too busy to become disadvantage information, at the same time causing some information to be idle, some information to be busy, resulting in load imbalance, which is not conducive to the pursuit of UI color interface Find the goal of target data integration [8]. Therefore, in order to achieve load balancing, this algorithm adds a load factor to the scheduling, calculates the load factor when a scheduling cycle is completed, and updates the control information accordingly, as shown in formula (1):

$$v = U_c / U_{sun} \tag{11}$$

Among them: v is the load factor, U_c is the completed task amount, and U_{sun} is all the tasks assigned to the resource.

To enhance the cooperation among targets is helpful to find the optimal solution. A single target is easy to converge to the local optimal, and the communication between targets mainly depends on the diffusion of information [9]. The amount of information is related to the distance of the information source. The information diffuses to the surrounding area with the radius of r as the center of O points of the information source. Assuming that the amount of information of O points is τ_{max}, the formula for calculating the total amount of information of any point 8 in the circular area is as follows (12):

$$\tau_p = (1 - L/r)\tau_{max} \tag{12}$$

In addition, suppose that target k has selected nodes a and b, and the distance between them is L_{ab}. For any node 9 adjacent to nodes a and b, if it is within the scope of target c diffusion information, the amount of information that target c diffuses to the nodes can be calculated.

Set $\tau_{max} = v \bullet \Delta \tau_{ab}^k$, $r = \psi \bullet L_{ab}$, $\psi = L_{ab}/L'$. Where: k is the previously calculated load factor, 14 is the amount of information about the path from target k to nodes a and b, and L' is the average distance from all nodes to target k. Then the total amount of information about the path from target k to nodes a, c and b, c is calculated as follows:

$$\Delta \tau_{ac}^k = \begin{cases} \frac{vQ}{l_k}\left(1 - \frac{L_{ac}L'}{L_{ac}^2}\right), L_{ac} & < r \\ 0 \end{cases} \tag{13}$$

$$\Delta \tau_{bc}^k = \begin{cases} \frac{vQ}{l_k}\left(1 - \frac{L_{bc}L'}{L_{bc}^2},\right) L_{bc} < r \\ 0 \end{cases} \tag{14}$$

Where: $\frac{vQ}{l_k}$ is the current pheromone strength, l_k is the distance of target k from the start node to the current node. In this way, formula (13)–(14) can calculate all the information of each target left in the relevant path in each cycle. Each activity of the target will not only affect the information of the path it passes through, but also affect the information of all the paths within the range of its adjacent information diffusion. Through such a local updating method of pheromone, the communication cooperation ability between the targets can be greatly improved, and the algorithm can be enhanced Because of the participation of load threshold coefficient, the load balance performance of the interface can be improved.

2.4   Optimal Path Multi-objective Equilibrium Integration

The ultimate goal of the optimal path multi-objective balanced integration is to simplify the process of the objective, improve the efficiency of implementation, and at the same time minimize the cost of the objective. Therefore, in the integration of the objective process, only considering from the whole can this objective be achieved, so it is necessary to adopt the global optimization. According to the requirements of the efficiency and expenditure of the target object, these two indicators are regarded as the final implementation target of the target integrated calculation [10]. At the same time, considering the credibility and reliability of the target object, the final model is established as follows:

$$MinF(P) = (T(P), C(P)) \tag{15}$$

The constraint conditions of path P subject to reliability and reputation are:

$$\mathrm{Re}\,p(P) \geq \mathrm{Re}\,p_0 \tag{16}$$

$T(P), C(P)$ is the calculation formula of each parameter, which is established by defining different target execution models. By taking the whole as the objective of optimization and adopting constraints, this algorithm can be applied to the calculation of N kinds of target set integration algorithms.

Taking genetic algorithm as an example, the integration process path of each goal is set as a gene chromosome [11–14]. Through gene operations such as cross mutation and recombination of chromosomes, the next generation of chromosomes will be generated. The new chromosome is closer to the set target value, and the next generation of chromosomes will be continuously generated through gene operation method, and finally the set target value will be achieved Convergence. Let the target population be medium gene population P_1 and limited population P_2, the algebra of chromosome evolution is T, and the final path result is P^*. The operation flow of gene evolution target is shown in Fig. 1.

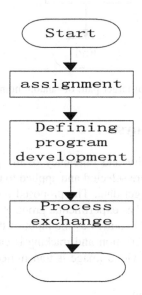

Fig. 1. Operation flow of target evolution

Each chromosome target evolution process needs an initial evolution path. In this calculation, a set of initial target paths is randomly generated, and the initial target paths are constrained by the set constraint algorithm to obtain an optimal target path generated by random numbers. The optimal target path method generated by random numbers is shown in Fig. 2.

According to the theory of gene evolution, in the same way, only each generation selects the highest quality target data to enter the next generation of gene genetic operation can get the optimal solution data, so it is necessary to judge the quality of the target data.

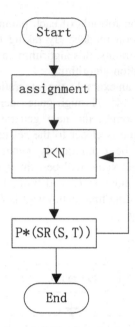

Fig. 2. Optimal target path of random number generation

3   Experiment and Analysis

In order to verify the reliability and efficiency of the proposed integration method, traditional detection methods are selected and applied to the test of dynamic integration method for detecting multi-target data. The proposed method is used as experimental group A and traditional method as experimental group B. Taking the video information collected by the front-end CCD camera as the input of the video image, the algorithm processing of moving target detection and tracking is carried out on the development board, and then the processed video image is transmitted to the large screen display.

3.1   Experiment Preparation

We collect CIF (352 × 288 resolution) images in YUV422 format to Da Vinci platform in real time, and start the moving target detection and tracking system. The first 50 frames are used to get the position of the moving target, and then the tracking and integration are started. Two integration methods are used to detect and integrate the target data sequence. Get and analyze the experimental test results.

3.2   Experimental Results and Analysis

Under the condition of keeping other experimental conditions unchanged, the two data integration methods are compared, and the target tracking performance during the inspection period is shown in the following table:

Table 1. Comparison of target tracking performance statistics

	Experimental B group (traditional methods)	Experimental A Group (Methods)
Frame rate	21	18
DSP load	24%	20%
ARM load	1%	0%
Time(s)	9	8

According to Table 1, the real-time performance index of the system when the moving target tracking integration algorithm is used, it can be seen that the integration method designed in this paper is fully in place in data integration, the occupancy rate of arm end is 0%, the load line of DSP is 20%, the system maintains reliable real-time performance, and achieves the ideal state of UI color interface operation. On the contrary, the traditional method of data integration is not in place; Under the same conditions, the time used is 1 s less than the traditional method, which improves the integration efficiency. It can be seen that the traditional integration method is not suitable for the requirements of UI color interface with large target data. Comprehensive experimental results show that the proposed multi-objective data integration method can perform data integration in the UI color interface with large target data.

4   Conclusion

Because traditional integration methods cannot meet the needs of UI color interfaces with large target data. The proposed multi-target data integration method can perform data integration in a UI color interface with larger target data, and has high practical applicability. This paper analyzes the multi-objective data integration method of UI color interface. Combined with the technical characteristics of UI color interface, the relevant algorithm is used to optimize parameters and improve the integration rate of multi-objective data. It is hoped that the multi-objective data integration method of UI color interface designed in this paper can provide theoretical basis for the application of UI color interface in China.

References

1. Zeng, J., Dou, L., Xin, B.: Multi-objective cooperative salvo attack against group target. J. Syst. Sci. Complex. **31**(1), 244–261 (2018)
2. Zhang, X., Tan, Y., Yang, Z.: Resource allocation optimization of equipment development task based on MOPSO algorithm. J. Syst. Eng. Electron. **30**(6), 1132–1143 (2019)
3. Gao, K., Cao, Z., Zhang, L., et al.: A review on swarm intelligence and evolutionary algorithms for solving flexible job shop scheduling problems. IEEE/CAA J. Automat. Sin. **6** (4), 904–916 (2019)
4. Cheng, S., Lei, X., Lu, H., et al.: Generalized pigeon-inspired optimization algorithms. Sci. China (Inf. Sci.) **62**(7), 120–130 (2019)

5. Hu, Y., Wang, J., Liang, J., et al.: A self-organizing multimodal multi-objective pigeon-inspired optimization algorithm. Sci. China (Inf. Sci.), **62**(7), 73–89 (2019)
6. Yan, L., Qu, B., Zhu, Y., et al.: Dynamic economic emission dispatch based on multi-objective pigeon-inspired optimization with double disturbance. Sci. China (Inf. Sci.) **62**(7), 108–119 (2019)
7. Yang, Yu., Gao, S., Wang, Y., et al.: Global optimum-based search differential evolution. IEEE/CAA J. Autom. Sin. **6**(2), 379–394 (2019)
8. Shuai, L., Gelan, Y.: Advanced Hybrid Information Processing, pp. 1–594. Springer, USA. https://doi.org/10.1007/978-3-030-36402-1
9. Liu, A., Deng, X., Ren, L., et al.: An inverse power generation mechanism based fruit fly algorithm for function optimization. J. Syst. Sci. Complex. **32**(2), 634–656 (2019)
10. Sun, J., Ling, B.: Software module clustering algorithm using probability selection. Wuhan Univ. J. Nat. Sci. **23**(2), 93–102 (2018)
11. Gong, D.W., Sun, J., Miao, Z.: A set-based genetic algorithm for interval many-objective optimization problems. IEEE Trans. Evol. Comput. **22**(99), 47–60 (2018)
12. Bradford, E., Schweidtmann, A.M., Lapkin, A.: Efficient multiobjective optimization employing Gaussian processes, spectral sampling and a genetic algorithm. J. Glob. Optim. **71**(2), 1–33 (2018)
13. Liu, S., Bai, W., Liu, G., et al.: Parallel fractal compression method for big video data. Complexity **2018**, 2016976 (2018). https://doi.org/10.1155/2018/2016976
14. Ben Elghali, S., Outbib, R., Benbouzid, M.: Selecting and optimal sizing of hybridized energy storage systems for tidal energy integration into power grid. J. Mod. Pow. Syst. Clean Energy **7**(1), 113–122 (2019)
15. Liu, S., Lu, M., Li, H., et al.: Prediction of gene expression patterns with generalized linear regression model. Front. Genet. **10**, 120 (2019)

The Application of Visualization of Internet of Things in Online Teaching of Mobile Interactive Interface Optimization

Feng Zhai$^{(\boxtimes)}$ and Ling-wei Zhu

Xi'an Eurasia University, Xi'an 710065, China
zhaifeng5431@sina.com

Abstract. The existing interactive interface of online teaching mobile terminal is not well used in the process of user experience. To optimize it, this paper puts forward the application of the visualization of the Internet of Things in the optimization of the interactive interface of online teaching mobile terminal. Utilizing the visualization technology advantages of the Internet of Things, we can improve the user's sense of use by optimizing the mobile interactive interface vision and human interaction design with the use module. Design simulation experiment compares the number of user choices before and after optimization to verify the validity of the design.

Keywords: Internet of Things · Online teaching · Mobile terminal · Interactive interface

1 Introduction

The innovation and development of mobile equipment and social network technology has caused great changes in the way people communicate, learn and play [1]. The use of mobile devices and social media enables everyone to maintain communication and communication with other members of society at anytime, anywhere, flexibly and conveniently [2]. The Internet of Things (IOT) is considered as the third typical application developed by the explosion of information after human beings entered the Internet era: the first typical application is the application of information, that is, search engine; the second typical application is the application of interpersonal relationships, that is, social networking sites; and the third is the application of things, that is, the Internet of Things. The Internet of Things (IOT) technology is designed to achieve a worldwide physical network for object awareness and interchange, enabling wider interconnection. With the rise of the Internet of Things, the development prospects of related industries are very promising [2]. At present, there are few studies on the experimental teaching of the Internet of Things. Most of the papers are about some technology and development of the Internet of Things itself. There are few studies on specific teaching tools and teaching methods. Although many colleges and universities in China have set up the related specialties of Internet of Things, they have not changed the teaching according to the characteristics of their disciplines. In the experimental teaching, the existing software and hardware are mostly used directly, and the actual teaching effect is not taken into account.

© ICST Institute for Computer Sciences, Social Informatics and Telecommunications Engineering 2021
Published by Springer Nature Switzerland AG 2021. All Rights Reserved
S. Liu and L. Xia (Eds.): ADHIP 2020, LNICST 348, pp. 255–265, 2021.
https://doi.org/10.1007/978-3-030-67874-6_24

In addition, through the investigation of the experimental teaching platforms used by domestic colleges and universities that have set up the specialty of Internet of Things, the drawbacks are also obvious: most of these experimental platforms provide very simple basic verification experiments, which are far from the actual application. The cost is relatively high, and the parts in the test box are highly customized, which damages one or more parts, so the whole test platform cannot be used anymore, and the daily maintenance of the equipment is very troublesome [4]. Existing systems have problems such as inconvenience in operation and maintenance, and do not use cloud technology. Therefore, it puts forward the application of the Internet of Things visualization in the optimization of the interactive interface of the mobile side of online teaching, setting up the instant interaction function between teachers and students, highlighting the guiding role of education teachers and the main role of students, supporting the upload of continuation points such as video and audio, so as to make the learning forms of students more diverse. The function of tripartite evaluation of students, teachers and business managers is set up to make the evaluation of students more comprehensive. The user experience of interface is smooth, easy to maintain and update, and it is suitable for school application. The experimental results show that the interface optimized by the proposed method is more popular with testers, and the selection rate is as high as 94.29%. And in the experimental process, the program runs well and the interface runs smoothly.

2 Optimize the Interactive Interface of Mobile Teaching on the Visualization of the Internet of Things

2.1 Design Optimization of Mobile Interactive Interface Based on Visualization

Interfaces are a very important part of human-machine interaction. From the point of view of design psychology, it can be divided into two main dimensions, affection and sensation, among which sensation includes touch and vision and hearing [5]. The design of interactive interface is a fusion of different disciplines, including cognitive psychology, design and language. The interactive interface is the most direct part of user contact. Design includes icons, background and overall interface effects, which affect the style of the entire interface design.

The development of graphical interface on mobile side has led to a new direction in the design of human-computer interaction. The development of interactive design and GUI design has evolved from a single functional requirement to a bridge to coordinate users and satisfy their emotional needs. Emphasizing the user's feeling and satisfying the user's emotional experience is the general trend of interactive design.

Complete the interface design based on feeling and emotion. Emotion is a user's subjective perception of the interface. Interface is the only channel for user perception, which is divided into sensory, emotional and cultural levels [6]. Traditional interactive design pays more attention to practical usability, ignoring emotional penetration will directly affect the friendly performance and usage evaluation of the interface. This requires us to pay more attention to the importance of emotion and sensory invocation in interface interaction design. Starting with pictures and graphic elements, we will use the tensive and expressive interface design to stimulate user's emotional feedback, which will increase the fun and pleasure, thus encouraging users to use it. The core part is always to focus on and understand the mindset of the target user. User-centered interface design needs to grasp user's psychology, immerse emotions in the design, adjust interface zoning and classification, and help users complete operation and decision-making in pleasure and novelty.

Applied visual design includes: font, color combination, icon, spacing, style uniformity, visual continuity. The readability and recognizability of text include readability, size, contrast between text color and background, and no interference from surrounding design elements. Consider the age group of a particular population, such as the elderly, and have tips to adjust the font size. Instead of using low-resolution pictures, use vector graphics for design rather than bitmaps. Follow the high resolution display design and scale down. Screen resolution is getting higher and higher, requiring scaling down from high resolution devices. Design drawings can be imported into mobile phone test corrections after design. The final result graph of the final design implementation on mobile devices must also undergo continuous iteration and upgrade.

Experience ease of use: Streamline the main information and operations, prioritization, the most important core functions and content to be reflected in the layout, so that there are obvious functional options in the operation; try to avoid introducing animation. The application loading process transitions from picture display to animation, which requires excessive natural panning. The first time a user opens an app, introducing animations is fun and lengthens the user's exposure to the app. It's worth waiting unless you can guarantee a short period of time and a fine and attractive design.

The button click design of the operation interface should have implication effect, dynamic conversion, concave and convex effect, and shadow effect, and ensure that the button design can ensure the click range to avoid mis-operation of keys. Use theme style, size, and color resources together to avoid redundancy.

Responsive design, also known as adaptive design, requires a unified visual effect across platforms and the unified visual effect of different mobile terminal screen sizes. By choosing the appropriate layout composition type, the visual display effect that mobile WEB needs to consider will be extended, the display of horizontal and vertical interfaces will also make APP design more difficult, and the corresponding screen size category will be expanded [7].

Mobile application interface design should avoid panning other systems causing obtrusiveness due to differences in application operating system versions and aesthetics and interaction. The design panels described in this section include application startup pages, size specifications for interface design, icon design and color design. This section describes the differences between different devices, especially the classification of screen size and density. The Android interface size is usually 480 * 800,720 * 1280,1080 * 1920. After multi-machine testing, 720 * 1280 is suitable for display, 1080 * 1920 is clear, and the size of the picture file after cutting is appropriate to reduce memory consumption. The basic Android interface is divided into navigation and status bars, as well as the main menu content area. Usually the base size chosen is 720 * 12800. Take the Android system as an example, and the basic composition of the interface diagram is shown in Fig. 1:

258   F. Zhai and L. Zhu

Fig. 1. Interface diagram of android application basic composition

Droid sans fallback is Google's default font on Android. The acceptable range of text resolution and comfort values can be analyzed from the survey of Baidu user experience in the following table, as shown in Table 1:

Table 1. Acceptance of font resolution for users

		Minimum value	Acceptability	Comfort value
High resolution	Long text	21px	24px	27px
	Short text	21px	24px	27px
	Notes	18px	18px	21px
Low resolution	Long text	14px	16px	18px–20px
	Short text	14px	14px	18px
	Notes	12px	12px	14px–16px

The screen density benchmark for the device is medium. As a result, adding benchmark icons increases generation to create a high-density version, and vice versa. Put the icon in the application's specific density resource directory. For example: (run under the default res/drawable/directory). Avoid filling in too much information or UI components on high-density pixel screens. When designing a UI for a high PPI (pixel per inch) screen, because more pixels are available, placing more information on the interface or UI components can lead to a preview of the design results, the interface clutter can affect the display of key functions on the interface. Screen density standard version size, as shown in Table 2:

Table 2. Screen density standard version size

Classification	Low density screen	Medium density screen	High density screen
Menu	36 × 36px	48 × 48px	72 × 72px
Status bar	24 × 24px	32 × 32px	48 × 48px
Label	24 × 24px	32 × 32px	48 × 48px
Dialogue	24 × 24px	32 × 32px	48 × 48px
List view	24 × 24px	32 × 32px	48 × 48px

The value of interface color is hexadecimal. Because there are 10 million different colors that can be distinguished by the naked eye, users pay much attention to color during the process of interface interaction. Color also has implications on users' psychology and culture. For example, red usually means warning, green means calm and healthy, etc. To reduce the complex memory of user interface information and actual operation, color information needs to be used efficiently to distinguish content classification from level [8]. This requires choosing appropriate colors for different interface scenarios, such as managing cool tone anomalies in the interface. Secondly, avoid more than three colors in the same interface in terms of matching style to avoid visual confusion. Finally, the contrast of colors should be emphasized clearly, such as using dark text on a light background, or using special colors to emphasize important information tips that users need to be aware of.

2.2 Human Computer Interaction Design Optimization

In the process of learning the boutique course, the learner will encounter difficult problems. The learner can go to the FAQ module to find solutions. If the FAQ module has no questions that the learner encounters or the learner is not satisfied with the presented answers, it is necessary to seek help from the teachers or other learners through the interpersonal communication tools provided to the learners through the elaborate course platform. Interpersonal interaction can be divided into synchronous interaction and asynchronous interaction according to the time of interactive feedback. Synchronous interaction is a real-time interaction in which the learner can get feedback immediately while asking a question [9]. Asynchronous interaction is a non-real-time interaction in which the learner takes some time to get feedback after asking a question. At present, the interaction provided by the boutique courses is mainly asynchronous, which means that the learner's questions need to wait for some time to get feedback. If the time interval is long, it will frustrate the motivation of the learner. Therefore, synchronous interaction should be combined with asynchronous interaction. Based on this, the structure of interpersonal interaction system is designed, as shown in Fig. 2:

Asynchronous interaction design: Because the learner is separated from the tutor and other learners in time and space when learning the boutique courses, it determines that the communication between people in the boutique courses must be mainly asynchronous interaction. At present, the asynchronous interactive tools provided by the boutique courses are relatively single, different learners are accustomed to using different interactive tools, and each interactive tool has its own unique functions.

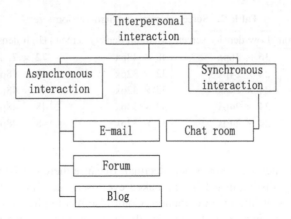

Fig. 2. Diagram of interpersonal interaction design

Therefore, boutique courses should provide a variety of interactive tools to enhance the communication between learners and teachers, and between learners and learners. Based on this, design three interactive ways: e-mail, forum and blog.

E-mail: E-mail is a kind of interaction often provided in the boutique courses. Its advantages are fast communication, convenience, low cost, and one-to-one non-real-time communication. Through e-mail, learners can send their homework or difficult questions to teachers for guidance and answers. Teachers can understand the learning situation of learners and give them guidance and encouragement. At the same time, e-mail reduces the barriers to traditional face-to-face communication and makes it easier for teachers to maintain contact with learners [10].

Forum: In the boutique course website, the course builder can establish a BBS site, or connect a BBS site, which is a commonly used interactive way to achieve one-to-one, one-to-many interaction [11]. In the forum, you can set up a variety of learning topics related to the content of the course, or you can set up a theme dedicated to emotional communication. The learner can choose the theme module of interest to learn [12]. If the learner has any questions or puzzles, you can post a post in the forum to ask for help, and the teacher or other learners can reply through the forum. At this time, as long as the online learners can see the reply information, which not only avoids repeated questions of the same problem, but also reduces the time for teachers to reply. By participating in the discussion in the forum and observing the learners' statements and comments, teachers can know what the learners care about and want to know, communicate with the learners in time, and stimulate the learners' enthusiasm for learning more easily [13]. Due to the variety of topics in the forum, you need to set up a search bar in the forum to help learners quickly find the topic or problem they need. At the same time, in each theme module of the forum, upload and download functions are provided to enable resource sharing among learners.

Blog: Blog, also known as web log, is a space for publishing personal thoughts, opinions, emotions and other content in the form of a journal. Its advantages are easy to create, easy to operate, instant updates, personalization, sharing and interaction, etc. Each student can have his or her own blog, as long as he or she logs in to relevant websites (such as Sina, NetEase) to register [14]. Moreover, blogs between learners can be shared with each other, and even with the authorization of the blogger, they can comment on a topic. In addition, in the blog, the learner can set up a blog circle according to their own interests and hobbies, that is, a virtual learning group. The learner can study and explore a certain problem or topic. This not only reduces the loneliness of the learner learning in the network, but also increases the emotional communication between the learners.

Synchronous interaction design should follow the principle of "asynchronous interaction as the main factor and synchronous interaction as the supplement" when designing the interpersonal interaction mode of the boutique course. Therefore, synchronous interaction design is also an indispensable part of the boutique course [15]. Synchronized interaction requires that teachers and learners in different areas be online simultaneously in order to communicate smoothly. Due to the limited time and energy of the teachers, it is impossible to be online frequently. This requires the teachers to inform the learners of the specific online time through the bulletin board, so that the learners can grasp the opportunity of "face-to-face" communication with the teachers. Provide online chat rooms for learners when designing synchronized interactions for boutique courses. Chat room is a real-time communication tool, which can achieve one-to-one, one-to-many interaction. Its advantage is the timeliness of interaction. In the chat room, the learner can consult the teacher and get feedback from the teacher immediately. The learner can also collaborate with other learners online to encourage each other, overcome learning difficulties and accomplish the construction of the meaning of knowledge together. When students communicate with teachers and other learners, they can use text, images, videos, audio, which makes the communication between them more vivid and interesting.

3 Simulation Experiments

3.1 Experimental Preparation

Design simulation experiment to analyze the application performance of the interactive interface optimization of online teaching mobile end in the visualization of the Internet of Things. By comparing the number of users using online teaching mobile end before and after optimization, the difference between before and after optimization is proved.

The test uses the method of static test to test the software function and user interface layout one by one. The test environment is an Android system4.0 and above. The test case design takes into account the complexity of the Android mobile phone brand, and has been tested with different models and system versions to ensure that the software can be compatible with the mainstream mobile phone types in the market, so as to ensure the use of different models and maximize the user group. Therefore, different brands and resolutions of mobile phones are used for testing, and the model configuration is as follows (Table 3):

Table 3. Test model configuration

Brand and model	Operating system	Resolution	PPI
Galaxy S3 (I9300)	Android OS4.0	2930 * 720	306
ZTE U930	Android OS4.0	960 * 540	256
Huawei Ascend P7	Android OS4.4	1920 * 1080	441
HTC One	Android OS4.1	1920 * 1082	469

1 Install test. The software download prompts the user that the current network environment is a WIFI or mobile network and asks if they want to install it. Software installation can choose its own path and be installed on a mobile phone or SD card. The new version installation can overwrite the old version, and the software application will not be abnormal, while the software can be uninstalled normally.

2 Performance test. Software usage means that there will be no Carton situation in the system, software application function is normal, each module works normally, and there is no unexpected exit. In the unstable network environment, the software application is normal, the network can still be used normally after disconnection, downloads such as courses will prompt the network disconnection information.

3 Interface test. The user interface is simple and generous, coordinated as a whole, consistent with the design prototype, and there is no interface scale imbalance. During the operation, pictures and text can be displayed normally without picture or text errors.

At present, the application function of language learning is basically developed, and the core part of application content basically meets the design requirements.

3.2 Comparison of Experimental Results

By comparing the selection of 350 groups of users, the experimental results of the two groups before and after optimization are shown in Fig. 3:

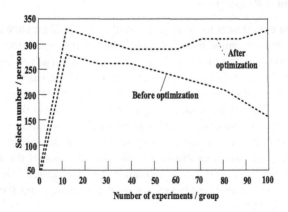

Fig. 3. Comparison of experimental results

According to the experimental results in Fig. 3, the optimized interactive interface of the online mobile terminal is more popular with the testers. The maximum number of users is 330, and the selection rate is 94.29%. The highest number of people who chose the learning interface before optimization was 270, and the selection rate was 77.14%. And in the process of the experiment, the software installation does not appear abnormal, can be unloaded normally. In the process of operation, each function module can run normally and the graphic interface display is normal.

By comparing user selectivity, we can see that most users prefer to choose an optimized interface. The limitations of optimizing the front interface design, the lack of structure and template design have been found in the user experience survey, which makes the overall layout single. The lack of visual and color design, as well as the absence of a GUI, results in a dull and lifeless interface. Although you can choose from Topics, the design section that should be attractive has become a toll file. The lack of practicality and user experience mainly takes the form of listening to pronunciation through a table.

Choose more courses to introduce knowledge as the main knowledge, but charge more, which makes the user experience extremely degraded. Lack of breakthrough and innovation. Compared with the traditional teaching mode, copying the book content to the mobile end of the Internet does not make better use of the enhanced advantages of mobile APP interface interaction. Inconsistent interface design. This is reflected in the picture selection and overall interface effect, as well as the visual color and GUI response. This also violates the user-centered interface design principles, the lack of research on picture selection leads to stale and dull overall interface senses, and the overall style rhythm depression makes learning fun decline.

The optimized interface takes the user as the core, and considers both sensory and emotional aspects. It uses cool tones and three colors to make the interface information and actual operation clearer and reduce the complex visual memory. Includes the use of text Icon backgrounds and contrast to make the interface simple, clear and interesting. Secondly, it is convenient to strengthen learners' autonomous learning and interaction. Interface interaction is guided by the establishment of effective hierarchical learning modules and enhanced autonomy in the form of learning tasks. In particular, a third-party sharing interaction link has been added to enhance the experience. By combining rich audio effects and interesting graphics, users can improve the efficiency of frag-mentation learning by downloading interesting Chinese topics and learning stage tests. Finally, the simplicity and clarity of the interface are the main features and advantages of this study. Both the oral listening learning interface and the HSK test interface have clear and concise menu management, and the humanized collection of wrong questions is also the highlight.

4   Conclusion

This paper presents an application of Internet of things visualization technology in the optimization of mobile interaction interface in network teaching. From the beginning of the project to the research requirements analysis to the preparatory work, the prototype design of dual channel Android client interaction and the design of server-side communication and data storage are realized. The main feature of this application interaction is that it can provide a good interface interaction and experience for Chinese learners with different experience backgrounds and learning purposes in the oral listening scene environment and HSK test simulation environment.

Through simulation experiments, the effectiveness of this method is analyzed from three aspects: installation test, performance test and interface test. The experimental results show that the selection rate of the optimized teaching interface and interaction mode is as high as 94.29%. However, due to the limitation of the technical level of myself and the research group, the next research will be carried out in the direction of file transmission optimization and server stability in the future.

References

1. Yang, B., Shu, H.C., Zhang, R.Y., et al.: Interactive teaching-learning optimization for VSC-HVDC systems. Kongzhi yu Juece/Control and Decision **34**(2), 325–334 (2019)
2. Maity, K., Mishra, H.: ANN modelling and Elitist teaching learning approach for multi-objective optimization of μ μ\upmu -EDM. J. Intell. Manuf. **29**(7), 1599–1616 (2018)
3. Zhang, Y.: Design and curriculum optimization of college english teaching model based on esp. Int. J. Eng. Model. **31**(1), 359–364 (2018)
4. Ran, M.: Research on the construction and optimization of multi english teaching model based on computer cloud technology. Int. J. Eng. Model. **31**(1), 228–235 (2018)
5. Hamzeh, M., Vahidi, B., Nematollahi, A.F.: Optimizing configuration of cyber network considering graph theory structure and teaching–learning-based optimization (GT-TLBO). IEEE Trans. Ind. Inform. **15**(4), 2083–2090 (2019)
6. Hassanzadeh, Y., Jafari-Bavil-Olyaei, A., Aalami, M.T., et al.: Experimental and numerical investigation of bridge pier scour estimation using ANFIS and teaching-learning-based optimization methods. Eng. Comput. **35**(3), 1103–1120 (2019)
7. Tiryaki, S., et al.: Performance evaluation of multiple adaptive regression splines, teaching–learning based optimization and conventional regression techniques in predicting mechanical properties of impregnated wood. Eur. J. Wood and Wood Prod. **77**(4), 645–659 (2019)
8. Sharma, G., Kumar, A.: Modified energy-efficient range-free localization using teaching-learning-based optimization for wireless sensor networks. IETE J. Res. **64**(1), 124–138 (2018)
9. Liu, S., Liu, D., Srivastava, G., Połap, D., Woźniak, M.: Overview and methods of correlation filter algorithms in object tracking. Complex Intell. Syst. 1–23 (2020). https://doi.org/10.1007/s40747-020-00161-4
10. Lu, M., Liu, S.: Nucleosome positioning based on generalized relative entropy. Soft. Comput. **23**(19), 9175–9188 (2018). https://doi.org/10.1007/s00500-018-3602-2
11. Pan, Z., Liu, S., Sangaiah, A.K., et al.: Visual attention feature (VAF): a novel strategy for visual tracking based on cloud platform in intelligent surveillance systems. J. Parallel Distrib. Comput. **120**, 182–194 (2018)

12. Fu, W., Liu, S., Srivastava, G.: Optimization of big data scheduling in social networks. Entropy **21**(9), 902 (2019)
13. Chen, Q., Yue, X., Plantaz, X., et al.: Viseq: visual analytics of learning sequence in massive open online courses. IEEE Trans. Visual Comput. Graphics **126**(8), 1622–1636 (2018)
14. Liu, S., Bai, W., Zeng, N., et al.: A fast fractal based compression for MRI images. IEEE Access **7**, 62412–62420 (2019)
15. Sigut, J., Castro, M., Arnay, R., et al.: OpenCV basics: a mobile application to support the teaching of computer vision concepts. IEEE Trans. Educ. **29**(3), 1–8 (2020)

Research on Feature Extraction Method of UAV Video Image Based on Target Tracking

Xin Zhang, Zhi-jun Liu, and Ming-fei Qu[✉]

College of Mechatronic Engineering, Beijing Polytechnic, Beijing 100176, China
Jameszlx@163.com, qmf4528@163.com

Abstract. In order to extract the key and useful features of the target in the UAV video image and strong marking ability, a feature extraction method for the UAV video image based on target tracking is proposed. The sparse beam method is used to adjust the splicing of UAV video images. Based on this, the pixel coordinates are obtained through the frame difference method to detect and locate the target. According to the target detection and positioning results, the video image of the target area is selected and preprocessed by the wavelet transform algorithm Target area video image, and extract the target area video image feature, through hierarchical particle filtering to achieve target tracking, to achieve the extraction of UAV video image feature. The experimental results show that: in the ORL database experiment, the average feature extraction percentage is 78.08%, and the average target tracking error is 1.16; in the COIL-20 database experiment, the average feature extraction percentage is 82.55%, and the average target tracking error is 1.20, which meets the needs of UAV video image feature extraction and target tracking.

Keywords: Target tracking · Drone · Video image · Feature · Extraction

1 Introduction

As a new aerial remote sensing platform, UAV has the characteristics of fast sailing, flexible operation and low cost. The digital cameras and digital cameras mounted on the aircraft can obtain high-resolution video images, and the processing of video images can satisfy the vast majority. The needs of most users in aerial photography and target monitoring [1]. However, the video image acquired by UAV has the characteristics of high altitude and small image amplitude, which can not reflect the overall situation of the camera area. In emergency rescue, it is often unable to meet the needs and applications of information. Therefore, it is necessary to use video image feature extraction method to provide the target information of camera area. Using wireless transmission technology, the video images collected by UAV can be downloaded in real time. Video image has the characteristics of small frame, low resolution, large amount of data and high redundancy. When aiming at a certain target, we can't get the complete information of the target quickly. Therefore, the feature extraction of UAV video image has important practical significance and application requirements.

S. Liu and L. Xia (Eds.): ADHIP 2020, LNICST 348, pp. 266–278, 2021.
https://doi.org/10.1007/978-3-030-67874-6_25

From the existing research results, feature extraction is one of the most fundamental problems in the field of pattern recognition, and extracting effective screening features is a prerequisite for solving target tracking. The research of feature extraction has two main purposes: one is to find the most discriminative description between targets, so as to distinguish different types of targets; the other is to compress the dimension of target data under certain circumstances [2]. According to whether it can be linearly separable, the feature extraction method can be divided into two kinds: one is linear feature extraction method, the other is non-linear feature extraction method. Among them, linear feature extraction methods include principal component analysis, independent component analysis, factor analysis, local preserving projection, linear discriminant analysis, local feature analysis and multi-dimensional scale analysis. The linear feature extraction method is easy to understand and easy to implement, and has been successfully applied to many fields such as face recognition, character recognition, speech recognition and target classification. In order to extract the key and useful features of the target in the UAV video image and strong marking ability, a feature extraction method for the UAV video image based on target tracking is proposed.

2   Research on Feature Extraction Method of UAV Video Image

2.1   Drone Video Image Stitching

The time difference of UAV video image acquisition results in gaps between video images. In order to obtain complete video image information of high quality, the UAV video image is spliced based on sparse beam adjustment [3].

Sparse beam adjustment is a method of minimizing the error between the measured value and the estimated value of the matching point based on the Levenberg-Marquardt algorithm using the irrelevance of the projection matrix. In this method, the sparse structure of the normal equation is used to reduce the computational complexity, and the optimal solution of the normal equation is obtained quickly, so that the error between the measured value and the estimated value of the matching point pair is minimized.

The sparse beam method adjustment can be used to globally optimize the spliced UAV video image, and minimize the conversion error of each video image to the reference plane. The UAV video stream splicing belongs to sequence video image splicing. If the video image is globally optimized, it can only be done at the end of aerial photography. In order to obtain high-quality splicing video image in real time, this study uses the center constraint to dynamically select the reference plane, and ensures that the spliced video image has the local optimal characteristics each time the reference plane is changed. Finally, the middle frame of the flight belt is taken as the final reference plane to complete the splicing of UAV video stream [4]. The specific splicing steps are as follows:

Step 1: Using UAV trajectory planning data to obtain the approximate range of the shooting area, determine the longitude and latitude of the first and last video images of each flight area, and convert them into geodetic coordinates to determine the geodetic coordinates of the middle position of the flight area.

Step 2: When the drone reaches the preset altitude and is flying at a constant speed, calculate the ground area of the video image during vertical shooting, and then roughly determine the video image to be extracted for a single flight strip according to the first and last geodetic coordinates of the flight strip and the given overlap Quantity n, recursively calculate $n/2$, and store the result of each calculation in vector v_1, then recursively calculate $(n/2 + n)/2$, store the result of each calculation in vector v_r, and transform the reference plane before and after the intermediate frame as P_{li} and P_{ri};

Step 3: according to the camera parameters, altitude and flight speed, calculate the sampling interval of the key frame, extract the key frame and correct the video image, and cut the corrected video image into the corrected image with the shortest width as the image width and the width height ratio of 4:3;

Step 4: Perform sequence stitching on the cropped corrected images. Before the middle frame, when stitching to the reference plane P_{li}, optimize the absolute homography matrix of each video image to the $i - 1$th reference video image and re-splice; when splicing to the middle frame video image $P_{n/2}$, the projection surface is fixed to the middle frame to continue splicing. When splicing to P_{ri}, the video image that has been spliced after the middle frame is optimized using the middle frame as the projection surface, so that each transformation is guaranteed The stitched video image before the reference plane is locally optimal and avoids the transfer of errors at both ends of the aircraft belt [5].

Through the above process, the UAV video image splicing is completed, which is ready for the following target detection and positioning.

2.2 Target Detection and Positioning

On the basis of the above video images, the pixel coordinates are obtained based on the frame difference method, and the target is detected and located.

Among the many target detection methods, considering the high speed of the frame difference method, this study selected this method as the basis of the pixel extraction algorithm. According to the result of the frame difference method, the update of the UAV video image is divided into pixel level and frame level. The former is used to detect small and slow changes; the latter is used to detect global and sudden changes. Different learning strategies and update speeds [6].

A dynamic feature matrix $D_{i,j}(t)$ is constructed to reflect the state of video image at time t, which is expressed as:

$$D_{i,j}(t) = \begin{cases} D_{i,j}(t-1) - 1 & S_{i,j}(t) = 0, D_{i,j}(t-1) \neq 0 \\ \lambda & S_{i,j}(t) \neq 0 \end{cases} \quad (1)$$

In formula (1), $S_{i,j}(t)$ represents a logical matrix; λ represents the pixel of frame λ.

The calculation formula of logical matrix $S_{i,j}(t)$ is:

$$S_{i,j}(t) = \begin{cases} 0 & |f_{i,j}(t) - f_{i,j}(t-\tau)| \leq T_s \\ 1 & otherwise \end{cases} \qquad (2)$$

In formula (2), $f_{i,j}(t)$ represents the gray value of pixel $p(i,j)$ at time t; τ represents the interval; T_s represents the threshold.

If the values of logic matrix $S_{i,j}(t)$ and continuous λ frame are all 0, it means that the gray value of corresponding pixel has little change in continuous λ frame. It is considered that there is no target or noise at this point in this period of time. Therefore, it can be considered that the gray value is the background gray value with great possibility. The gray value can be used to update the corresponding background point. The updating formula is:

$$B_{i,j}(t) = \alpha \cdot f_{i,j}(t) + (1-\alpha) \cdot B_{i,j}(t-1) \qquad (3)$$

In Eq. (3), $B_{i,j}(t)$ represents the background point update value; α represents the update coefficient.

If the value of logical matrix $S_{i,j}(t)$ is 1 or it can't guarantee that the continuous frame λ is 0, the gray value of the corresponding point is considered to be unstable, and the gray value of the point should not be collected. According to the above analysis, matrix $D_{i,j}(t)$ stores the changes of the corresponding pixels in the video image space at time t. if the value is 0, it belongs to the pixels that have not changed the gray value of the continuous λ frame; if the value is not 0, it belongs to the pixels that have changed in the continuous λ frame. The larger the value, the closer the change is to the current frame [7].

The difference between the real-time gray-scale matrix $f_{i,j}(t)$ and the background matrix $B_{i,j}(t)$ and the thresholding will obtain the binarized video image data, which is expressed as:

$$F_{i,j}(t) = \begin{cases} 1 & |f_{i,j}(t) - B_{i,j}(t-\tau)| \leq T_F \\ 0 & otherwise \end{cases} \qquad (4)$$

In Eq. (4), T_F represents the threshold parameter.

Binary video image data a usually contains $F_{i,j}(t)$ lot of noise - unreal target area, which is called false detection target, needs to be filtered. Because the specific location of the target can not be determined, the convolution of filtering is carried out in the whole video image space. Excessive filtering may cause the target area to be destroyed by the convolution operation and the serious blurring of the edges. Even if a target splits into multiple small targets or multiple targets merge into a target phenomenon, these results will be It causes difficulties for follow-up [8]. Therefore, the goal of filtering should not be to eliminate all noises, but to protect the target from being eroded or blurred as much as possible while filtering. Therefore, this study adopts a relatively conservative filtering process. The main steps are as follows: in order to reduce the amount of calculation, $F_{i,j}(t)$ is sub sampled twice; after the sub sampled video image is processed by morphological filtering once, the video image matrix $I(t)$

is obtained, and the pixel coordinates with the value of 1 are extracted to form a data set composed of n pixel coordinates: $X(t) = \{x_1(t), x_2(t), \cdots, x_n(t)\}$.

Calculate the initial clustering of $X(t)$, and merge the clusters repeatedly under the guidance of the criterion function J_{nn} until the criterion function is the smallest. The target is calibrated according to the final stable clusters. The calibration position used is the cluster centroid coordinate, which completes the detection and positioning of the target.

2.3 Feature Extraction of Video Image in Target Area

Based on the above target detection and positioning results, the target region video image is selected, the target region video image is preprocessed by the wavelet transform algorithm, and the target region video image feature is extracted.

Wavelet transform is widely used in video image processing because of its good properties of multi-scale and multi-resolution. In video image processing, wavelet transform has perfect reconstruction power, that is, it will not lose the original information, nor generate redundant information. Wavelet transform can easily obtain the frame information and detailed information of the original video image after wavelet decomposition. After decomposing the video image, the two-dimensional wavelet transform will produce four sub bands, namely LL, LH, HL and HH. Among them, LL represents horizontal and vertical low frequency information; LH represents horizontal low frequency and vertical high frequency information; HL represents horizontal high frequency and vertical low frequency information; HH represents horizontal high frequency and vertical high frequency information. The low frequency part corresponds to the average gray level in a video image, and reflects the smooth part of the video image. The high level part is the gray level which changes faster and faster in the video image, corresponding to the edge, detail and noise in the video image. Therefore, the processing of the low frequency part of the video image will not lose or change the details and edge information of the video image, nor will it change the noise in the video image [9, 10].

The steps of video image preprocessing based on wavelet transform algorithm are as follows:

Step 1: Acquire the video image of the target area, record it as $f(x, y)$, and use db8 wavelet to decompose $f(x, y)$;

Step 2: In the four sub bands LL, LH, HL and HH, LL sub bands are selected for histogram equalization;

Step 3: Set an appropriate threshold and use wavelet threshold denoising technology to suppress noise in high-frequency video images;

Step 4: The processed LL low-frequency video image and other high-frequency video image are reconstructed by wavelet, and the video image $f_1(x, y)$ is obtained;

Step 5: Then perform histogram equalization processing on the video image $f_1(x, y)$ to obtain the final video image $g(x, y)$.

Based on the video image $g(x, y)$ of the target area obtained in the above steps, Vectorize it to form a vector χ of $mn \times 1$ connected end to end, as shown in Fig. 1.

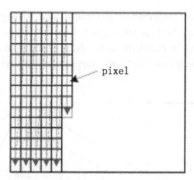

Fig. 1. Vectorization of video image in target area

Then the average vector of M target area video image is:

$$\mu = \frac{1}{M}\sum_{i=1}^{M}\chi_i \tag{5}$$

Then the covariance matrix of the video image in the target area is expressed as:

$$C = \frac{1}{M}\sum_{i=1}^{M}(\chi_i - \mu)(\chi_i - \mu)^T = \frac{1}{M}XX^T \tag{6}$$

In formula (6), XX^T is the simplified formulation of $\sum_{i=1}^{M}(\chi_i - \mu)(\chi_i - \mu)^T$, by solving formula (6), the eigenvalues and eigenvectors of video images in the target area can be obtained.

The feature vectors are sorted according to the size of the feature value. The larger the feature value is, the better the feature vector can reflect the video image features of the target area, and the size of the feature value decreases exponentially [11, 12]. The video image of the target area corresponding to the feature vector is called a feature sub-video image, that is, a feature face. The more blurred the feature face, the less information it contains. The feature sub video image is used to reconstruct the video image in the target area. It can be seen that a large amount of information of the original video image in the target area can be recovered with fewer features. Then the feature vector obtained is representative and recorded as $X\eta$.

2.4 Feature-Based Target Tracking

Based on the video image feature vector of the target area obtained above, the target tracking is realized by hierarchical particle filter.

Target tracking can be understood as adaptively "dragging" the window as the target moves, so that the tracked target is always in the window, and the size and angle of the window are adjusted in real time according to the size and posture of the target. In the video image coordinate of UAV, the target tracking window is shown in Fig. 2.

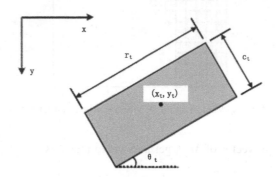

Fig. 2. Schematic diagram of target tracking window

As shown in Fig. 2, (x_t, y_t) represents the target coordinate; r_t represents the length of the window; c_t represents the width of the window; θ_t represents the angle between the tracking window and the coordinate axis.

The steps of target tracking algorithm based on hierarchical particle filtering are as follows:

Step 1: set the tracking window manually, select the target to be tracked, take the center of the window as the initial position of the target, the angle between the long axis of the window and the axis as the initial direction of the target, and the size of the window as the initial size of the target. Based on this, the initial value of the target state can be determined as $s_0 = [x_0, y_0, r_0, c_0, \theta_0]^T$. And extract the video image in the window, and use the process of Sect. 2.3 to find the feature vector;

Step 2: initialization of particle filter. The initial state of the particle filter is obtained by initializing N position particle with the position information in the initial value vector of the target state: $\{x_0^i, 1/N_s\}_{i=1}^{N_s}$;

Step 3: Importance sampling. At time $t > 0$, N_s particles are re-sampled according to the particle group of the previous two frames and the target state transition model to obtain the predicted value of the filter: $\{x_t^i, 1/N_s\}_{i=1}^{N_s}$;

Step 4: Similarity measurement. Calculate freely to the left according to the derivation process of the hierarchical particle structure, and finally get the target state $s_t^k = [x_t^k, y_t^k, r_t^k, c_t^k, \theta_t^k]^T$ corresponding to the k th first-layer particle.

Step 5: target state estimation and target vector update. Firstly, the state value of the target at the moment is estimated according to the state value of N_s target: $E(s_t) = \frac{1}{N_s} \sum_{i=1}^{N_s} s_t^i \cdot \omega_t^i$, where $s_t^i \cdot \omega_t^i$ represents the state value of the first layer particle; then, the corresponding video image is grabbed according to the target area corresponding to $E(s_t)$ and the feature vector $p_{E(s_t)}$ is calculated, and the similarity between it and the original target is compared, and the target vector is updated or not according to this. If it is updated, it will be used as a new target prototype feature;

Step 6: Determine whether the target disappears. If it does not disappear, return to step three; otherwise, exit.

Through the above process, the feature extraction of UAV video image based on target tracking is realized, which helps the application of UAV.

3 Performance Analysis of Feature Extraction Method for UAV Video Image

This research will verify the effectiveness of the UAV video image feature extraction method based on target tracking on ORL database and COIL-20 database. In the experiment, two random partition methods are selected for each database video image set to verify the feasibility of the proposed method. In the experiment, the nearest neighbor classifier is used. The experimental software environment is matlab 7.0a. All the experimental results are executed ten times Then take the average.

3.1 Simulation Experiment on ORL Database

The information of ORL database is not described in detail due to space limitation. In this experiment, all UAV video images are cropped to the size of 64 × 64 dimensions, and then each video image is divided into 4 blocks, row and column are divided into 16 blocks in total, then the size of each sub block video image is 16 × 16.

In this experiment, the ORL database is used as the input UAV video image set, and 10 video images belonging to the same target are used as a class. For the first time, 5 video images in the database are randomly selected as training video images, and the remaining 5 video images as a test video image, the training video image set has 200 images, and the test video image set has 200 images, referred to as 5train for short. In the second random extraction of 3 video images in the database as training video images and the remaining 7 video images as test video images, the training video image set has 120 images and the test video image set has 280 images, referred to as 3train for short.

Through experiments, the percentage of feature extraction of the proposed method is shown in Table 1.

Table 1. Feature extraction percentage analysis

Number of goals	5train	3train
1	85.10%	85.46%
2	86.44%	80.12%
3	82.13%	82.25%
4	85.01%	80.00%
5	86.66%	79.45%
6	85.03%	78.11%
7	85.55%	76.23%
8	85.00%	75.00%
9	85.49%	75.00%
10	85.11%	74.22%
Average value	78.08%	

It can be seen from Table 1 that at 5train, the proposed method is more robust, and the feature extraction percentage remains almost unchanged, while at 3train, as the number of targets increases, the feature extraction percentage decreases rapidly. Through calculation, the average feature extraction percentage is 78.08%, which satisfies the requirements for feature extraction of UAV video images.

The experimental results of target tracking error are shown in Table 2.

Table 2. Target tracking error analysis

Number of experiments	5train	3train
10	1.2	0.7
20	1.1	1.0
30	1.0	1.1
40	0.9	1.1
50	1.0	1.2
60	1.1	1.3
70	1.2	1.5
80	1.0	1.6
90	1.0	1.6
100	0.8	1.8
Average value	1.16	

As shown in the data in Table 2, at 5train, the proposed method error is small and almost unchanged; at 3train, the proposed method error gradually increases. The average value of target tracking error is 1.16, which meets the requirements of target tracking.

3.2   Simulation Experiment on COIL-20 Database

Due to space limitation, the information of COIL-20 database is not described in detail [13]. In this experiment, in COIL-20 database, we select 24 video images under 15O rotation of UAV, and there are 480 video images in total. If the image size is 64 × 64, the video image is divided into 16 blocks, and the row and column are all 4 blocks, then the video image size of each sub block is 16 × 16, so as to prepare for the experiment.

The COIL-20 database is used as the input video image set, and 24 video images belonging to the same target are used as a category. For the first time, 12 video images in the database are randomly selected as training video images, and the remaining 12 video images are used as test videos. For video, the training video image set has 240 images, and the test video image has 240 images, referred to as 12train. In the second random extraction of 9 images as training video images, and the remaining 15 video images as test video images, the training video image set has 180 images, and the test video image set has 300 images, referred to as 9train.

In the case that the dependent variable is the number of targets, the feature extraction percentage of the proposed method is obtained through the simulation comparison experiment, as shown in Table 3.

Table 3. Feature extraction percentage analysis

Target quantity	12train	9train
1	84.12%	85.02%
2	84.00%	83.12%
3	84.02%	81.00%
4	84.13%	80.94%
5	84.44%	80.50%
6	85.00%	80.44%
7	85.01%	80.00%
8	84.49%	79.45%
9	83.39%	78.51%
10	85.46%	78.00%
Average value	82.55%	

It can be seen from Table 3 that the proposed method is relatively robust at 12 train, and the feature extraction percentage almost remains unchanged, while at 9 train, with the increase of the number of targets, the feature extraction percentage drops rapidly. The average percentage of feature extraction is 82.55%, which meets the requirements of UAV video image feature extraction.

The experimental results of target tracking error are shown in Table 2.

As shown in Table 4, at 12train, the proposed method error is small and almost unchanged; at 9train, the proposed method error gradually increases. Through calculation, the average value of target tracking error is 1.20, which meets the requirements of target tracking.

Table 4. Target tracking error analysis

Experiments	12train	9train
10	1.0	1.2
20	1.1	1.1
30	0.8	1.1
40	0.9	1.6
50	1.0	1.2
60	0.9	1.4
70	0.9	1.5
80	1.0	1.9
90	1.0	1.6
100	0.8	2.0
Average value	1.20	

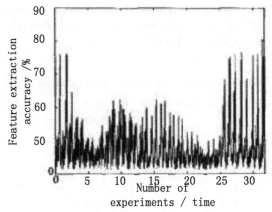

(a) The accuracy of image feature extraction based on this method

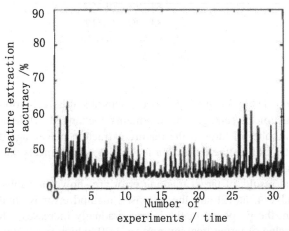

(b) Accuracy of image feature extraction based on traditional methods

Fig. 3. Comparison of video image feature extraction accuracy of UAV

In order to further verify the effectiveness of this method, the proposed method and the traditional method of UAV video image feature extraction accuracy are compared and analyzed, the comparison results are shown in Fig. 3.

According to Fig. 3, the accuracy of UAV video image feature extraction in this paper is up to 80%, while that of traditional method is only 65%. The accuracy of UAV video image feature extraction in this paper is higher than that of traditional method.

4 Empirical Conclusion

This paper proposes a feature extraction method for UAV video images based on target tracking, and the experimental results show that: under the ORL database experiment, the average feature extraction percentage is 78.08%, and the average target tracking error is 1.16; under the COIL-20 database experiment, the average feature extraction percentage is 82.55%, and the average target tracking error is 1.20, which can meet the needs of UAV video image feature extraction and target tracking.

Acknowledgements. The application of UAV spray in the city pest control service (2019H033-KQ)

References

1. Chen, Y., Liu, J., Pei, J., et al.: The risk factors that can increase possibility of mandibular canal wall damage in adult: a cone-beam computed tomography (CBCT) study in a Chinese population. Med. Sci. Monitor **24**(2), 26–36 (2018)
2. Azimi, S.M., Britz, D., Engstler, M., et al.: Advanced steel microstructural classification by deep learning methods. Sci. Rep. **8**(1), 2128 (2018)
3. Ballerini, L., Lovreglio, R., Valdés Hernández, M.C., et al.: Perivascular spaces segmentation in brain MRI using optimal 3D filtering. Sci. Rep. **8**(1), 2132–2132 (2018)
4. Lewis, A.G., Schriefers, H., Bastiaansen, M., et al.: Assessing the utility of frequency tagging for tracking memory-based reactivation of word representations. Sci. Rep. **8**(1), 7897 (2018)
5. Liu, S., Liu, D., Srivastava, G., et al.: Overview and methods of correlation filter algorithms in object tracking. Complex Intell. Syst. (2020). https://doi.org/10.1007/s40747-020-00161-4
6. Hilchey, M.D., Rajsic, J., Huffman, G., et al.: Dissociating orienting biases from integration effects with eye movements. Psychol. Sci. **29**(3), 328–339 (2018)
7. Jiang, H., Qu, P., Wang, J.W., et al.: Effect of NF-κB inhibitor on Toll-like receptor 4 expression in left ventricular myocardium in two-kidney-one-clip hypertensive rats. Eur. Rev. Med. Pharmacol. Sci. **22**(10), 3224–3233 (2018)
8. Rina, L.D.N., Walburg, K.V., Martin, V.H.P., et al.: Retinal pigment epithelial cells control early mycobacterium tuberculosis infection via interferon signaling. Investigative Opthalmol. Vis. Sci. **59**(3), 1384–1395 (2018)

9. Harris, H., Sagi, D.: Visual learning with reduced adaptation is eccentricity-specific. Sci. Rep. **8**(1), 608 (2018)
10. Liu, S., Liu, G., Zhou, H.: A robust parallel object tracking method for illumination variations. Mob. Netw. Appl. **24**(1), 5–17 (2019)
11. Liu, S., Lu, M., Li, H., et al.: Prediction of gene expression patterns with generalized linear regression model. Front. Genet. **10**, 120 (2019)
12. Barrea, A., Delhaye, B.P., Lefèvre, P., et al.: Perception of partial slips under tangential loading of the fingertip. Sci. Rep. **8**(1), 7032–7032 (2018)
13. Lu, M., Liu, S.: Nucleosome positioning based on generalized relative entropy. Soft. Comput. **23**(19), 9175–9188 (2018). https://doi.org/10.1007/s00500-018-3602-2

Automatic Recognition of Tea Bud Image Based on Support Vector Machine

Wang Li[1], Rong Chen[1], and Yuan-yuan Gao[2(✉)]

[1] College of Big Data, TongRen University, Tongren 554300, China
[2] Changsha Medical College, Changsha 410219, China
ijnm98760@126.com

Abstract. The existing recognition method of tea shoots is only to judge the single color or shape features, resulting in low recognition accuracy. Therefore, an automatic recognition method of tea shoots image based on support vector machine is designed. In this method, two kinds of image features, color and shape texture, are extracted from the tea bud image for discrimination. The RGB model is used to extract color features, and LBP/C operator is used to extract the shape and texture features of the bud. The extracted features are used as the feature vectors of the training samples, and support vector machine model training is carried out to obtain the support vector machine classifier, and the tea bud image is recognized. The experimental results show that the recognition rate, recall rate and comprehensive evaluation index of the method are higher than those of the traditional method, which proves that the method has high recognition accuracy and improves the recognition efficiency.

Keywords: Support vector machine · Image recognition · Feature extraction

1 Introduction

China is an important tea producing and selling country in the world. Tea culture is deeply loved by people all over the world. Due to the huge production of tea, the use of mechanical picking instead of manual picking can speed up the picking efficiency and alleviate the shortage of labor, but on the one hand, this method is lack of selectivity and will pick the old leaves, young leaves and buds together, but the efficiency of manual picking is low, the cost is high, and the phenomenon of hard work sometimes occurs in the peak period of tea picking. Therefore, it is necessary to develop an efficient and selective method of tea bud recognition to realize the classification production of Longjing tea. One of the key technologies is to study the automatic detection and recognition of Longjing tea buds.

S. Liu and L. Xia (Eds.): ADHIP 2020, LNICST 348, pp. 279–290, 2021.
https://doi.org/10.1007/978-3-030-67874-6_26

Support vector machine (SVM) is a very popular algorithm in the field of artificial intelligence. In 1995, Vapnik et al. Proposed a classification algorithm based on statistical learning theory to realize structural risk minimization [1]. Its main principle is to construct an optimal linear hyperplane in the sample space, to maximize the distance between the two closest samples on both sides of the plane, and to obtain a good generalization ability. The support vector in the algorithm refers to the training points which are closest to the classification decision surface and the most difficult to identify in the training set samples. Whether the distance between these points and the plane has reached the maximum is the standard for SVM to reach the optimal classification. With the rapid development of machine learning methods in the 1990 s, SVM has been widely used in biology, handwriting, text recognition and other fields [2]. Some domestic scholars have carried out multi-agent research on tea image grade recognition technology. The recognition method is simple and fast, but when the image quality is low and the number of pixels is small, it can not accurately identify tea buds. Therefore, this paper proposes an automatic recognition method of tea bud image based on support vector machine.

2   Extracting Image Feature Parameters of Tea Buds

2.1   Tea Shoots Image Collection

Because there are many types of tea, this article selects Longjing tea as the research object. The image of Longjing tea is collected into a computer, and the image is digitized using imaging technology and analog-to-digital conversion technology. With the continuous development of computer technology and microelectronic technology, especially the rapid improvement of the performance of the charge-coupler (CCD), the imaging quality is good, and the performance is stable, which is widely used. The tea images obtained in this article were taken with a digital camera at an angle of approximately 45° to the horizontal ground. When the camera shooting angle is 0° (i.e. horizontal shooting), the contour features of tea shoots obtained are the clearest and the most complete. However, due to the different growth position of tea shoots, this shooting method will appear serious occlusion phenomenon, which is not convenient for the subsequent processing of the shoot image. When 900 is used to shoot vertically and downward, the outline of the tender leaves is clear, but the shape of the tender buds is not easy to identify. In this case, the tender buds tend to converge into a nearly circular area, which can not be distinguished from the young leaves, which is not conducive to subsequent processing. When the shooting angle is 45°, the outline of tea shoots and leaves is clear, and the occlusion phenomenon is less [3], which is easy to distinguish from the background of old leaves, as shown in the following figure:

(1)

(2)

(3)

Fig. 1. Tea sample image

The shooting angle of Fig. (1) in the above figure is 0°, that of Fig. (2) is 45°, and that of Fig. (3) is 90°. When sampling, close shot mode is adopted, flash is turned off, and direct sunlight is avoided. The number of buds in each frame is 1−4. When the number of buds exceeds 4, the occlusion in Longjing tea pictures is serious. At the same time, there are some incomplete tea bud areas, which will bring inconvenience to subsequent processing.

2.2   Extract Color Features of Buds

The RGB color space is the most used color model in color image processing. This is due to the standardization of the three primary colors. Today, almost all colors can be represented by the linear combination of the three components R, G, and B [4, 5]. The original image of tea samples obtained in this experiment is based on RGB color space, so the preferred color feature is RGB. According to the principle of RGB three primary colors, each color can be weighted by red, green and blue primary colors. RGB model is based on Cartesian system, as shown below:

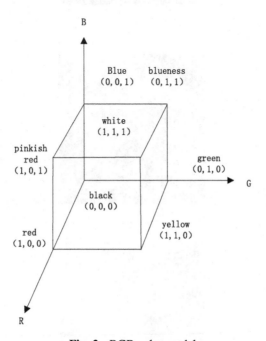

Fig. 2. RGB color model

In the picture above, R, G, and B are located at three corners respectively. Black is at the origin, and white is at the furthest distance from the origin. The gray level is distributed along the line of black and white. Normalize and count the color characteristics of 400 tea bud samples [6]. In view of the space limitation of the article, only part of the data is listed, as shown in the following table (Table 1):

Table 1. Color characteristic data of tea shoots

Serial number	R	G	B
1	167.5	168.7	91.5
2	143	166	67.8
3	88.6	109.7	39
4	118.3	131.8	67.8

(*continued*)

Table 1. (*continued*)

Serial number	R	G	B
5	73.8	84.9	36.7
6	113.8	119.7	70.9
7	135.8	116.8	58.3
8	129.8	133	59
9	68.6	75.9	40.8
10	106.8	113.5	58

Although the RGB color model is widely used and ideal for hardware implementation. However, from the perspective of visual psychology, the human eye's description of colored objects is not based on the RGB primary colors, but on hue, saturation, and brightness (HSI). Hue describes the characteristics of pure color, saturation measures the degree of pure color diluted by white light, brightness is the key parameter to describe color perception. The HSI model is suitable for image processing and explaining the local characteristics of objects, and can effectively separate color information and achromatic information of an image. In addition, with the improvement of hardware level, the conversion of color space can be effectively accelerated by hardware [7, 8], PC and workstations have available modules to convert video or RGB images to HSI images in real time. Based on the above analysis, this paper selects a total of 9 color features R, G, B, H, S, I, L, u, v to train the classifier model. Color feature selection adopts the method of averaging the pixels of the tea bud area. Each tea image is adjusted to the same size (width: 100 pixels, height: 150 pixels), and RGB is converted into HSI, Luv color space, H, S, I, L, u, v6 color feature values are unified in the range of 0−255.

2.3   Texture Feature Extraction of Buds

By observing the image, we can know that the new leaf (including the bud) area is above the upper edge of the old leaf, so we use the edge contour extraction, first extract the upper edge contour of the old leaf. The extraction method is as follows: firstly, the coordinate of the lower left corner of the image is (0,0) Secondly, the upper left corner of the image is scanned line by line until the upper edge of the old leaf blade is detected, and the coordinate value is recorded. The method to determine the edge of old leaves is as follows:

$$B(i) = \begin{cases} j & (T_{bw}(j,i) = 0) \\ 0 & (T_{bw}(j,i) = 1) \end{cases} \tag{1}$$

In the above formula, $B(i)$ is the vertical coordinate value of the edge of column i of the record. If there is no edge pixel in this column, the record is zero; $T_{bw}(j,i)$ represents the pixel value of row j and column i in the figure, 1 represents the white background, 0 Indicates black leaves.

The texture of tea belongs to natural texture, which is analyzed by statistical method. Among the methods of texture analysis based on statistics, LBP (local binary patterns) is widely concerned because of its simple calculation and strong applicability. LBP can describe the spatial structure of local texture, but does not emphasize the contrast information of texture, so it often combines the contrast information (LBP/C) as the texture feature descriptor [9]. The calculation of C descriptor only needs a part of LBP template, and the specific calculation method of LBP/C operator is shown in the figure below:

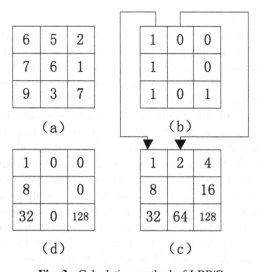

Fig. 3. Calculation method of LBP/C

In the 3 × 3 neighborhood with gray level 6 as the center, binarize the neighborhood pixels with the center of the neighborhood as the min value to obtain (b) in the above figure, and then compare the above figure with (a). The corresponding binary template in (b) can be used to find the mean difference. Each pixel of the original image will get an image of the same size after LBP operation, as shown in the following figure:

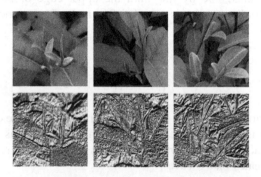

Fig. 4. Obtained texture image

The first row is the original tea shoot image, and the second row is the corresponding texture image. Four mutually uncorrelated and easy-to-calculate features were extracted from it, namely energy, entropy, moment of inertia and correlation, which were counted by LBP/C. At this point, the extraction of image feature parameters of tea buds is completed.

3 Research on Automatic Recognition of Tea Bud Image Based on Support Vector Machine

Support vector machine is a machine learning method based on statistical learning theory, which adopts the principle of structural risk minimization, and has the advantages of small sample, non-linear and "avoiding digital disaster" [10]. Support vector machine can be used to solve linear and nonlinear problems. Its main principle is to find the optimal classification surface for classification. The nonlinear problem can be solved by introducing kernel function into high dimensional space. The linear support vector classifier used in this paper is based on the maximum interval method. The maximum interval method transforms the problem of finding the optimal classification surface into the problem of finding the maximum classification interval. By using Lagrange multiplier method and introducing dual function, the optimization problem is transformed into a quadratic linear programming problem, and the characteristics of training samples are extracted. The training SVM model is used as the feature vector of the training samples, and the training model, i.e. the classification device, is obtained. First determine the number of classifiers, using the SVM algorithm, the goal is to divide the sample into 2 categories, this experiment only needs to train a classifier; determine the feature vector of the training sample, after extracting the image features above, obtain the training sample, and obtain the color feature $C_{i,j}$ With the texture feature $T_{i,j}$, as the training sample feature vector $F(C_{i,j}, T_{i,j})$, j represents the category, i represents the i th pixel in the j th category. The color features of each pixel in the R, G, and B channels are extracted as:

$$C_{i,j} = \left(C_{i,j}^R, C_{i,j}^G, C_{i,j}^B \right) \tag{2}$$

The texture characteristics of pixels at energy $E(n)$, entropy $H(n)$, moment of inertia $I(n)$ and correlation $C(n)$ are:

$$T_{i,j} = \left(T_{i,j}^E, T_{i,j}^H, T_{i,j}^I, T_{i,j}^C \right) \tag{3}$$

In the training process, use the SVM toolkit that comes with Matlab and use the following model model = svmtrain (TrainLabel, TrainData) to complete the training. Among them: TrainLabel is the category label, TrainData is the training sample data, and the extracted feature vector $F(C_{i,j}, T_{i,j})$ Import and conduct model training, that is, build a classifier.

After subtracting the pixels of R = B = G = 0 (background color), the color features and texture feature values of the remaining pixels are used as test samples, and imported into the classifier trained in the previous step for recognition, and the data is divided into the following on the basis of minimizing the error function. According to the predetermined number of classes, each sample in the image is classified according to its distance from the center of each class, and the two sets of pixel points identified and output are respectively marked. According to the test result of the previous step, the pixel set of the same mark is output again to the image, and the area of the pixel set RGB feature value of the tea shoots in the output image is:

$$\begin{cases} R_{\text{tender}} \in [75, 116] \\ G_{\text{tender}} \in [54, 133] \\ B_{\text{tender}} \in [39, 128] \end{cases} \tag{4}$$

Through RGB discrimination of the tea in the image, the tea shoots identified and output are as follows:

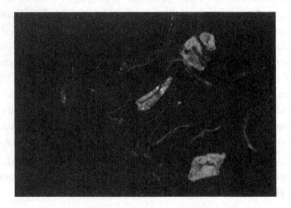

Fig. 5. Tea shoots output

So far, the automatic identification of tea buds has been completed

4   Experiment

In order to verify the effectiveness of the method in this paper, a comparative experiment needs to be designed, using the K-Means method, Grabcut method and the method in this paper to conduct experiments.

4.1   Experimental Design

Under the operating environment of the Windows7 operating system with a CPU of 3.2 GHz, 4 G memory, and 500 G hard disk, the method proposed in this article was tested using MATLAB_R2013b software. In order to avoid the accidentalness of the experimental process as much as possible, and to better verify the effectiveness and practicability of the experimental method in this article, the same equipment was used to shoot under different light sources, different angles and different environments. Six of them are selected as samples, as shown in Fig. 6:

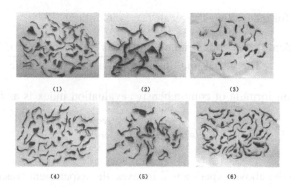

Fig. 6. Test image in the experiment

The above picture, Fig. (1) It is collected under indoor incandescent lamps at night, Fig. (2) is collected under natural light during the day, Fig. (3) is collected under indoor incandescent lamps during the day, Fig. (4) (5) is collected under indoor natural light during the day, and Fig. (6) is night For flash collection. Before the experimental detection, the image is pre-processed (enhanced). The total number of tea particles in each figure is 49, 16, 31, 51, 34, and 59 respectively, and the corresponding number of tender tea particles is 21, 2, 9, 12, 5, 18 grains, the experiment is to deal with the selected 6 images (image enhancement and unified background color ($R = B = G = 0$) processing, to obtain the color features and texture feature values of the image as a test sample), and then use the method proposed in this article to identify, you can get two types of pixel sets of sprouts and non-sprouts, and finally output the sprout image for the sprout pixel set according to the results as shown below (Fig. 7):

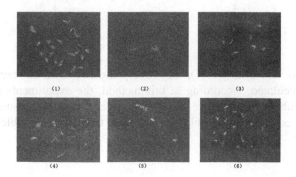

Fig. 7. The results of this method for the identification of tea buds

In order to better illustrate the effectiveness and feasibility of the experimental method, the recognition rate P, recall rate R, and comprehensive evaluation index F_1 are used to evaluate. The formula of recognition rate P is as follows:

$$P = \frac{\text{Identify the correct number of buds in the number of buds}}{\text{Total number of buds identified}} \tag{5}$$

The formula for calculating the recall rate is as follows:

$$R = \frac{\text{Identify the correct number of buds in the number of buds}}{\text{The actual total number of buds in the sample}} \tag{6}$$

The calculation formula of comprehensive evaluation index is as follows:

$$F_1 = \frac{2}{\frac{1}{P} + \frac{1}{R}} = \frac{2 \times P \times R}{P + R} \tag{7}$$

According to the above experimental indexes, the experiment was carried out, and the results were analyzed.

4.2 Statistical Results Analysis

Carry out the experiment according to the above process, and the statistical results of this article are shown in the following table (Table 2):

Table 2. Experimental results in this paper

Image	Total grain number of tea	Number of buds of tea	Identify the number of tea bud particles	Correctly identify the number of tea bud particles
1	49	20	21	18
2	16	2	1	1
3	31	10	9	9
4	51	13	14	13
5	34	4	3	3
6	59	17	17	16

According to formulas (5)–(7), the three-phase index of the experiment in this paper can be calculated. According to this method, the experiments using K-Means method and Grabcut method are used respectively, and the results of the three experiments are counted. The results are shown in the following table:

Table 3. Comparison of experimental results of three methods

Methods	Recognition rate P/%	Recall rate R/%	F_1/%
K-Means	73.4	81.4	78.6
Grabcut	88.9	77.4	83.9
Method in this paper	95.5	84.7	88.3

According to the data in Table 3, the recognition rate of K-means method is the lowest among the three methods, and the recall rate of grabcut method is the lowest among the three methods. However, the recognition rate of this method can reach 95.5%, the recall rate can reach 84.7%, and the highest comprehensive evaluation rate is 88.3%. It shows that the recognition rate, recall rate and comprehensive evaluation of the method in this paper are high.

In order to further verify the effectiveness of the method in this paper, k-means method and grabcut method for tea bud image recognition time were compared and analyzed. The comparison results are shown in Fig. 8.

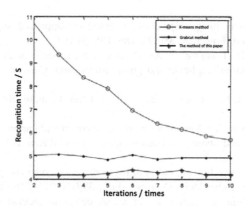

Fig. 8. Comparison results of tea bud image recognition time

According to Fig. 8, the recognition time of tea bud image in this method is within 4.5 s, which is shorter than that of K-means method and grabcut method, which indicates that the recognition efficiency of tea bud image can be improved by using this method.

5 Conclusion

In view of the color difference of tea shoots in Longjing tea image, combining digital image processing technology and application machine learning, this paper proposes an automatic recognition method of tea shoots image based on support vector machine, which makes full use of the characteristics of buds to improve the image recognition rate, and the automatic selection of training samples also provides good features for support vector machine to avoid the subjective selection by hand Sex. The experimental results show that the method selected in this paper is feasible for the recognition of tea shoots in the tea image, and the recognition effect is better than the two traditional methods, achieving the expected experimental effect. However, there are still some shortcomings in this method. When collecting tea images, there are differences in the orientation of tea particles and the shooting light, so the clustering features that affect the selection of the experimental process are not perfect. The next step is to consider the corresponding reflective features in combination with the next research. In addition, this experiment is only conducted on the limited image categories, and the recognition accuracy also needs to be further improved Raise.

Fund Projects. Youth science and technology talent growth project of Guizhou Provincial Education Department (Qian Education KY [2019]181, Qian Education KY [2016]298), Science and technology cooperation project of Guizhou Provincial Science and Technology Department (Qian Science LH [2016] 7288).

References

1. Meng, W.: Ultrasonic image segmentation based on improved support vector machine algorithm. J. Biomed. Eng. Res. **38**(02), 186–195 (2019)
2. Tingting, S., Xue, H., Huaqing, L.: An identification method of indicator diagram based on LIBSVM fusion fourier amplitude and phase information. Comput. Meas. Control **26**(10), 240–245 (2018)
3. Jun-yu, L., Yi-xuan, L., Cong-wei, Z., et al.: J. Chin. Comput. Syst. **40**(06), 1330–1335 (2019)
4. Tongxiao, Y., Caijun, Y.: Research on hydrometeor classification of convective weather based on SVM by dual linear polarization radar. Torrential Rain Disasters **38**(04), 297–302 (2019)
5. Xiaoxiao, S., et al.: Detection algorithm of tea tender buds under complex background based on deep learning. J. Hebei Univ. (Nat. Sci. Ed.), **39**(02), 211–216 (2019)
6. Zhiwei, L., Xiaochun, Z., Kehui, Z., et al.: A detection method of low/zero fault and contamination fault of insulator string based on infrared image feature and BP neural network. Insulators Surge Arresters **03**, 204–211 (2019)
7. Yuchen, L., Jianqing, W.: Image recognition of TCM decoction pieces based on HOG-LBP. Chin. J. Inf. Tradit. Chin. Med. **26**(04), 106–110 (2019)
8. Liu, S., Pan, Z., Cheng, X.: A novel fast fractal image compression method based on distance clustering in high dimensional sphere surface. Fractals **25**(4), 1740004 (2017)
9. Pengliang, P., Guojun, W., Fangmei, Z., et al.: Evaluation of leaf-damaged severity from tea geometrid by geometric morphometrics. J. Henan Agric. Sci. **47**(05), 62–68 (2019)
10. Jingming, N., Shuhuan, L., Yujie, W., et al.: Hyperspectral imaging for quantitative quality prediction model in digital blending of congou black tea. Food Sci. **40**(04), 318–323 (2019)

Automatic Color Image Segmentation Based on Visual Characteristics in Cloud Computing

Jia Wang[1(✉)] and Jie Gao[2]

[1] Mechanical Engineering College, Yunnan Open University,
Kunming 650500, China
zigler.wang@163.com
[2] Nantong Polytechnic College, Nantong 226002, China

Abstract. Aiming at the problems existing in traditional color image segmentation methods, namely, image noise and image quality are poor, a color image automatic segmentation method based on visual characteristics is proposed. The method first analyzes the human visual characteristics, then uses the weighted average method to grayscale the color image, then uses the histogram equalization method to enhance the image, and then detects the edge of the image through the binary wavelet, and finally in the image. Image segmentation based on edge detection. The results show that compared with the traditional image segmentation method, the segmented color image of this method has a SNR of 5.3 dB, less noise and improved image quality.

Keywords: Visual characteristics · Color image · Segmentation method

1 Introduction

Image segmentation is a major problem in image processing and an important part of the field of computer vision and pattern recognition [1].Image segmentation is the first step in image processing in image engineering. The result of image segmentation directly affects the subsequent image processing. In the past forty years, black and white images have been widely used, so there are many studies on black and white image algorithms. However, with the advancement of technology and the reduction of various hardware costs, color images have achieved undoubted protagonist status in the new century. Color images are in line with human visual characteristics and rich color information. It plays an important role in daily life and technology applications. However, because color images are widely used in a short period of time, color image processing methods and research are not as good as the system specifications for gray images, resulting in color image processing effects that are not as good as black and black. The image is white, noisy, and the image segmentation quality is poor. Aiming at the above problems, a method of automatic segmentation of color images based on visual characteristics under cloud computing is proposed. The method is mainly divided into four steps: color image graying, image enhancement, image edge detection, image segmentation [2].

S. Liu and L. Xia (Eds.): ADHIP 2020, LNICST 348, pp. 291–300, 2021.
https://doi.org/10.1007/978-3-030-67874-6_27

Finally, it is verified that the image signal-to-noise ratio of the segmentation method is much higher than that of the traditional image segmentation, and the image segmentation quality is improved, which solves the problems of the traditional method.

2 Automatic Color Image Segmentation Method Based on Visual Characteristics

Image segmentation is an important pre-processing process in the early stage of image processing. Its definition is to divide a complete image into several different connected domains. The pixel characteristics in each region are similar, that is to say, the region is homogeneous and heterogeneous compared with other adjacent regions. Image segmentation is the first step in graphic analysis and pattern recognition. In the system of image analysis and pattern recognition, image segmentation is an important and necessary part. It is the most difficult part in the process of image processing and determines the quality of the final result of image processing [3].

2.1 Visual Characteristics

Nowadays, there are many applications in color space. RGB model is the most widely used model for pictures taken by digital cameras. Human eyes rely on three types of cone cells for color sensing, namely red cone cells, green cone cells and blue cone cells. Among them, the human eye has the weakest sensitivity to red, followed by blue, and the strongest sensitivity is green. In other words, in a digital image with the same 8bit depth of 8 bits, the human eye can perceive the color with the most difference as green. RGB model can be described by Cartesian coordinate system[4]. In this model with a cube shape, the transformation from black to white is expressed as a zero point to a diagonal point, and each one-dimensional coordinate axis represents the transformation of the brightness of the color from small to large. Taking a digital image with a 8bit depth of 8 bits as an example, when the origin is R = O, G = O, B = O, the color represented at this time is black. The diagonal points of its cube are R = 255, G = 255, B = 255, and the color represented by this point is white. According to this model, each color image can be decomposed into three independent planes for calculation [5].

Now, according to the different sensitivities of human eyes to the common red, green and blue colors, automatic color image segmentation is carried out. The specific flow is shown in Fig. 1.

As can be seen from Fig. 1, the automatic color image segmentation method based on visual characteristics is mainly divided into the following steps: firstly, the original color image is input into a computer, then the color image is grayed out, then image enhancement and image edge detection are carried out, and finally image segmentation is realized.

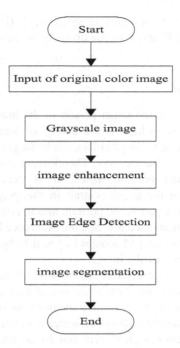

Fig. 1. Flow of automatic color image segmentation method

2.2   Color Image Grayscale

Color images are composed of multiple colors. If contour features are extracted directly from color images, the extraction effect will be blurred due to the influence of colors, so it needs to be grayed out [6].

Image graying refers to a series of processing of color images with multiple colors to produce corresponding gray images containing only black, white and different shades of gray. The pixels of each color in a color image are determined by the components of R(red), G(glass) and B(blue), and the value range of each component is [0,255]. However, all the different pixel points in the grayed color image become the same value, which is called the gray value of the pixel point, and the gray value ∈ [0,255]. Generally, there are four methods to gray color images: component method, maximum value method, average value method and weighted average method. Among them, the weighted average method is the most commonly used, it is based on the importance and other indicators, the components of the three colors are weighted average with different weights, the formula is as follows:

$$Grayg(x,y) = 0.299R(x,y) + 0.578G(x,y) + 0.114B(x,y) \tag{1}$$

In the formula, $Grayg(x,y)$ is a gray image; $R(x,y)$, $G(x,y)$ and $B(x,y)$ represent the component values of R, G and B of the pixels coordinated (x,y) in the gray image respectively.

2.3   Image Enhancement

After the image is grayed out, the target person in the image will be weakened to a certain extent. Therefore, according to the theory of human visual characteristics, human eyes are more sensitive to bright colors, while the graying of the image weakens the sensitive characteristics of human eyes. Therefore, in order to ensure the quality of the automatically generated portrait sketch, it is necessary to perform image enhancement processing on the target portrait in the grayed-out image, that is, to compensate for the weakened part in the portrait, and to compensate the sharpness and contrast of the portrait. Portraits increase, and the difference between the features of the target and other objects in the image becomes larger, thereby enhancing the computer's interpretation and recognition of the image.

In order to make the image clearer, the purpose of image enhancement can be achieved by adjusting the proportion of brightness of each part. This method is called histogram adjustment. Histogram adjustment methods include histogram equalization and histogram matching. The histogram of an image is based on statistical principle and reflects the relationship between gray level and its occurrence probability. Usually, when performing histogram processing, the histogram normalization is generally performed first, and the formula is as shown in (2):

$$P(r_k) = n_k/N, (k = 0, 1, \ldots, 255) \tag{2}$$

In the formula, r_k is the k-level gray value; r_k is the normalized number of pixels whose gray value is r_k; n_k is the number of pixels whose gray level is r_k; N is the number of pixels in the image.

Histogram equalization, also called histogram equalization, combines the histogram of the image with the gray level corresponding to the histogram, and then maps the corresponding gray level to a new gray level through a certain mapping, so that The transformed image becomes a uniform probability density distribution. This transformation can not only expand the effective gray level of pixels, but also improve the frequency of occurrence of each gray value, thus improving the contrast of the image and achieving the purpose of image enhancement.

The following calculation steps for histogram equalization are given:

1) calculating a histogram of an image;
2) normalizing the histogram of the image;
3) calculating a new gray value of the image;
4) replacing the corresponding gray value with the gray value calculated in step 3 in the original image to form a new image.

The implementation process of histogram equalization image enhancement method is relatively simple. After equalization, each gray level of the new image is more balanced and closer to our desired ideal result. When histogram equalization is used to enhance an image with a relatively small gray scale range, the gray scale range of the image can be effectively expanded, and the information of the new image obtained after equalization becomes clearer and more useful. When histogram equalization is jointly applied to image edge detection methods, the histogram equalization image enhancement method enlarges the gray difference between various factors of the image and enhances the gray contrast of the image so as to detect useful detail edges of the image. Therefore, this method has great practical effect and value in the process of image edge detection [7].

2.4   Image Edge Detection

The most basic feature of an image is the edge. The so-called edge refers to the set of pixels with contrast changes in the gray intensity of surrounding pixels. It usually exists between the target and the background, the target and the target, the region and the region, and the primitive and the primitive. Therefore, it is the most important feature on which image segmentation depends, and it is also the important information source of texture features and the basis of shape analysis.

Edge detection is one of the most classical research topics in the field of image processing and computer vision. It has a long research history and has achieved great results. There are many methods of edge detection, but in summary, they all use differential operators to find out the abrupt change points of gray scale. Among the existing edge detection algorithms, the commonly used algorithms are primary differentiation, secondary differentiation, template operation, surface fitting, etc. In recent years, with the deepening of wavelet research, the application of binary wavelet in edge detection is also increasing [8].

When the image is decomposed by dyadic wavelet, the noise of the image is decomposed into high-frequency sub-images, so the low-frequency sub-images obtained by the original image by dyadic wavelet decomposition can effectively remove the noise when performing edge detection. In order to effectively remove noise and realize edge detection in more details of the image, this paper proposes an image edge detection method based on binary wavelet transform, i.e. the low-frequency sub-image obtained by binary wavelet decomposition of the original image is enhanced by histogram equalization, and then edge detection is performed by the modulus maximum point method of binary wavelet transform. These processes are implemented by programming with Matlab7.04. The specific steps are as follows:

Step 1: The quadratic spline function is selected as the wavelet function, and the original image is subjected to one-layer binary wavelet decomposition to obtain a low-frequency sub-image of the original image.

Step 2: The low frequency sub-image is enhanced by histogram equalization.

Step 3: Edge detection is carried out on the enhanced low-frequency sub-image by adopting a dyadic wavelet transform modulus maximum point method to obtain an edge image, namely, the modulus value and the amplitude angle of the dyadic wavelet transform are obtained first, and then the local maximum point of the modulus value along the amplitude angle direction is obtained from the modulus value and the amplitude angle of the dyadic wavelet transform. The positions of these dyadic wavelet transform modulus maxima points give the edge of the image.

2.5 Image Segmentation

Image segmentation based on edge detection is a typical method. Human vision is very sensitive to the edge of the image. In general, when observing objects with edges, the first thing people perceive is the edge. In theory, the definition of edge is: the abrupt position of data such as structure or gray value, which is called edge. It is the cut-off of one part and the start of another part. According to this visual imaging characteristic of human beings, image segmentation can be performed [9, 10].

Edge detection is a very important step in image processing technology. When detecting the edge of the target, firstly roughly detect its contour points, then connect the detected contour points according to a certain principle, and test and link the missing contour points, at the same time remove the wrong boundary points. However, there is noise in real target signals, and their edges are also composed of many different types of edges and their blurred parts, so the processing process is quite complicated, so edge detection is also a difficulty in image processing technology[11, 12].

Image segmentation refers to the process of subdividing an image into multiple image sub-regions. At present, the commonly used segmentation methods are based on threshold, region, edge and specific theory. Since the image edge extraction has been completed in the above section, in order to reduce the workflow. This section performs image segmentation based on the above image edge detection. The specific flow is shown in Fig. 2.

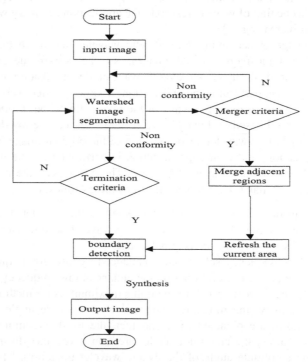

Fig. 2. Image segmentation implementation flow

3 Simulation Test

For color image segmentation algorithms, the quality of segmentation results depends on many factors, such as consistency, spatial compressibility, continuity, smoothness, and so on. The use of a single metric does not include all of the factors, so the quality of the segmentation should be based on whether it is used for evaluation in a particular application area.

The main solution of this research is that the traditional image segmentation technology has a lot of noise, which leads to the problem of poor image quality after segmentation. Therefore, the simulation results are mainly used to verify the segmentation effect of this method and traditional methods. Figure 3 below shows the image to be segmented (the yellow area is the target to be extracted).

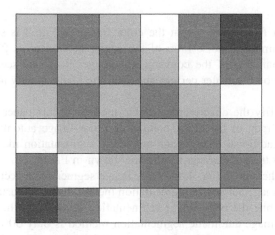

Fig. 3. Image to be segmented

Now, the yellow area in Fig. 3 is segmented with this method and the conventional method, and the signal-to-noise ratio after image segmentation is shown in Table 1 and Table 2 below.

Table 1. The SNR/DB after image segmentation of this method

Number of iterations	This method
1	23.5
2	21.4
3	22.8
4	23.4
5	22.7
6	20.6
7	22.5
8	22.9
9	23.5
10	27.1

Table 2. SNR after image segmentation by traditional method/db

Number of iterations	Traditional method
1	18.2
2	11.2
3	11.2
4	15.6
5	14.7
6	15.6
7	19.2
8	19.4
9	19.8
10	17.0

It can be seen from Table 1 that the color image in Fig. 3 is segmented by the method, and the image-to-noise ratio of the segmented image is 27.1 dB, which is 10.1 dB higher than that of the conventional image. It can be seen that the image segmentation method has better performance and the image quality after segmentation is higher.

In order to verify the effectiveness of the method in this paper, the color image segmentation accuracy of the cloud computing color image automatic segmentation method and the traditional color image automatic segmentation method is compared and analyzed, and the comparison results are shown in Fig. 4.

According to the data in Fig. 4, the color image segmentation accuracy of the cloud computing color image automatic segmentation method based on visualization features can reach 80%, while the color image segmentation accuracy of the traditional cloud computing color image automatic segmentation method is only 50%. Based on visualization The color image segmentation accuracy of the feature-based cloud computing color image automatic segmentation method is higher than that of the traditional color image automatic segmentation method, which shows that the color image segmentation effect of the cloud computing color image automatic segmentation method based on visual features proposed in this paper is better it is good.

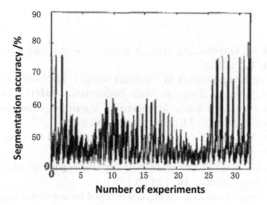

(a)Segmentation accuracy of automatic segmentation method of color image based on Visualization

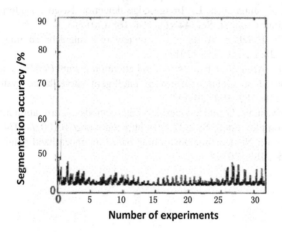

(b)The segmentation precision of traditional cloud computing color image automatic segmentation method

Fig. 4. Comparison of image segmentation accuracy

4 Conclusion

In summary, in view of the problems existing in color image segmentation, the color image segmentation technology based on human visual features is studied. The research process includes image graying, image enhancement, image edge detection and image segmentation. It is verified that the method solves the problem of the traditional color image segmentation technology, that is, the signal-to-noise ratio is improved, which lays a foundation for the development of color image processing technology. However, this study does not study the mechanism of human visual segmentation, so the method needs to be further improved. Subsequent in-depth research should be conducted on edge detection and segmentation in order to propose new reference methods.

References

1. Hongya, Y., Jingxiu, Z., Guanhua, X., et al.: A survey of color image segmentation methods. Softw. Guide **17**(4), 1–5 (2018)
2. Anonymous. Explore the application of computer image segmentation algorithm based on visual characteristics. Comput. Program. Skills Maintenance, **401**(11), 144–154 (2018)
3. Marlowe. Research on graphic image segmentation algorithms based on visual characteristics. Comput. Knowl. Technol. **14**(17), 222–223 (2018)
4. Jie, Z., Hongxia, P., Mingjun, T.: An image perception method for crop diseases based on machine vision features. Agric. Technol. **37**(18), 11–13 (2017)
5. Liu, S., Lu, M., Li, H., et al.: Prediction of gene expression patterns with generalized linear regression model. Front. Genetics **10**, 120 (2019)
6. Yuelin, G.L., et al.: Image enhancement algorithm based on histogram segmentation coupled with clipping control equalization. Comput. Eng. Des. (2) 465–469 (2017)
7. Xinchun, L., Shidong, C., Moyan, Z., et al.: Edge detection of contrast images based on local histogram correlation. Chin. J. Image Graph. **5**(9), 750–754 (2018)
8. Zhiguo, Z., Qian, Z., Jingchuan, L.: Image edge detection based on interpolation wavelet tower decomposition. Comput. Sci. **44**(s1), 164–168 (2017)
9. Feng, J., Qing, G., Huizhen, H., et al.: A review of content-based image segmentation methods. J Softw. **28**(1), 160–183 (2017)
10. Zheng, P., Shuai, L., Arun, S., Khan, M.: Visual attention feature (VAF): a novel strategy for visual tracking based on cloud platform in intelligent surveillance systems. J. Parallel Distrib. Comput. **120**, 182–194 (2018)
11. Liu, S., Liu, D., Srivastava, G., et al.: Overview and methods of correlation filter algorithms in object tracking. Complex Intell. Syst. (2020). https://doi.org/10.1007/s40747-020-00161-4
12. Mengye, L., Shuai, L.: Nucleosome positioning based on generalized relative entropy. Soft. Comput. **23**, 9175–9188 (2019)

Research on Moving Target Behavior Recognition Method Based on Deep Convolutional Neural Network

Jian-fang Liu[1], Hao Zheng[1], and He Peng[2(✉)]

[1] Department of Computer and Software, Pingdingshan University,
Pingdingshan 467000, China
babygongzuo@126.com
[2] Center of Engineering Practice Training, Tianjin Polytechnic University,
Tianjin 300387, China
wwzzmmhhhh@126.com

Abstract. In order to solve the problem that the average recognition degree of moving target line is low by the traditional method of moving target behavior recognition. Therefore, a motion recognition method based on deep convolutional neural network is proposed. Construct a deep convolutional neural network target model, and use the model to design the basic unit of the network. The returned unit is calculated to the standard density map by the set unit, and the moving target position is determined by the local maximum method to realize the moving target behavior recognition. The experimental results show that The experimental results of the multi-parameter SICNN256 model are slightly better than other model structures. And the average recognition rate and the recognition rate of the moving target behavior recognition method based on deep convolutional neural network are higher than the traditional method, which proves its effectiveness. Since a single target is more frequent than multiple recognitions and there is no target similar recognition, similar target error detection cannot be excluded.

Keywords: Convolutional neural network · Moving target · Recognition · Depth

1 Introduction

The moving target recognition means that the computer simulates an eye to retrieve the target object of interest in the image. The recognition of the moving target is to complete the judgment of the target category and the calibration of the location of the target, which is a basic visual processing task for one, but it is very difficult for the computer [1]. An image is converted into a group after it is entered into the computer. The binary value of 0–255, the computer should abstract the high-level semantic information of the target category from this set of data, and determine the location of the target. The target will show different degrees of deformation due to the influence of angle of view, illumination, occlusion between objects and self-occlusion, noise, etc., which increases the difficulty of recognition of moving targets. Although there are

S. Liu and L. Xia (Eds.): ADHIP 2020, LNICST 348, pp. 301–312, 2021.
https://doi.org/10.1007/978-3-030-67874-6_28

many difficulties in moving target recognition, it is the first step for the computer to "see the world" to handle advanced visual tasks [2]. Therefore, moving target recognition is of great significance in the field of computer vision and practical applications. Moving target recognition, also known as target extraction, combines the segmentation and recognition of the target to achieve the purpose of finding the target and identifying the target in the image. The speed and efficiency of moving target recognition is a very important evaluation criterion for the recognition system. Especially in complex scenes, when multiple targets are identified and processed, the target recognition ability becomes more important [3]. The research focuses on the research and development of moving target recognition methods based on deep convolutional neural networks. Through the analysis and comparison of these research work, the status quo of the development of moving target recognition is summarized, and some forward-looking research directions in moving target recognition are proposed.

2   Design of Moving Target Behavior Recognition Method Based on Deep Convolutional Neural Network

2.1   Deep Convolutional Neural Network Target Model

The moving target recognition method based on the deep convolutional neural network extracts the target features through the convolutional neural network. If the convolutional neural network is too shallow, its recognition ability is often inferior to that of ordinary SVM and boosting; If the convolutional neural network is deep, a large amount of data is needed for training, otherwise there will be over-fitting in the learning [4].

The moving target behavior recognition method of deep convolutional neural network has powerful characterization and modeling capabilities [5]. Through supervised, semi-supervised or unsupervised training methods, it is possible to learn the feature representation of the target layer by layer and automatically, and realize the abstraction and description of the object hierarchy. The moving target recognition process based on the deep convolutional neural network is shown in Fig. 1. The input image is subjected to pre-measurement, standardization, etc. Then the image is input into the deep convolutional neural network model. The convolutional neural network learns the target feature and the location of the target from a large amount of input data, and finally determines the category by softmax and other methods [6].

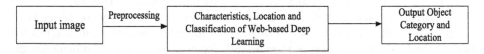

Fig. 1. Moving target detection framework based on deep convolution neural network

The advantage of the moving object recognition method based on deep convolutional neural network is that it can learn features from a large amount of data, and the learned features are robust and have a strong generalization ability, which is very important for moving target recognition.

2.2   Network Basic Unit Design

The basic unit setting of the network is an important part of the method. By increasing the width and depth of the network, the network structure adjusts the parameters required for training and mitigates over-fitting problems [7]. In surveillance video or pictures, because of the distance, the height of the lens and the difference in the target, it usually causes a large difference in the size of the moving target. Therefore, it is necessary to extract features using three convolution layers of different convolution kernel sizes. The basic unit structure is shown in Fig. 2.

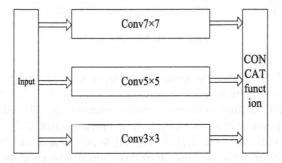

Fig. 2. Network basic unit

The input feature maps are extracted by three convolution layers of different convolution kernel sizes, and the obtained feature maps are stitched by Concat function to obtain a new feature map [8]. In general, the near-point target diameter is about 10, and the far point is about 5. Considering that the pooling layer is used for down-sampling operation, the convolution kernel size is selected by 7×7, 5×5, and 3×3 for feature extraction. Among them, the 7×7, 5×5 and 3×3 convolution kernels respectively extract the features of the moving targets of the near, middle and far levels, and the extracted feature maps are stitched by Concat function to obtain the feature map of the layer [9].

To use the convolutional neural network to identify the moving target behavior, first create a model and set the parameters of each layer of the deep convolutional neural network target model. The model includes a data layer, a feature extraction layer, an activation layer, a loss layer, and the like. In order to extract multiple features faster and better, the Inception structure is used as the basic structural unit of the network to extract different scale features. And after the convolutional layer, the largest pooling layer is added to downsample, and the multi-convolution kernel features at more scales are obtained [10]. Therefore, two models of convolutional neural network model based on serial Inception structure (Serial-Inceptions Based Convolutional Neural Network, SICNN) and convolutional neural network model based on composite

Inception structure (Multi-Inceptions Based Convolutional Neural Network, MICNN) are proposed. The SICNN model and parameters are shown in Fig. 3.

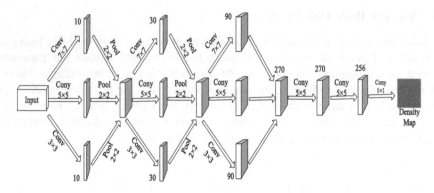

Fig. 3. SICNN_256 model

In the SICNN_56 model, the input data is input into the first Inception structure, and the convolution kernels are convolutional layers of 7×7, 5×5, and 3×3, respectively. And 10 feature maps are exported respectively. The 30 feature maps are integrated into 30 new feature maps by Concat function after 2X2 and the maximum pooling process with step size 2 [11]. The 30 feature maps and the first Inception structure process enter the second and third Inception structures, respectively, to obtain 90 and 270 feature maps, respectively. In this process, for example, the input image of pixel 480×640 is extracted by three convolution kernels of 7×7, 5×5, and 3×3 at three scales of 480×640, 240×320, and 120×160, respectively. It extracts information from different dimensions at different scales and obtains richer features. Then, the feature is further extracted by two convolution layers of 5×5 convolution kernel size; Finally, a 120×160 size density map is output through a 1×1 convolution kernel size convolutional layer. The MICNN_56 model and parameters are shown in Fig. 4.

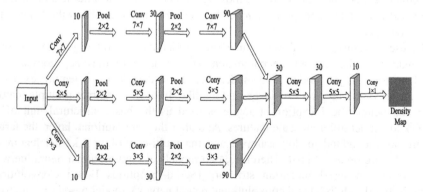

Fig. 4. MICNN_56 model

In the MICNN_56 model, there is only one Inception structure, and the depth is increased while keeping the width of the Inception structure unchanged [12]. At the same time, in order to eliminate the influence of parameters on the network structure, the number of convolution kernels in the MICNN_256 model is the same as that of the SICNN_56 model. The input data is input into the Inception structure, and is divided into three branches for operation. The first branch passes through the convolution layer with a convolution kernel size of 7 × 7, and outputs 10 feature maps. The feature map is subjected to a maximum pooling process of 2 × 2 and a step size of 2, and then a convolution layer having a convolution kernel size of 7 × 7 is output, and 30 feature maps are output [13]. The feature map is subjected to a maximum pooling process of 2 × 2 and a step size of 2, and then a convolutional layer with a convolution kernel size of 7 × 7 is output, and 90 feature maps are output. The second and third branches are similar to the first branch, but the convolution kernels are 5 × 5 and 3 × 3, respectively, and the three branches output 90 characteristic maps. In addition, in order to further explore the influence of convolution kernel parameters on the network, the structure and parameters of another set of experimental models SICNN_10 and MICNN_10 are designed when the network structure is unchanged and the number of convolution kernels in the network is changed. They are shown in Fig. 5 and Fig. 6, respectively.

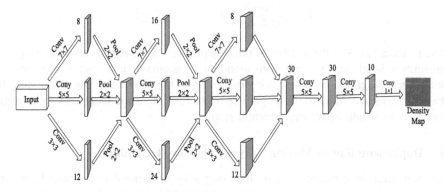

Fig. 5. SICNN_10 model

The activation layer activates the input data, that is, a function transformation, which is performed on an element-by-element basis. Commonly used activation functions are Sigm oid function, Tanh function, ReLU function, etc. The most commonly used function is ReLU function.

A linear rectification function, also known as a modified linear unit, usually refers to a nonlinear function represented by the ramp function $f(x) = \max(0, x)$ and its variants.

The loss layer calculates the loss value by calculating the difference between the training sample output and the real sample value. At present, there are three main Loss layer loss functions: Sigmoid, Softmax and Euclidean. Where Sigmoid is mainly used

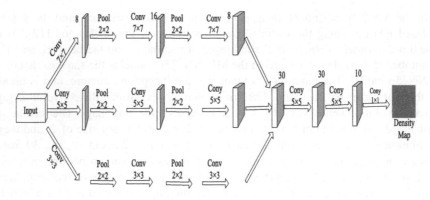

Fig. 6. MICNN_10 model

for the two-classification problem, Softmax can be used for the multi-classification problem, and Euclidean is the commonly used loss function for linear regression. The designed network structure model uses the Euclidean loss function as the Loss layer to return the density map of the network output to the standard density map. The loss function is:

$$L = \frac{1}{2N} \sum_{i=1}^{N} \left\| F(X_i) - D(X_i) \right\|_2^2 \qquad (1)$$

In the formula (1), N is the number of training pictures, X_i represents an input image, D represents a label density map corresponding to standard data, and F represents a density map generated by a network structure; L is the calculated loss value, and the network judges according to the magnitude of the loss value and feedbacks the relevant parameters to obtain better experimental results.

2.3 Implementation of Moving Target Behavior Recognition

After obtaining the estimated density map output by the convolutional neural network model, the local maximum value is extracted from the estimated density map, and the extracted local maximum is the position where the target of the image is likely to be located. In the process of target behavior recognition, the density map needs to be denoised, and the target should be located within a certain range according to the target size. Because the size of the target is different due to the influence of the height of the lens and the distance from the target to the lens, the positioning should be selected according to certain classification measures; Due to the complexity of the background and the degree of light and darkness, the numerical values of the target points in the estimated density map obtained by learning are not the same. It is necessary to use a histogram to set a threshold to remove the background and remove the wrong target point to perform the target positioning. The specific operation process is shown in Fig. 7.

Fig. 7. The flow chart of motion target behavior recognition algorithm

Degree histogram, calculate the pixel value size P with the highest proportion, and set a certain threshold value T, the pixel point whose pixel value $F_k(x, y)$ is smaller than $P + T$ in the estimated density map is defaulted to the image background, and the interference term is removed, and is set to 0, and the estimated density map $D_k(x, y)$ of the background is obtained, as shown in Eq. 2:

$$D_k(x,y) = \begin{cases} 0, & F_k(x,y) < P + T \\ F_k(x,y), & F_k(x,y) \geq P + T \end{cases} \qquad (2)$$

Selecting a moderately sized sliding window $M_k(x, y)$ selects the target point by local maximum in the estimated density map $D_k(x, y)$ of the removed background. The maximum point is set to 0, and the estimated position map $R_k(x, y)$ of the target is obtained, as shown in Eq. 3.

$$R_k \; x \; y = \begin{cases} 0, \; D_k \; x \; y < \max \; M_k \; x \; y \\ 255 \; D_k \; x \; y = \max \; M_k \; x \; y \end{cases} \qquad (3)$$

The target point obtained by using the local maximum value sometimes has the same value of two similar points while retaining the information of the two position points, thereby causing adhesion, resulting in a re-inspection when the target is marked. In order to avoid such things, the obtained target center point position coordinate map will eliminate some wrong target points according to the law of two points, in order to obtain better recognition results. First, input the labeled coordinates and estimated coordinates of the test set; secondly, calculate the distance between each point in the estimated coordinate data and each point in the labeled coordinate data; Again, find the nearest point in the estimated coordinates by the labeled coordinates, and set the nearest distance threshold as the search range; finally, retrieve the target.

3   Simulation Experiment

3.1   Experiment Data

The algorithm is validated using the Mall data set. The Mall dataset contains different target densities and lighting conditions and is widely used in target counting work. The dataset uses a surveillance camera to collect data. It is a continuous video sequence. The video sequence in the dataset consists of 2000 frames of 640×480 color images, which are tagged with more than 60,000 moving targets. A total of four experiments were performed, including SICNN_256 model, SICNN_10 model, MICNN_256

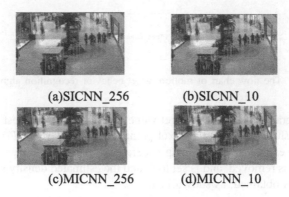

(a)SICNN_256   (b)SICNN_10

(c)MICNN_256   (d)MICNN_10

Fig. 8. Recognition results

model, and MICNN_10 model. Each model targeted the test images with 27 real targets, and the results are shown in Fig. 8.

Figure 8(a) shows the target recognition result of the SICNN_256 model, in which 31 target targets are estimated, 23 targets are correctly identified, 8 targets are misidentified, and 4 targets are missing. Figure 8(b) shows the target recognition result of the SICNN_10 model, in which 32 targets are estimated, 22 targets are correctly identified, 10 targets are misidentified, and 5 targets are missing. Figure 8(c) shows the target recognition result of the MICNN_256 model, in which 36 targets are estimated, 23 targets are correctly identified, 13 targets are misidentified, and 5 targets are missing. Figure 8(d) shows the target recognition result of the MICNN_10 model, in which 31 targets are estimated, 21 targets are correctly identified, 10 targets are misidentified, and 6 targets are missing.

3.2   Average Literacy Rate and Comparison Rate

In this experiment, the training time of the less parameter network is 0.39 s/5 times, and the training time of the network with more parameters is 1.68 s/5 times.

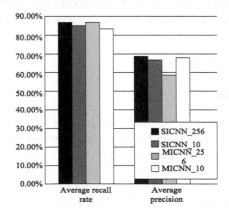

Fig. 9. Comparison of average recall rate and precision rate

SICNN_256 model, SICNN_10 model, MICNN_256 model, MICNN_10 model 4 groups of experiments on the average recognition accuracy and time-consuming situation of 400 test set pictures under the same number of iterations, as shown in Fig. 9.

Through comparison between different experiments, it can be seen from Fig. 9 that the experimental results of the multi-parameter SICNN256 model are slightly better than those of other model structures. However, the model structure has a large amount of computation and takes a long time, and performance and timeliness cannot be simultaneously provided.

At the same time, in the test set, select some pictures to compare the recognition results, and compare the results of the full rate and the identification rate parameters, as shown in Fig. 10, respectively:

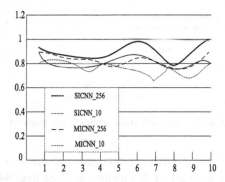

Fig. 10. Test set picture recognition rate

Based on the comprehensive experimental data, it can be concluded that the omission recognition caused by the occlusion situation in the test picture and the decrease in the recognition rate caused by the excessive number of samples are the main reasons for the poor experimental results. It can be seen from Fig. 10 that the missed detection of the sample mostly occurs in the area with more obstructions or edges, and the misdetection of the sample mostly occurs on the same target body or the like. Since the individual is replaced by a single head as the recognition target, although the recognition number is accelerated, since there is no further recognition judgment. Therefore, the single target is recognized more frequently than many times, and since there is no target similar recognition, the similar target false detection cannot be excluded.

3.3 Comparative Experiment

In order to verify the effectiveness of the proposed motion target behavior recognition method based on deep convolution neural network, a comparative experiment was conducted. The method based on spatiotemporal semantic information and the method based on intelligent video analysis were selected as the experimental comparison methods. The recognition accuracy and recognition time were selected as the experimental indicators. The results are as follows.

(1) Recognition Accuracy

Three methods are selected to test the recognition accuracy, and the results are shown in Fig. 11.

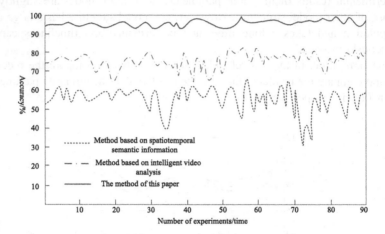

Fig. 11. Identification accuracy comparison

Analysis of Fig. 11 shows that, compared with the experimental comparison method, the recognition accuracy of this method is more than 94%, which shows that the method can accurately identify the behavior of moving targets.

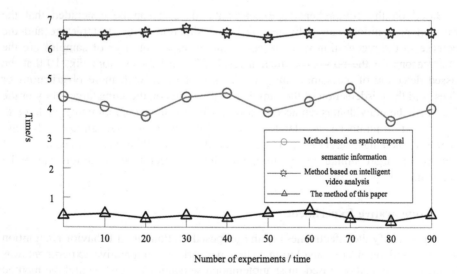

Fig. 12. Identification time comparison

(2) Identification Time

Three methods are selected to test the recognition time, and the results are shown in Fig. 12.

Analysis of Fig. 12 shows that the recognition time of this method is always below 0.9 s, which is far lower than the experimental comparison method, which shows that the method can realize fast recognition of moving target behavior.

4 Conclusion

In order to solve the problem that the average recognition degree of moving target line is low in the traditional method of moving target behavior recognition, a motion recognition method based on deep convolution neural network is proposed. The method mainly uses the deep convolution neural network target model to realize the behavior recognition of moving objects. The experimental results show that the method has excellent performance and can be further applied in practice.

Acknowledgments. This work is supported by Henan provincial department of science and technology planning project social development (NO.182102310040), and Pingdingshan University Youth Scientific Research Fund Project (PXY-QNJJ-2018005).

References

1. Liu, Z., Huang, J., Feng, X.: Constructing behavior recognition model of multiscale depth convolution neural network. Opt. Precis. Eng. **25**(3), 799–805 (2017)
2. Tang, Z., Zhang, K., Li, C., et al.: Motion imagination classification based on deep convolution neural network and its application in brain-controlled exoskeleton. J. Comput. Sci. **40**(6), 1367–1378 (2017)
3. Zhou, T., Xu, Y., Zheng, W.: Classification and recognition of tomato main organs based on deep convolution neural network. J. Agri. Eng. **33**(15), 219–226 (2017)
4. Li, C., Qin, P., Zhang, J.: Research on image denoising based on deep convolution neural network. Comput. Eng. **43**(3), 253–260 (2017)
5. Wang, Z., Min, H., Zhu, Q.: Optical flow detection of moving objects based on deep convolution neural network. Optoelectron. Eng. **45**(8), 1–9 (2018)
6. Liu, S., Liu, D., Srivastava, G., et al.: Overview and methods of correlation filter algorithms in object tracking. Complex Intell. Syst. (2020). https://doi.org/10.1007/s40747-020-00161-4
7. Yuan, G., Tang., Y., Han., W., et al.: Vehicle type recognition method based on deep convolution neural network. J. Zhejiang Univ. Eng. Edn. **17**(4), 12–25 (2018)
8. Liu, Z., Ho, D., Xu, X., et al.: Moving target indication using deep convolutional neural network. IEEE Access **6**, 1 (2018)

9. Liu, S., Lu, M., Li, H., et al.: Prediction of gene expression patterns with generalized linear regression model. Front. Genetics **10**, 120 (2019)
10. Zhang, Y., Li, J., Guo, Y., et al.: Vehicle driving behavior recognition based on multi-view convolutional neural network (MV-CNN) with joint data augmentation. IEEE Trans. Veh. Technol. **68**(5), 1 (2019)
11. Shuai, L., Gelan, Y.: Advanced Hybrid Information Processing, pp. 1–594. Springer, New York (2019). https://doi.org/10.1007/978-3-030-36402-1
12. Fei, G., Teng, H., Jinping, S., et al.: A new algorithm of SAR image target recognition based on improved deep convolutional neural network. Cogn. Comput. **17**(6), 1–16 (2018)
13. Liu, G., Liu, S., Khan., M., et al.: Object tracking in vary lighting conditions for fog based intelligent surveillance of public spaces. IEEE Access **6**, 29283–29296 (2018)

Design of 3D Image Feature Point Detection System Based on Artificial Intelligence

He Peng(✉)

Center of Engineering Practice Training,
Tianjin Polytechnic University, Tianjin 300387, China
wwzzmmhhhh@126.com

Abstract. Aiming at the problems of low efficiency and accuracy in the traditional 3D image feature point detection system, an efficient 3D image feature point detection system based on artificial intelligence is designed. Firstly, the whole frame of the system is designed. Then the hardware system is designed, including the development board, peripheral equipment and interface, basic engineering reconstruction and feature point detection unit. Then the software system is designed, including image collection module, image feature point display module. Image feature point processing module, image feature point extraction module, image feature point description module, and using the combination of hardware system and software system to achieve three based on artificial intelligence Dimension image feature point detection system. Finally, the effectiveness of the 3D image feature point detection system based on artificial intelligence is verified by experiments, and the detection efficiency and accuracy are much higher than the traditional methods. This study lays a foundation for the further study of images.

Keywords: Artificial intelligence · 3d image · Feature points · Efficient detection

1 Introduction

Artificial intelligence has entered the life of modern people. In many fields of artificial intelligence, image detection is the foundation stone, especially in the field of Internet of things [1]. Image is a special data format. If we want to obtain different connotations according to different images, we need to use certain image detection technology. With the wide application of 3D image in daily life, 3D image feature point detection has gradually become a hot topic. Feature point detection of 3D image is an important part of computer vision application. With the rapid development of information technology, people rely more and more on information, how to obtain more useful information becomes more and more important. The most intuitive way to get information is to observe and recognize the outside world through the human eye, and according to statistics, it is possible to obtain external information through human vision, which is enough to show that the human visual system can image through the eyes. And then sense the three-dimensional objects in life. This ability to obtain 3D image information from two-dimensional scene images is a visual function that people have been trying to give

© ICST Institute for Computer Sciences, Social Informatics and Telecommunications Engineering 2021
Published by Springer Nature Switzerland AG 2021. All Rights Reserved
S. Liu and L. Xia (Eds.): ADHIP 2020, LNICST 348, pp. 313–323, 2021.
https://doi.org/10.1007/978-3-030-67874-6_29

computers and intelligent machines. It can replace human to complete the acquisition and processing of external scene information. Another way to obtain information is to obtain external information indirectly from the image. Photography can record all kinds of image information, that is, the process of exposing the sensitive medium inside the camera by shooting the reflected light of the scene, or "painting with light" Therefore, the general imaging method can only obtain the plane image of the scene [2]. If the three-dimensional shape information of the object scene in the image is to be obtained, the scanning mode can only be used, and the time taken by the process is longer, so that not only the real-time property obtained by the image information is affected, but also the definition of the imaging picture is affected. The commonly used 3D image feature point detection methods include feature detection method based on image edge information, corner information detection method and various interest operators. However, the traditional detection methods all have the problems of large computation and long time consuming. In order to solve these problems, a three-dimensional image feature point detection system based on artificial intelligence is designed.

2   System Overall Frame Design

When constructing the whole frame of the 3D image feature point detection system based on artificial intelligence, four modules are designed, which are shown in Fig. 1.

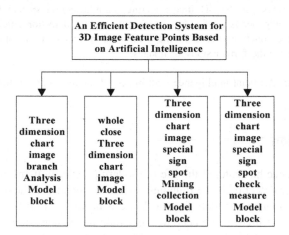

Fig. 1. Overall frame construction

For 3D image analysis module, its main function is to collect and analyze 3D image resources. The purpose of integrating 3D image module is to recombine the digital signal output from the system, generate 3D image and then complete the feature point detection of 3D image [3]. The feature point acquisition module of three-dimensional image is to extract and collect the feature point of three-dimensional image. And the three-dimensional image feature point detection module is used for realizing the high-efficiency detection system of the three-dimensional image feature point based on the artificial intelligence by detecting the three-dimensional image feature point.

3   Hardware System Design

3.1   Development Boards

The hardware platform used in the 3D image feature point detection system based on artificial intelligence is the PYNQ-Z1 development board, which integrates the 7000 series ZYNQ chip of Syringes Company. It can realize the heterogeneous processing of ARM and FPGA so as to realize the efficient detection of 3D image feature points [4]. The PYNQ development board is an artificial intelligence development board developed by Dizzelon and Syringes. It is simple in shape and equipped with ZYNQ artificial intelligence chip. It simplifies and improves the design of software and hardware related to APSOC. Because it integrates features that other ZYNQ development boards do not have, based on the heterogeneity of ARM and FPGA: It can utilize Python, an efficient language with rich third party libraries, and apply the API command based on Python language to the control of FPGA. It can detect feature points by controlling FPGA. Its integrated Jupyter Notebooks framework compiler tool has high visibility and is very suitable for the development process of detection system [5]. In addition, the embedded platform can effectively collect the feature points of 3D images, and the feature points can be analyzed by integrating the collected feature points. Based on the advantages of PYNQ and the requirement of high efficiency detection of 3D image feature points based on artificial intelligence, The PYNQ development board is selected as the hardware system of the 3D image feature point detection system based on artificial intelligence [6]. In addition, the PYNQ-Z1 board can be started from the JTAG, Quad-SP flash memory and the microSD card. There is a digital-to-analog converter on the chip, and the on-chip system and peripheral interfaces are shown in Fig. 2.

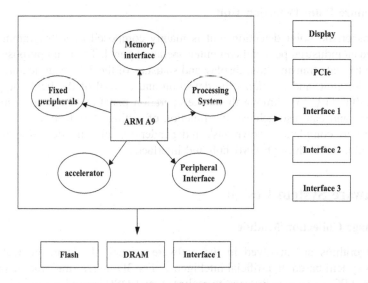

Fig. 2. System and peripheral interface

3.2   Peripherals and Interfaces

ZYNQ PL can support multiple types of external protocols and interfaces. PYNQ-Z1 has two Pmod ports and a ArduINo interface for connecting peripherals directly to ZYNQ PL, allowing peripheral devices to be controlled by hardware. Other peripherals can be connected to the above port through adapters or breadboards [7].

Although the FPGA architecture is reconfigurable and used to achieve high performance in hardware, the design of FPGA is a task that requires in-depth knowledge of hardware engineering and expertise. OVERLAY is a programmable/configurable FPGA design. Extend user application from ZYNQ processing system to programmable logic [8]. Software programmers can use OVERLAY, in a similar way to the software library to run some 3D image feature point processing functions on the FPGA architecture, such as edge detection, threshold, etc. [9].

3.3   Reconstruction of Basic Engineering

Since all the configurations required for the pre-defined ZYNQ SoC PS are provided on the basic OVERLAY, and the PS-PL interface is defined, the hardware OVERLAY, which creates a custom PYNQ, can use the existing Base OVERLAY as the starting point to add the required feature IP core to achieve the desired function [10, 11]. The connection information of four AXI interface information is as follows: GP0 has 13 interfaces, which is in charge of LEDS GPIO, Buttons GPIO, BRAM, Video HPD, Video GPIO, Video Dynamic Clock, Video VDMA, Audio, RGB LEDS and HP0 is mainly responsible for Video VDMA function; GP1 and HP2 are responsible for PMOD and Arduino. Of which, basic workers The program allocates a total of 512 megabytes of space for memory in advance, and HP0,HP2 shares this address space.

3.4   Feature Point Detection Unit

The characteristic point detection unit is mainly composed of a programmable controller and an industrial personal computer operating panel. The main purpose of IPC is to realize the data transmission, display and statistics of the 3D image feature points of the detection unit, and to realize the detection and control of the field equipment by using the IPC [12, 13]. Among them, the operation panel can be used to modify the parameters, and can also achieve the function of setting. The feature point detection unit needs to complete its own task independently, and the detection results are transmitted to CPU through USB hub and interface.

4   Software System Design

4.1   Image Collection Module

Many algorithms are involved in the software system of 3D image feature point detection system based on artificial intelligence. For 3D image stratigraphic processing algorithm, FPGA is generally used to realize it, and DSP is used for high-level image processing algorithm. The system allocates the algorithm to each processing unit, and then realizes the task processing by pipeline, so that the efficiency of the system can be further improved. The main function of image collection card is to convert the microscopic image signal which has been converted into analog video signal in CCD camera device to digital signal direction. Use an image collection card that meets your own performance and price, depending on your work needs. In the software system of 3D image feature point detection system based on artificial intelligence, multimedia 3D image collection card is mainly used to realize 3D image analog preference A/D transform, digital signal storage and so on. And computer to achieve text, image, graphics and voice artificial intelligence integrated processing.

4.2   Image Feature Point Display Module

Image feature point display module can effectively collect and display 3D image feature points and capture 3D images. The capture refers to storing the scene image in memory, then copying the image into the buffer, then storing the data and BMP file format in the image buffer to the hard disk to show that it has been called. First of all, we initialize the acquisition, start the image acquisition process, and turn off the image to screen function in real time, so as to prepare for the image to memory acquisition. The parameters of the image resolution are set, and the input and output windows are determined, then the window is set by using the corresponding functions. After the setup is completed, the image is called into the memory function, and the resource allocation of the file information is realized in the memory, and the file information header is set. After the installation, BMP file information and image data information will be written to the empty file. If the function is executed successfully, BMP file will be written to disk and 3D image capture is completed.

4.3   Image Feature Point Processing Module

Image feature point processing module mainly deals with 3D image feature points, including gray processing, threshold segmentation, median filtering, edge detection, thinning, coordinates and distance calculation. Grayscale processing means that it is difficult to collect 32 true color images, so it is necessary to transform them into 256-color grayscale images. Threshold segmentation means that there are gray noise points in the process of collecting 3D image feature points, which can be removed by threshold method. Median filter refers to the use of 3×3 modules, a point as the center, if there is only this pixel value in the module region, then this point is noise, the pixel value will be changed to remove it. Edge detection refers to the use of functional Gauss algorithm to remove noise and then implement edge detection because noise points have a certain effect on edge detection. Thinning refers to the realization of thinning the image, effectively highlighting the geometric features of the image and reducing the amount of redundant information. Coordinate and distance are calculated by Hough algorithm, and the distance between two circles is calculated.

4.4   Image Feature Point Extraction Module

ORB descriptor is an improved algorithm of BRIEF algorithm proposed at the International Conference of computer Vision, which has the characteristics of invariance and anti-noise of 3D image rotation. The algorithm also combines the calculation flow of corner detection operator algorithm, which is recognized to be faster at present, and is a new matching algorithm. When extracting feature points, the algorithm gives the feature points a direction information after corner detection, which can keep the rotation invariance of 3D images. When matching feature points, in order to overcome the sensitivity to noise, the pixel block is used to determine the gray value in the pixel block, and the corresponding feature descriptor is generated, which improves the attack ability of the algorithm against noise, and has rotation invariance. The algorithm is a combination of operators and descriptors, which combines the advantages of the two algorithms. Extracting feature points is also an interesting part of the extraction image. The general feature points include boundary points, spots and corners. These points not only contain a lot of information of the image, but also determine the attributes of the image. Using the FAST feature detection operator to extract the corner information of the image, this is a fast feature extraction method, this method is mainly divided into two steps: The first step is to select and determine the gray value of the peripheral pixels compared with the points to be detected, and the second step to compare the gray values between the detection points and the peripheral pixels, and to determine the corner points according to the preset conditions. First, the pixel points should be selected, and the distribution of the peripheral pixel values determines whether the correct image feature points can be extracted. If we take a pixel point P in the image and judge whether point P is a feature point, you need to create a circle with radius r = 3 centered on a point P as a template graph, which is a discrete circle. The generation of discrete circle needs to use the algorithm of drawing circle in computer graphics, and through multiple iterations and traversing 16 pixels. The selection process of 16 pixels around the 3D image feature points is as follows:

Initial conditions: radius rn 3 D ~ (1) ~ (-1) ~ (-1) ~ (-1).
Termination condition: $X > y$.
Iterative steps: starting point $(XY) = (0nr)$
IF($x<y$),IF($d<0$),d=d+2x+3;

ELSE (dapd 2 (x-y) 5Cy -X);
END.

Where x is the x-axis coordinate value of generating the peripheral pixel point, y is the y axis coordinate value of generating the peripheral pixel point, and d is the constant changing threshold value.

4.5   Image Feature Point Description Module

The concept of gray moment is used to calculate the corner direction of feature points. The determination of corner direction is mainly divided into two parts [14, 15]. Firstly, the gray balance point of the small pixel block with the feature point as the center of 7×7 is determined, that is, the centroid point; The offset between the centroid point and the feature point, that is, the primary direction of the corner point, is then calculated. In the circular neighborhood radius r, the formula for calculating the offset m is as follows:

$$m_{pq} = \sum_{x,y} x^p y^q I(x, y) \tag{1}$$

Where the $I(x, y)$ table does not have a gradation value at the coordinates of (x, y), and p and q are the order number. The centroid point C can be obtained by using the following formula:

$$C = \left(\frac{m_{01}}{m_{10}}, \frac{m_{01}}{m_{00}} \right) \tag{2}$$

If the characteristic point is P, the vector PC is the main direction of the angle point, and the main direction of the ER angle point can be obtained by using the following formula:

$$\theta = \arctan\left(\frac{m_{01}}{m_{10}}\right) = \arctan\left(\frac{\sum_{x,y} y I(x, y)}{\sum_{x,y} x I(x, y)}\right) \tag{3}$$

5   Experimental Results and Discussion

In the process of experiment, a 3D image is used as the experimental object to detect the feature points of the 3D image. The connection between the PC machine and the PYNQ development platform is carried out by using the HDMI to USB connection line, and then the PYNQ development platform and the display screen are connected by the same

Fig. 3. Experimental sample

connection wire. After completing the environment connection, we load the original 3D image on the Jupyter Notebook, and the 3D image loaded is 720 or 1280 pixels.

In the experiment, part of the action screenshot in a 3D animation is selected as the experimental sample through the 3D image simulation system, as shown in Fig. 3.

The feature points are extracted from the 3D image by using the 3D image feature point detection system based on artificial intelligence and the traditional 3D image feature point detection system. The extraction results are shown in Fig. 4.

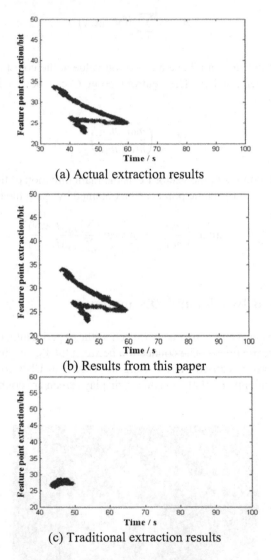

(a) Actual extraction results

(b) Results from this paper

(c) Traditional extraction results

Fig. 4. Three-dimensional image feature point extraction results of two systems

According to Fig. 4, the extraction results of 3D image feature points in this system are consistent with the actual three-dimensional image feature points extraction results, while the traditional system's three-dimensional image feature point extraction results are quite different from the actual three-dimensional image feature point extraction results, which shows that the system in this paper can accurately extract three-dimensional image feature points.

In order to verify the effectiveness of the system in this paper, the efficiency of feature point detection in 3D image of this system and traditional system is compared and analyzed. The comparison results are shown in Fig. 5.

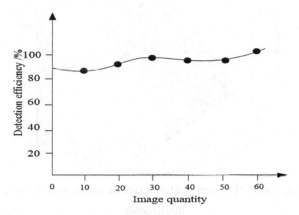

(a)Detection efficiency of 3D image feature point detection system based on Artificial Intelligence

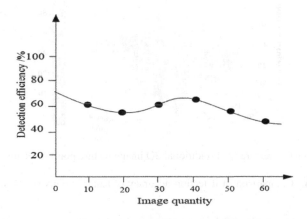

(b)Traditional 3D image feature point detection system

Fig. 5. Comparison of detection efficiency of 3D image feature point detection

Figure 5 shows that the high efficiency of 3D image feature point detection based on artificial intelligence is higher than that of traditional 3D image feature point detection system.

In order to further verify the effectiveness of the method in this paper, the detection accuracy of three-dimensional image feature points based on artificial intelligence and traditional three-dimensional image feature point detection system is compared and analyzed, and the comparison results are shown in Fig. 6.

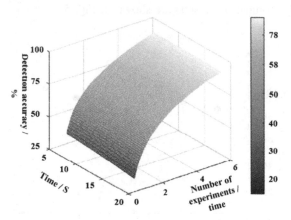

(a)Detection accuracy of 3D image feature point detection system based on Artificial Intelligence

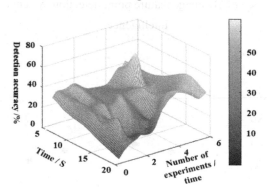

(b)Detection accuracy of traditional 3D image feature point detection system

Fig. 6. Comparison of detection accuracy of feature points in 3D image

According to Fig. 6, the 3D image feature point detection accuracy of the 3D image feature point detection system based on artificial intelligence can reach up to 95%, which is higher than that of the traditional 3D image feature point detection system.

6 Concluding Remarks

The traditional 3D image feature point detection system has the problems of low detection accuracy and long detection time, which leads to the poor detection effect of 3D image feature points. Therefore, this paper designs a 3D image feature point detection system based on artificial intelligence. Through the design of system hardware and software, the design of efficient detection system of 3D image feature points based on artificial intelligence is successfully realized. Experimental results show that the detection accuracy of the system is high, which improves the detection efficiency and detection effect, which proves that the system has good application value.

References

1. Zhang, H.: Design and implementation of artificial intelligence image detection system based on Internet of Things. Comput. Measur. Control **25**(22), 215–218 (2017)
2. Cui, Y.: Design of artificial intelligence image detection system based on Internet of Things. J. Liaoning Univ. Sci. Technol. **23**(24), 222 (2018)
3. Jin, Z., Cao, J., Zhang, Y., et al.: Novel visual and statistical image features for microblogs news verification. IEEE Trans. Multimedia **19**(3), 598–608 (2017)
4. Slave tree: Design of embedded image recognition and detection system based on ARM. Comput. Measur. Control **25**(29), 136–138 (2017)
5. Tien, N.D., Danh, P.T., Rae, B.N., et al.: Combining deep and handcrafted image features for presentation attack detection in face recognition systems using visible-light camera sensors. Sensors **18**(3), 699–702 (2018)
6. Gao, Q., Wang, J., Ma, X., et al.: CSI-based device-free wireless localization and activity recognition using radio image features. IEEE Trans. Veh. Technol. **66**(11), 10346–10356 (2017)
7. Testolin, A., Stoianov, I., Zorzi, M.: Letter perception emerges from unsupervised deep learning and recycling of natural image features. Nat. Hum. Behav. **1**(9), 657–664 (2017)
8. Eichinger, P., Alberts, E., Delbridge, C., et al.: Diffusion tensor image features predict IDH genotype in newly diagnosed WHO grade II/III gliomas. Sci. Rep. **7**(1), 13396 (2017)
9. Fu, W., Liu, S., Srivastava, G.: Optimization of big data scheduling in social networks. Entropy **21**(9), 902 (2019)
10. Jaccard, N., Szita, N., Griffin, L.D.: Trainable segmentation of phase contrast microscopy images based on local basic image features histograms. Comput. Methods Biomech. Biomed. Eng. Imaging Visualiz. **5**(5), 359–367 (2017)
11. Wan, T., Cao, J., Chen, J., et al.: Automated grading of breast cancer histopathology using cascaded ensemble with combination of multi-level image features. Neurocomputing **229**, 34–44 (2017)
12. Liu, S., Pan, Z., Cheng, X.: A novel fast fractal image compression method based on distance clustering in high dimensional sphere surface. Fractals **25**(4), 1740004 (2017)
13. Lu, Mengye, Liu, Shuai: Nucleosome positioning based on generalized relative entropy. Soft. Comput. **23**(19), 9175–9188 (2018). https://doi.org/10.1007/s00500-018-3602-2
14. Johnson, P.B., Young, L.A., Lamichhane, N., et al.: Quantitative imaging: correlating image features with the segmentation accuracy of PET based tumor contours in the lung. Radiotherapy Oncol. **123**, 257–262 (2017). S0167814017301044
15. Shuai, L., Weiling, B., Nianyin, Z., et al.: A fast fractal based compression for MRI images. IEEE Access **7**, 62412–62420 (2019)

An Optimal Tracking Method for Moving Trajectory of Rigid-Flexible Coupled Manipulator Based on Large Data Analysis

Yang Fu-Jian[1] and Wei Tao[2([⊠])]

[1] Guilin University of Aerospace Technology, Guilin 541004, China
aklpl23456@163.com
[2] Northwestern Polytechnical University, Xi'an 710072, China
whai7863@163.com

Abstract. The manipulator has dynamic characteristics, and the trajectory tracking system of the manipulator has non-holonomic constraints and various uncertainties, which makes tracking control of the mobile manipulator more difficult. There is a big error in tracking a rigid flexible coupling manipulator with a single neural network. A new method for trajectory optimization tracking of a rigid-flexible coupled manipulator based on big data analysis is proposed. This method takes neural network as the research object, introduces fuzzy control into neural network, optimizes a single neural network, forms a composite method of fuzzy neural network, and uses a hybrid method to track the trajectory of the manipulator. Experimental results show that the tracking error of this method is less than 0.035 rad, which improves the tracking efficiency and improves the tracking accuracy. The method can complete the operation faster and more accurately according to the predetermined trajectory, and has higher practical applicability.

Keywords: Rigid-flexible coupling · Manipulator · Moving trajectory · Tracking method · Fuzzy nerve

1 Introduction

With the progress of science and technology, mechanized production gradually replaces manual production, completes all kinds of dangerous work that human beings can not directly contact, saves production costs greatly for enterprises, improves production efficiency, and its application in industry and other fields is of great practical significance. In the manipulator control, the most difficult part is how to accurately control the arm in accordance with the predetermined trajectory, which is also one of the basic tasks of robot control [1]. At present, there are many control methods, such as feedforward compensation control, sliding mode variable structure control, adaptive control, robust control, neural network control, iterative learning control, inversion control and so on. However, with the development of the manipulator industry, these methods have been unable to meet the requirements of accurate control in various environments. This paper takes the neural network as the research object. This method has great advantages in solving the non-linearity and uncertainties in the process of

© ICST Institute for Computer Sciences, Social Informatics and Telecommunications Engineering 2021
Published by Springer Nature Switzerland AG 2021. All Rights Reserved
S. Liu and L. Xia (Eds.): ADHIP 2020, LNICST 348, pp. 324–334, 2021.
https://doi.org/10.1007/978-3-030-67874-6_30

modeling. However, this method needs a lot of time training, and it is difficult to meet the requirements of high control accuracy.

Therefore, in order to solve the above problems, this study introduces the fuzzy control into the neural network, and uses the fuzzy neural network to optimize the tracking of the rigid-flexible coupling manipulator trajectory [2]. Firstly, the kinematics and dynamics model of the manipulator is established, and then the desired trajectory of the manipulator is tracked by designing corresponding control methods.Experiments show that the tracking error of the proposed method is much smaller than that of the single neural network method, and it can complete the operation more quickly and accurately according to the predetermined trajectory.

2 An Optimal Tracking Method for the Moving Trajectory of a Rigid-Flexible Coupled Manipulator

The goal of trajectory tracking is to reach the desired position according to the given trajectory. The whole process is to change the angle and velocity of the manipulator joint according to the output torque of the controller. Mobile manipulator is a highly nonlinear and strongly coupled system. Mobile platform and manipulator have different dynamic characteristics. In addition, the system also has non-holonomic constraints and various uncertainties, which make the tracking control of mobile manipulator more difficult. Multi-mobile manipulators cooperate to accomplish a target task, which is a hot research direction in recent years [3].

2.1 Mathematical Model of Manipulator

The trajectory tracking of the most common six-joint manipulator is studied in this paper. The model part of the manipulator is mainly divided into two aspects, kinematics and dynamics. Kinematics is the research object of path planning and trajectory control of manipulator, and dynamics is to calculate the control moment input of space robot according to the planned path and trajectory [4]. The kinematics model establishes the mathematical relationship between the position, velocity and other states of the end of the robot and the position, velocity (or angular velocity) of each link (degree of freedom), which is an intuitive description of the motion. The dynamic model reflects the relationship between the driving moment of each joint and the position and velocity of the system. A complete robot control system needs to study not only kinematics model but also dynamics model. The establishment of model in this chapter is the basis of research [5].

(1) Kinematics Modeling

The manipulator consists of a series of joints and connecting rods, and each joint is assigned a reference coordinate system. The transformation from one coordinate system to the next coordinate system is the state transformation from one joint to the next. By combining all the transformations from the base to the last joint, the total transformation matrix of the manipulator can be obtained.

Figure 1 shows two connecting rods and three joints (either moving or rotating).

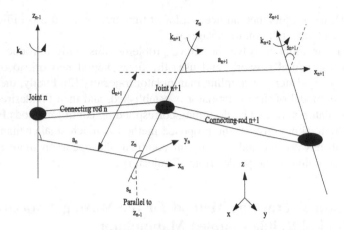

Fig. 1. Robot arm.

As can be seen from Fig. 1, the connecting rod n is between joint n and joint n + 1, and the connecting rod n + 1 is between joint n + 1 and joint n + 2. A reference coordinate system is specified for each joint, i.e. an X-axis and a z-axis.Since the y-axis is perpendicular to the x-axis and the z-axis, the y-axis can be determined by specifying the x-axis and the z-axis, so it is not necessary to specify the y-axis, nor does the D-H method need the y-axis.The coordinates of joint n + 1 are expressed in xn-z n coordinate system. An represents the common vertical length between Z n-1 an d z (if the z-axis of two joints is parallel, there are countless common vertical lines. If the z-axis of two adjacent joints is related, the common vertical line is zero). Angle K represents the rotation angle around the z-axis, D represents the distance between the adjacent two common vertical lines on z-axis, an d angle s represents the angle between adjacent z-axis, also known as joint torsion angle.Generally, only K and D are joint variables [6]. Usually, a local coordinate system can be transformed to the next local coordinate system by the following four steps.

1) Firstly, the rotation angle kn + 1 around the Zn axis makes the xn axis and the xn + 1 axis parallel to each other.
2) Translating the distance of dn + 1 along the Zn axis to make the xn axis and the xn + 1 axis collinear;
3) Translating an + 1 distance along the xn axis to make the origin of the xn axis coincide with that of the xn + 1 axis.
4) Finally, the Z-N axis is rotated around the x-n + 1 axis to align the Z-N with the z-n + 1 axis. At this time, the N coordinate system and the N + 1 coordinate system are identical, one coordinate system is transformed to the next coordinate system.

If there are multiple coordinate systems to be transformed, the above steps can be repeated between the two coordinate systems. For the manipulator, starting from the reference coordinate system, it can be converted to the base, then to the first joint, then to the second joint, then to the n joint, until the end of the actuator. Thus the coordinate transformation of the whole manipulator joint is completed, and the total coordinate transformation matrix, namely kinematics equation, is obtained. Since each transformation is relative to the current coordinate system, the transformation matrix A can be obtained by multiplying four matrices representing four motions by right, and the following results can be obtained:

$$Q_n = Rot(z, k_n) \cdot Trans(z, d_n) \cdot Trans(a_n, x) \cdot Rot(x, s_n) \tag{1}$$

Among them, Q_n represents transformation matrix of coordinate system; Rot represents rotation matrix; Trans represents translation transformation.

From the base of the manipulator to the end-effector, the transformation matrix of each joint can be obtained by defining each joint. The total transformation between the base and the end effector is as follows:

$$^R Q_m =^R Q_1^1 Q_2^2 Q_3^3 Q_4 \ldots Q_n \tag{2}$$

Among them, R represents the reference coordinate system, m represents the coordinate system where the end effector is located, and n represents the number of joints of the manipulator.

(2) Dynamics Modeling

Dynamics of manipulator mainly analyses and studies the interaction between joint moment and motion, and then expresses the relationship between them through mathematical model, and deeply understands the meaning of its mathematical and physical model, which will improve the tracking effect of the system to a certain extent. Lagrange function method and Newton-Euler method are commonly used to establish the mathematical model of manipulator [7]. Among them, Newton-Euler method is a method to analyze the force between the joints of the manipulator using dynamic equilibrium analysis method. This method is easy to understand, but it needs to analyze the relationship between various forces. When the complexity of the system is high and the number of forces between the systems is large, the analysis method is cumbersome and unsuitable for engineering application. Lagrange function considers the overall energy distribution of the current system, and then calculates the system dynamics equation by mathematical calculation. Although the understanding process is abstract, the calculation process is relatively simple and convenient, which is suitable for engineering application. In this chapter, Lagrange function is used to analyze the whole process of dynamic model of manipulator in detail, considering the constraints in the actual process.

Lagrange function:

$$L(q, \dot{q}) = H - P \tag{3}$$

Among them, q represents the rotation angle of each joint of the manipulator in the joint space; $\dot{q}$ is the angular velocity of each joint of the manipulator in the joint space; H is the total kinetic energy of the system, P is the total potential energy of the system, and they can be expressed in any convenient coordinate system.

Lagrange equation:

$$L_i' = \frac{1}{t}\frac{L}{\dot{q}_i} - \frac{L}{q_i} (i = 1, 2, \ldots, j) \tag{4}$$

Among them: L' is the force or moment acting on the first joint; J is the degree of freedom, that is, the number of connecting rods.

The kinetic energy H of the system is defined as the sum of the translational and rotational kinetic energies of the connecting rods. Its expression is as follows (5). Potential energy P is defined as the sum of the gravitational potential energies of the connecting rods. Its expression is shown in formula (6):

$$H = \sum_{i=1}^{j} \left(\frac{1}{2}b_i c_i^2 + \frac{1}{2}j_i \dot{r}_i \times \dot{r}_i \right) \tag{5}$$

$$P = \sum_{i=1}^{j} \left(j_i g_i \times r_{iy} \right) \tag{6}$$

Among them, b_i is the central inertial tensor of the connecting rod; c_i is the angular velocity of the connecting rod in the inertial coordinate system; $\dot{r}$ is the linear velocity of the connecting rod in the inertial coordinate system; r_{iy} is the component along the y-axis in the inertial coordinate system.

The total kinetic energy and total potential energy equation of the connecting rod of the manipulator are introduced into the Lagrange function as shown in Eq. (3), and then the partial derivative is introduced into Eq. (4). Finally, the dynamic equation of the manipulator is obtained as follows:

$$F(q)\dot{q} + W(q, \dot{q})\dot{q} + U(q) = L' \tag{7}$$

Among them, $F(q)$ is the inertia matrix of the system, $W(q, \dot{q})$ is the centrifugal force and the Gothic force matrix of the system, $U(q)$ is the gravity term matrix, q is the current joint angle vector, and L' is the driving moment output by the controller on the joints of the controlled manipulator.

2.2   Manipulator Trajectory Tracking Control Strategy

In recent decades, artificial intelligence methods and theories represented by fuzzy logic, neural network and evolutionary computation have been applied to the control of robotic arms, especially neural networks. In the application of manipulator control, almost all neural network models and learning algorithm application examples can be found. Because the neural network has the ability to learn in the control process, in order to make up for the lack of prior knowledge, so there are great advantages in solving the nonlinear and uncertain problems in the process of modeling. However, due to the lack of systematic and standardized methods to construct neural networks, and the fact that neural networks need a lot of time to train, have no clear physical significance, and are difficult to prove the stability of the system, many problems exist, which restrict the popularization and application of neural network control in the field of joint manipulator trajectory tracking control. So many experts and scholars have improved it a lot [8].

Fuzzy control draws lessons from the fuzziness of human thinking and uses control experience to realize control. It is the earliest form of intelligent control. Because the control system design does not require precise mathematical model, many effective membership functions are relatively simple, the required rule base is not very complex, robust and many other advantages, fuzzy control has been widely used. In the field of manipulator control, people seldom apply fuzzy control directly to manipulator control, but combine fuzzy algorithm with other control methods to construct composite control system. With the development of neural network technology, people begin to combine fuzzy control with neural network, and use fuzzy method, membership function and fuzzy rules to adjust the node function and connection weight of neural network.Fuzzy method is used to adjust and optimize the structure of the neural network. This not only makes the fuzzy control adaptive, but also makes the neural network have the ability of fuzzy reasoning. The weights of the network have a clear meaning of fuzzy logic [9].

2.3 Structure of Fuzzy Neural Network

In this paper, a four-layer Mamdani-type fuzzy neural network (FNN) is used to approximate the nonlinear link of the manipulator system. The center and width of the fuzzy basis function are variable. The structure of the FNN is shown in Fig. 2.

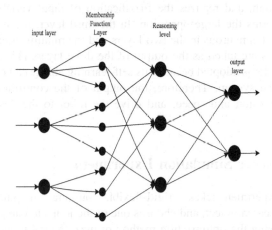

Fig. 2. Structure of fuzzy neural network.

It consists of input layer, membership function layer, reasoning layer and output layer. In the input layer, the data samples are input into a feedforward neural network, and the shape of the activation function of the network represents the shape of the fuzzy membership function. Activation functions can be linear or exponential, which represent triangular and Gaussian membership functions, respectively. The adjustment of the width and center position of the activation function can be obtained by adjusting the weights and deviations of the input layer of the network. The number of neurons in the input layer represents the number of fuzzy markers, which can be selected as needed. The membership function layer and inference layer belong to the middle layer of the neural network, which realizes the fuzzification and fuzzy reasoning of the input. The fourth layer is the output layer, which realizes the de-fuzzification of the control quantity [10].

The combination of fuzzy and neural network has both learning ability and reasoning ability. It provides an effective method to solve the high non-linearity, coupling and unmodeled uncertainty in robot control. The proposed controller consists of two parts: a fuzzy neural network (F'NN) controller and a CMAC controller.

FNN controllers need to be trained with samples, that is, supervised learning. This process is carried out offline, and then applied to the corresponding system. The training samples are taken from the system based on computational moment control. The system achieves better control performance with lower accuracy requirements. But in fact, the system can not be accurately modeled, and the existence of uncertainty is inevitable. The training data can not reflect the structural or non-structural uncertainties in the robot model, so the real-time performance of these data can not be guaranteed. In order to overcome this shortcoming, the self-learning ability of CMAC controller and the advantages of fast convergence are used to compensate the control error online [11, 12].

The first layer of FNN controller is the fuzzification layer. Each node of FNN controller corresponds to a linguistic variable, completes the calculation of an input membership function, and realizes the fuzzification of input variables [13, 14]. The second layer performs the large operation. In the third layer, the minimization of all outputs of the first ten neurons in the two layers and the minimization of all outputs of the last ten neurons are taken as the output of the three layers [15, 16].

CMAC controller is adopted because its self-learning ability can compensate for the control error of the system. Therefore, the output of the controller must reflect the change of joint position and speed, and drive the robot to the desired position and speed.

3   Tracking Error Simulation Experiment

The simulation experiment takes a rigid-flexible coupling manipulator of a six-joint machine as the research object, and chooses one of the joints to carry out the trajectory tracking test by using the optimization method of fuzzy neural trajectory tracking and the single neural network trajectory tracking method. The length of the joint arm is 6.73 m, the mass is 1.8 kg, and the simulation is carried out by MATLAB software. Figure 3 shows the experimental environment. The expectation given in Fig. 4 below is shown as follows. Track the trajectory curve.

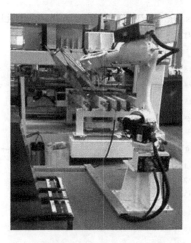

Fig. 3. Experimental environment.

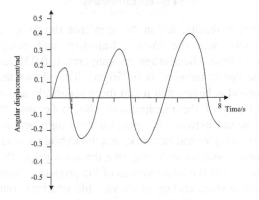

Fig. 4. Expected tracking trajectory curve.

Figure 5 below shows the tracking trajectory curve of the joint given by the fuzzy neural network and the neural network.

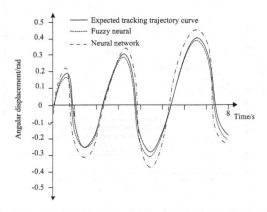

Fig. 5. Tracking trajectory curve given by fuzzy neural network and neural network.

Figure 6 below shows the error curve between the track curve given by the two methods and the desired track curve.

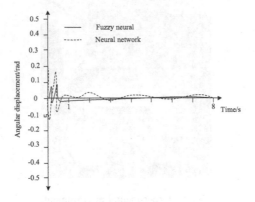

Fig. 6. Error curve.

From the simulation results, it can be seen that the fuzzy neural network can quickly and accurately plan the motion trajectory and complete the operation requirements in a short time. The average tracking error is 0.021 rad, while the average tracking error of the neural network is 0.056 rad. This shows that the tracking performance of this method is better, and it can more accurately and effectively complete the tracking control task of the manipulator. This is because the fuzzy control is introduced into the neural network to optimize the single neural network to form a compound method of fuzzy neural network, and the hybrid method is used to track the trajectory of the manipulator, so as to improve the accuracy of the tracking results.

In order to further verify the effectiveness of the proposed method, the feedforward compensation control method, sliding mode variable structure control method and the method in this paper are compared by taking the trajectory tracking time as the index. The results are shown in Fig. 7.

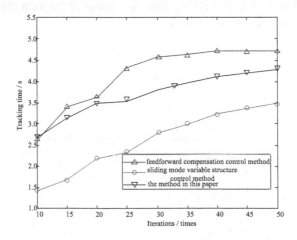

Fig. 7. Comparison of tracking time of different methods.

It can be seen from Fig. 7 that when the rigid flexible coupling manipulator trajectory is tracked by our method, the time spent is always less than 3.5 s, while the feedforward compensation control method, The tracking time of sliding mode variable structure control method is significantly higher than that of the method in this paper, which shows that the tracking efficiency of this method is higher, and it can effectively track the moving trajectory of the manipulator in a short time.

4 Conclusion

In summary, the development of computer information technology promotes the mechanized production. In the mechanized production, the application of the manipulator is more and more widely. Manipulator replaces human manual labor, and the production efficiency has been greatly improved. However, the manipulator also has a shortcoming. The elastic deformation of the rigid flexible manipulator will cause the manipulator operation deviation and vibration problems. In this context, an optimal tracking method for rigid-flexible coupling manipulator is studied. This method solves the problems in using single neural network to track the moving trajectory of manipulator by using fuzzy control algorithm, optimizes single neural network, provides a reference for the control of manipulator and improves the production efficiency of mechanization.

Acknowledgements. Guangxi University Young and Middle-aged Teachers Research Fundamental Ability Enhancement Project:Optimization of Vibration Problems of Thin Plates with Concentrated Mass Based on Modal Analysis(Code: KY2016YB530)

References

1. Jing, J., Song, C., Hongda, L.: Research of manipulator trajectory tracking based on gravity compensation. Trans. Shenyang Ligong Univ. **35**(2), 5–9 (2016)
2. Jiayu, S., Jianwei, M., Haitao, Z., et al.: Optimal trajectory tracking control for robot manipulator. Comput. Simul. **33**(12), 338–341 (2016)
3. Wa, Y., Wei, Q.: Research on trajectory tracking control of mechanical arm. Agric. Equipment Veh. Eng. **55**(10), 34–37 (2017)
4. Junqiang, L., Yanding, W., Guoping, L., et al.: Optimal trajectory planning of a flexible manipulator for its vibration suppression using genetic algorithm. J. Vib. and Shock **35**(11), 1–6 (2016)
5. Li, A.: Composite control of trajectory tracking and vibration suppression of one-link flexible manipulator. Mach. Tool Hydraulics **45**(21), 21–31 (2017)
6. Wang, F., Chao, Z., Li, H., et al.: Trajectory tracking control of robot manipulator based on wavelet neural network and fuzzy sliding mode. Comput. Simul. **34**(11), 353–359 (2017)
7. Yijun, G., Li, Y., Jianming, Xu.: The trajectory tracking control of manipulator based on fuzzy ADRC technology. J. Shaanxi Normal Univ. (Natural Science Edition) **45**(2), 42–48 (2017)
8. Guimaraes, J.S., Almeida, B.R.D., Tofoli, F.L., et al.: Three-phase grid-connected wecs with mechanical power control. IEEE Trans. Sustain. Energy **9**(4), 1508–1517 (2018)

9. Liu, S., Liu, D., Srivastava, G., et al.: Overview and methods of correlation filter algorithms in object tracking. Complex Intell. Syst. 1–23 (2020). https://doi.org/10.1007/s40747-020-00161-4
10. Kalani, H., Akbarzadeh, A., Nabavi, S., Moghimi, S.: Dynamic modeling and CPG-based trajectory generation for a masticatory rehab robot. Intell. Serv. Robot. 11(2), 187–205 (2018). https://doi.org/10.1007/s11370-017-0245-6
11. Liu, S., Liu, G., Zhou, H.: A robust parallel object tracking method for illumination variations. Mob. Netw. Appl. 24(1), 5–17 (2018). https://doi.org/10.1007/s11036-018-1134-8
12. Liu, S., Pan, Z., Cheng, X.: A novel fast fractal image compression method based on distance clustering in high dimensional sphere surface. Fractals 25(4), 1740004 (2017)
13. Kohl, N., Miikkulainen, R.: Evolving neural networks for strategic decision-making problems. Ann. Surg. Oncol. 22(3), 326–337 (2018)
14. Lu, D., Popuri, K., Ding, G.W., et al.: Multimodal and multiscale deep neural networks for the early diagnosis of alzheimer's disease using structural MR and FDG-PET images. Sci. Reports 8(1), 5697 (2018)
15. Lau, K., Leung, Y.Y., Poon, C., et al.: 630 development of an endoscopic surgical robotic system and from bench to animal studies. Gastrointest. Endosc. 85(5), 90–91 (2017)
16. Kitagami, H., Nonoyama, K., Yasuda, A., et al.: Technique of totally robotic delta-shaped anastomosis in distal gastrectomy. J. Minim. Access Surg. 13(3), 215–218 (2017)

Fast Recognition of Multi-combination Target Features in Motion Image Based on Large Data Analysis

Tao Wei[(✉)]

Northwestern Polytechnical University, Xi'an 710072, China
whai7863@163.com

Abstract. In order to overcome the low efficiency of traditional recognition technology, a fast recognition method of multi-combination features of moving images based on large data analysis is proposed. Based on feature extraction of multi-combination target, denoising of moving image and determination of Boolean correlation coefficient, fast recognition of multi-combination target feature of moving image under large data analysis is realized. The experimental data show that the proposed recognition method can not only effectively improve the efficiency of traditional recognition technology, but also make the recognition result more stable, and enhance the adaptability and flexibility of image recognition technology.

Keywords: Large data · Motion image · Target feature · Recognition method

1 Introduction

It is of great significance to classify multi-combination targets of moving images and recognize them in the practice of national defense and modernization construction. In this paper, a recognition method of multi-combination target features in moving images is proposed. Because there are many kinds of target features, the appropriate feature set should be selected according to the empirical evidence of different kinds of typical performance. If the selected feature set can not obtain the effective information of the target to be identified, the feature information should be re-selected. There are many traditional image classification methods, such as statistical classification method, The K-means algorithm, K-Means algorithm, CART classification and regression tree algorithm, but these methods are powerless for high-dimensional and massive classification and recognition problems. In this paper, the PSO algorithm of large data neural network is used for noise reduction, and the fast recognition of multi-combination target features in moving images is realized by the calculation of Boolean correlation.

S. Liu and L. Xia (Eds.): ADHIP 2020, LNICST 348, pp. 335–344, 2021.
https://doi.org/10.1007/978-3-030-67874-6_31

2 Fast Recognition of Multi-combination Target Features in Motion Image Based on Large Data Analysis

2.1 Feature Extraction of Multi-combination Objects in Image

The structure of the image affects the result of feature extraction of multi-combination targets. Most of the methods of feature extraction of multi-combination targets are based on the surface and local information of the image. The feature information extraction method based on multi-color features uses multi-color features of the image, divides the image according to the color features, and then uses discrete statistical processing fitting technology to reconnect the segmented regions, completing the process of extracting the feature information of the image.

Firstly, the multi-color features of the image are classified. In the Lab color channel, the brightness, color, texture and other color features are extracted, and then the color features of the image are segmented. Secondly, the segmented image blocks are represented by color feature difference:

$$x^2(g,h) = \frac{1}{2}\sum_i \frac{(g(i) - g(h))^2}{g(i) + g(h)} \tag{1}$$

In the formula, g and h represent the color features of the recognized image. After calculation, the feature information of the image is extracted. The method of feature extraction of multi-combination targets in images is greatly influenced by the intensity of illumination. It is also necessary to use the edge detection algorithm based on mathematical morphology to extract the secondary feature information of images. Firstly, the geometric model of mathematical morphology recognition is constructed to obtain the geometric feature information of the image. Then, feature information matching is carried out based on mathematical morphology extraction method.

In the least squares format, $z(k) = h^\tau(k)\theta + n(k)$ and $\theta = [a_1, a_2, \ldots a_{n_a}, b_1, b_2, \ldots, b_{n_b}]^\tau$ are the parameters to be estimated. $h(k) = [-z(k-1), \ldots, -z(k-n_a), u(k-1), \ldots, u(k-n_b)]^\tau$, for $k = 1, 2, \ldots L$ (L is the length of the data). Construct a system of linear equations, written in $z_L(k) = H_L(k)\theta + n_L(k)$;

$$Z_L = \begin{bmatrix} z(1) \\ z(2) \\ \vdots \\ z(L) \end{bmatrix}, H_L = \begin{bmatrix} h^\tau(1) \\ h^\tau(2) \\ \vdots \\ h^\tau(L) \end{bmatrix}, n_L = \begin{bmatrix} n(1) \\ n(2) \\ \vdots \\ n(L) \end{bmatrix} \tag{2}$$

According to the least square method, the parameters of the algorithm are estimated as follows: $\hat{\theta}LS = (H_L^\tau H_L)^{-1}H_L^\tau Z_L$.

The steps of image multi-combination target feature are as follows:

Step 1: Initialize $W(0) = 0$; $P(0) = \sigma^{-1}I$, where I is the unit matrix;

Step 2: Update $n = 1, 2, \ldots$ Calculation;

Update the gain vector: $g(n) = P(n-1)X(n)/[\lambda + X^T(n)P(n-1)X(n)]$ Extraction: $y(n) = W^T(n-1)X(n)$;

Error estimation: $e(n) = d(n) - y(n)$;

Update the weight vector: $W(n) = W(n-1) + g(n)e(n)$;

Update the inverse matrix: $P(n) = \lambda^{-1}[P(n-1) - g(n)X^T(n)P(n-1)]$. Among them, $P(n)$ is the inverse of the autocorrelation matrix $P_{xx}(n)$, constant λ is the forgetting factor, and $0 < \lambda < 1$.

In summary, according to data acquisition and generation, take $d(n)$, $X(n)$; initialization of parameters; extraction and processing of multi-combination target features of images; conversion of images into digital images; HMT transformation of recognized images. After HMT transformation of images, geometric correspondence can be established between recognized images and digital images. Finally, the task of extracting information from the target image is completed.

2.2   Noise Reduction Processing of Motion Image

In motion image denoising, PSO algorithm based on large data neural network is used to denoise. Big data neural network is a multi-layer network with one-way propagation. It has three layers: input layer, hidden layer and output layer [1], which are divided into forward and backward propagation. The weights of each layer are adjusted by forward and backward propagation errors. The process of weights adjustment is the training process of the neural network, which reduces the output errors and is used for noise reduction. This process is cyclical until the termination condition is reached. Figure 1 is a schematic diagram of denoising processing of large data neural network.

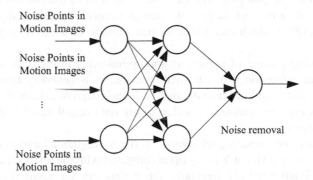

Fig. 1. Large data neural network denoising processing

In order to achieve image denoising of multiple combined targets in moving images, 15 nodes are selected in the input layer, including 12 time points data and 3 Influence Factors normalized and quantified data.

Write it down as xk = (xk1, xk2,... XKH,... Xk15) the output layer has one node, which is represented by empirical formula $m = \sqrt{n+l} + a$, M represents the number of nodes in the hidden layer, n represents the number of nodes in the input layer, l represents the number of nodes in the representative layer, and a takes the constant between 1 and 10. Considering the accuracy of the results, the number of nodes is 5, 6 and 7, and the number of nodes in hidden layer is 6 [2].

The transfer function between the hidden layer and the output layer is bipolar S-type function $f(x) = \frac{2}{1-e^{-x}} - 1 = \frac{1-e^{-x}}{1+e^{-x}}$. The weight matrix between the input layer and the hidden layer is expressed by v. The weight matrix from the hidden layer to the input layer is expressed by W. For the hidden layer, there are $y_i = f(net_j)$, $j = 1, 2, \ldots, w$, $net_i = \sum_{i=0}^{n} v_{ij}x_i$, $j = 1, 2, \ldots, w$. For the output layer, there are $O_k = f(net_k)$, $j = 1, 2, \ldots, m$, $net_j = \sum_{i=0}^{n} v_{jk}x_i$, $j = 1, 2, \ldots, w$.

Suppose the position Y and velocity V of the i particle are $Y_i = (Y_{i1}, Y_{i2}, \ldots, Y_{in})$, $V_i = (V_{i1}, V_{i2}, \ldots, V_{in})$, respectively. The historical optimal solution of the i particle is $P_i = (P_{i1}, P_{i2}, \ldots, P_{in})$, the optimal solution of the particle swarm is $P_m = (P_{m1}, P_{m2}, \ldots, P_{mn})$, and the velocity and position of the particle are updated as follows [3, 4]:

$$\begin{cases} V_{in}(t+1) = wv_{in}(t) + c_1r_1(P_{in} - x_{id}(t)) + c_2r_2(P_{mn} - x_{in}(t)) \\ Y_{in}(t+1) = Y_{in}(t) - V_{in}(t+1) \end{cases} \tag{3}$$

In the formula, W is called inertia factor, the range of values is [0.4, 0.9]; c1, C2 is noise factor, the value is 2. R1 and R2 are random numbers distributed in [0,1].

The PSO algorithm of large data neural network is used to automatically adjust the parameters of image extraction to adapt to the statistical characteristics of unknown signals and noises or changing with time, so as to achieve the optimal extraction.

The PSO algorithm of large data neural network is essentially an adaptive extraction algorithm which can adjust its transmission characteristics to achieve the optimal.

In the running process of particle swarm optimization in big data neural network, adaptive digital extraction with adjustable parameters is generally called FIR digital adaptive extraction, also known as dot matrix digital adaptive extraction. On this basis, the particle swarm optimization algorithm of big data neural network can be divided into two processes.

Firstly, after the input signal adjusts x (n) to digital adaptive signal through parameters, the output signal Y (n), y (n) is compared with the reference signal D (n) to get the error signal e (n) [5]. Secondly, the parameters are adjusted by an adaptive algorithm and the values of X (n) and E (n). The input signal x (n) is weighted to the digital adaptive output signal Y (n). The adaptive algorithm adjusts the extraction weight coefficient to minimize the error signal e (n) between the output y (n) and the adaptive extraction expected response D (n).

The PSO algorithm coefficients of large data neural networks are controlled by error signals, and are automatically adjusted according to the value of E (n) and the adaptive algorithm. By adjusting the weight coefficient, the mean square error between the adaptive extracted output signal Y (n) and the expected response signal D (n) is minimized, or $e^2(n)$.

$$\hat{\nabla}(n) = \frac{\partial[e^2(n)]}{\partial w(n)} = -2e(n)x(n) \tag{4}$$

This instantaneous estimation method is unbiased because its expected value $E[\hat{\nabla}(n)]$ equals vector $\nabla(n)$. Therefore, according to the relationship between the variation of coefficient vectors and the direction of gradient vector estimation extracted by PSO algorithm of large data neural network, we can first write the formula of PSO algorithm of large data neural network as follows:

$$\hat{w}(n+1) = \hat{w}(n) + \frac{1}{2}\mu[-\hat{\nabla}(n)] = \hat{w}(n) + \mu e(n)x(n) \tag{5}$$

By substituting the formulas e(n) = d(n)–y(n) and e(n) = d(n)–wHx(n) into the formulas above, we can get:

$$\begin{aligned}\hat{w}(n+1) &= \hat{w}(n) + \mu x(n)[d(n) - \hat{w}^H(n)x(n)] \\ &= [I - \mu x(n)x^H(n)]\hat{w}(n) + \mu x(n)d(n)\end{aligned} \tag{6}$$

The denoising of moving image is realized.

2.3   Fast Recognition of Target Characteristics

In order to realize the fast recognition of image target, statistical analysis should be carried out from the Angle of moving image data station, and then the fast recognition function of target should be completed by summarizing and processing data set, eliminating mixed data and calculating data.

The fast recognition algorithms for target features can be divided into the following categories:

The K-means algorithm, K-Means calculation theory, Support Vector Machines, The Apriori algorithm, Boolean association rule frequent iteration calculation theory, and Adaboost iteration calculation theory.

Based on the theory of strategy tree and Boolean association rule frequent itemset, a motion image guidance algorithm is constructed. It is applied to fast recognition of target features in moving images.

Boolean association rule frequent itemset computing theory is a big data computing theory of probability correlation statistics. Based on the statistical method of frequent itemsets of association rules published by James Boole in statistical, and the continuous optimization of the algorithm of large data of association statistics by mathematicians represented by Egstrom, the Boolean association rule frequent itemset computing theory is established [6]. Boolean Association Rules Frequent Item Set Computing Theory is a mathematical large data computing method suitable for data statistical recognition and mining [7]. Let there be a set of target characteristic variables A = {A1, A2,... An}, where the data of variable set A satisfies a certain recognition trend relation D and the local variables conform to the correlation probability distribution R, then a recognition trend network can be formed. The sketch diagram of the recognition trend network is shown in Fig. 2 [8].

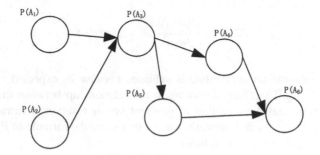

Fig. 2. Identifying trend network diagram

In Fig. 2, A1 and A2 represent target features to quickly identify user behavior, D represents causality, R represents correlation probability distribution, and A3 represents self-Media recognition trend. At the same time, derivatives can be calculated according to operation A3, and derivatives of identification can be inferred to calculate A4 and A5, and conclusion A6 can be drawn. Its derivative calculation satisfies the following formula [9, 10]:

$$C = \sigma_o \gamma_i + W_0(\partial^2 q_k dx)/S \tag{7}$$

From formula (7), the value of Boolean correlation coefficient will directly affect the accuracy of user behavior estimation results A4 and A5. At the same time, the response time of accurate Boolean correlation coefficient will directly affect the speed of large data analysis. The selection of Boolean correlation coefficient is related to causality D and probability distribution R, which satisfies the relationship in Table 1.

Table 1. Selection of boolean correlation coefficient

Causality D	Probability distribution R	Boolean coefficient S
[0.00,0.30]	[0.00,0.50]	[0.00,0.45]
[0.30,0.50]	[0.00,0.50]	[0.45,0.62]
[0.50,0.80]	[0.50,1.00]	[0.62,0.83]
[0.80,1.00]	[0.50,1.00]	[0.83,1.00]

In order to obtain the Boolean correlation coefficient quickly, the causality D and the probability distribution R are determined firstly, then the range of the Boolean correlation coefficient is determined according to Table 1, and then the specific value of the Boolean correlation coefficient is determined according to Egstrom function, which greatly saves the time of obtaining the Boolean correlation coefficient directly by Egstrom calculation.

Based on the Boolean correlation coefficient, the threshold function is used to compute the target features. In the image with an indivisible target feature subset, when the distance between the two target feature subsets is long enough, the two subsets can be segmented by using the linear function relationship. However, when the two subsets are close to each other, the Kmean ++ algorithm is used to discretize the extracted image, and then the importance SGA of the target eigenvalues of the two subsets is calculated. Finally, the attributes with small importance of the target eigenvalues are selected as the image segmentation points for segmentation. However, when the two subsets are close to each other, the Kmean ++ algorithm is used to discretize the extracted image, and then the importance SGA of the target eigenvalues of the two subsets is calculated. Finally, the attributes with small importance of the target eigenvalues are selected as the image segmentation points for segmentation [11, 12].

Then the undivided image is segmented from the important points by the same method, and the decision tree is generated. After two segmentation, two recognition results will be produced. Thirdly, Kmean ++ algorithm is used to discretize the original image, randomly select points on the recognized image, and then form a discrete sequence, which is segmented several times, and the recognition results are processed by discrete statistics, finally the image recognition is completed.

3   Simulation Experiment and Result Analysis

In order to verify the practical application performance of the above moving image multi-combination target feature recognition method based on big data analysis, the following simulation experiment is designed for verification.

In the simulation experiment, the recognition effect of the traditional image recognition technology and the moving image multi-combination target feature recognition method based on big data analysis is compared. The experimental data are from the ImageNet data set. In order to ensure the accuracy of simulation test, many simulation tests are carried out, and the data generated by the tests are displayed in the same data graph.

3.1   Experimental Data Setting

In order to ensure the accuracy of the simulation test process and results and set the test parameters, 120 multi-combination target images of different kinds of moving images were browsed by two image recognition technologies. Among them, there are 30 moving images of people, 30 moving images of cars, 30 moving images of aircraft and 30 moving images of animals. Five images are randomly extracted from each image classification, totaling 20 images. Two different image recognition techniques are used to identify and classify them.

3.2   Analysis of Experimental Results

The experimental results are counted into Table 2. It can be seen from Table 2 that the method in this paper can effectively improve the recognition rate of images, and the number of effective recognition is significantly higher than that of traditional methods. Moreover, the fast recognition method based on big data analysis for multiple target features of moving images is more efficient in recognizing animal moving images and aircraft moving images.

Table 2. Experimental results of two recognition techniques

Technology category	Image class	Number of valid identification	Effective recognition rate
Traditional image recognition technology	Character motion images	4	80%
	Animal motion images	3	60%
	Aircraft motion image	1	20%
	Automatic generation of portraits	5	100%
Image recognition technology based on artificial neural network	Character motion images	4	80%
	Animal motion images	4	80%
	Aircraft motion image	4	80%
	Automatic generation of portraits	5	100%

In the experimental process, if the number of random sampling is small, the deviation of test results will be caused. After many times of extraction and recognition experiments, the effective recognition efficiency of the two recognition technologies has changed. The recognition result deviation of the two recognition technologies is shown in Fig. 3.

From the analysis of Fig. 3, it can be seen that the stability of the recognition method designed in this paper is higher than that of the traditional image recognition technology, and the deviation of the recognition result is not only small but also stable.

To sum up, the fast recognition method of moving image multi-combination target features designed in this paper based on big data analysis can not only effectively

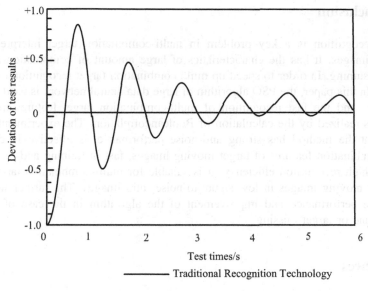

(a)Traditional Recognition Technology

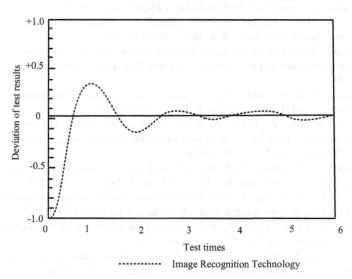

(b)Image Recognition Technology

Fig. 3. Comparison of deviation between two recognition techniques

improve the recognition efficiency and accuracy, but also make the recognition results more stable and enhance the adaptability and flexibility of the image recognition technology.

4 Conclusion

Target recognition is a key problem in multi-combination target interpretation of moving images. It has the characteristics of large amount of computation and long time-consuming. In order to speed up multi-combination target recognition of moving images. In this paper, the PSO algorithm of large data neural network is used for noise reduction, and the fast recognition of multi-combination target features in moving images is realized by the calculation of Booleancorrelation. The experimental results show that the method has strong anti-noise performance, is effective in extracting multi-combination features of target moving images, fast in training and recognition, and has high recognition efficiency. It is suitable for multi-combination target recognition of moving images in low signal-to-noise ratio images. The further work is to study the performance and improvement of the algorithm in the case of complex background or target missing.

References

1. Jing, T.: Precise prediction of monthly electricity consumption based on multi-dimensional feature analysis. Power Syst. Protect. Control **45**(16), 145–150 (2017)
2. Shaodong, P.: Mining method model and application of online learning behavior research in big data era. Audiovisual Educ. Res. **40**(1), 70–79 (2017)
3. Anonymous.: Simulation of intelligent data acquisition, operation and maintenance Mining Model for power consumption information under large data. Comput. Simul. **35**(10), 412–415 (2018)
4. Anonymous.: Prediction and analysis of electric heating power consumption based on big data platform. Appl. Electron. Technol. **44**(11), 67–69 (2018)
5. Qianyi, C., Wenbin, Z., Jiekai, P., et al.: Research on condition monitoring and fault handling methods of intelligent distribution network based on large data analysis. Modern Electron. Technol. **41**(4), 105–108 (2018)
6. Anonymous.: Performance evaluation of wind turbines based on mutual information association of large wind data. Sci. Technol. Eng. 18(29), 216–220 (2018)
7. Liu, S., Liu, D., Srivastava, G., et al.: Overview and methods of correlation filter algorithms in object tracking. Complex Intell. Syst. (2020) https://doi.org/10.1007/s40747-020-00161-4
8. Xiansheng, Y., Lei, J., Xiong, P., et al.: Research on anomaly detection methods based on large data. Comput. Eng. Sci. **23**(07), 38–44 (2018)
9. Zheng, P., Shuai, L., Arun, S., Khan, M.: Visual attention feature (VAF): A novel strategy for visual tracking based on cloud platform in intelligent surveillance systems. J. Parallel Distr. Comput. **120**, 182–194 (2018)
10. Liu, S., Bai, W., Zeng, N., Wang, S.: A fast fractal based compression for MRI Images. IEEE Access **7**, 62412–62420 (2019)
11. Jing, G., Xiaojun, Z.: A summary of learning analysis based on big data. Audiovis. Educ. China **12**(1), 121–130 (2017)
12. Mengye, L., Shuai, L.: Nucleosome positioning based on generalized relative entropy. Soft. Comput. **23**, 9175–9188 (2019)

Research on Accurate Communication Method of Spatial Scene Visual Information Based on Big Data Analysis

Wen-da Xie[1]([⊠]) and Jia-ju Gong[2]

[1] Computer Engineering Technical College (Artificial Intelligence College), Guangdong Polytechnic of Science and Technology, Zhuhai 519090, China
xiewenda5844@163.com
[2] Electronic and Information Technology, Jiangmen Polytechnic, Jiangmen 529030, China

Abstract. Aiming at the problem that the adaptive convergence of noise iteration performance is not fast enough in the traditional spatial scene visual information communication method, a method for accurately transmitting spatial scene visual information based on big data analysis is proposed. The digital image acquisition of the spatial scene is realized by signal filtering processing, then the spatial scene image is transcoded, intercepted and preprocessed, and the spatial scene image distortion correction, image smoothing processing and image segmentation are performed. Finally, through the scrial communication method, the visual information of the space scene is accurately transmitted. The experimental results show that the noise iteration performance of the spatial scene visual information accurate communication method based on big data analysis can quickly and adaptively converge compared with the traditional spatial scene visual information transmission method.

Keywords: Big data analysis · Space scene · Visual information · Accurate communication

1 Introduction

A basic spatial scene visual information accurate communication process, including five elements: communicator, recipient, information, media and feedback. The communicator, also known as the source of information, refers to the initiator of the dissemination of this behaviour, and also refers to the person who sends the message in the way of the main action for others. Then in the ordinary social communication behaviour, the communicator can be either an individual or an organization or a group. The recipient, also known as the sink, is the responder and receiver of the information, and is also the audience of the space scene, the target of the communicator [1]. The term "object" does not mean that the recipient is a passive being. On the contrary, it can influence the communicator through feedback activities. Similarly, the recipient can be an individual or an organization or group. In the general communication activities, the communicator and the recipient are not fixed roles, and the conversion of the two roles can be alternated. Information refers to a group of interrelated

S. Liu and L. Xia (Eds.): ADHIP 2020, LNICST 348, pp. 345–354, 2021.
https://doi.org/10.1007/978-3-030-67874-6_32

and meaningful symbols that can completely express a certain meaning. The medium is also known as the channel, means or tool of communication. It is the "porter" of information and the link that links various factors in the process of communication [2]. Feedback refers to the respondent's reaction or feedback to the received information, that is, the respondent's reaction to the communicator as described above. In the process of accurately conveying visual information in space scenes, participation or intervention is not only the attributes of the information itself, but also the meaning of the information communicator, the meaning of the information receiver and even the meaning of the information communication medium. Big data analysis studies the accurate communication method of visual information in spatial scenes, which can accurately convey visual information by accurately processing images.

2 Accurate Communication Method of Spatial Scene Visual Information Based on Big Data Analysis

2.1 Space Scene Digital Image Acquisition

Spatial scene digital filtering refers to the process of signal filtering processing, using discrete data analysis techniques and finite precision algorithms to achieve discrete-time linear time-invariant information [3]. Due to the imperfection of the imaging system, the transmission medium and the recording device during the image acquisition process of the space scene, the digital image is contaminated by various noises in different stages during its formation and transmission recording. Spatial scene image filtering is to suppress the noise of the target scene image under the condition of retaining the detailed features of the space scene image. It is an indispensable operation in the spatial scene image preprocessing, and the processing effect will be directly followed by the effectiveness and reliability of image processing and analysis in space scenes [4].

If the contour of the low-frequency component in the digital space scene image is too blurred, the edge extraction results at the contour will be lost, so that the gradient vectors at the lost edge are all turned to the missing point [5]. In this regard, we must first analyze the characteristics of the initial gradient field and GVF field characteristics at the missing edge in the spatial scene image. In order to remove noise, the following energy model is proposed:

$$E_{cp} = \iint \mu |\nabla v|^2 + |W(x,y) \bullet \nabla f|^2 dx dy \tag{1}$$

In the formula, W represents the corresponding map of the missing point, $v(x,y) = [u(x,y), v(x,y)]$ represents the new vector field [6], which is called the edge preservation GVF (Corner Preserving GVF, CP-GVF) field, ∇f represents the low-frequency edge image f of the gradient space scene image, and μ represents the weighted parameter, There are two types in the energy function formula. The first type $\mu |\nabla v|^2$ represents the transition of the vector field V in the coordinate (x,y). The smaller the change of V, the slower the vector field; The second type $|W(x,y) \bullet \nabla f|^2 |V - W(x,y) \bullet \nabla f|^2$ represents the degree of difference between V and $W(x, y) \bullet \nabla f$. The smaller the degree of difference, the closer V and $W(x, y) \bullet \nabla f$ are. So the GVF field is an extended gradient vector field.

Through the processing method of the GVF vector field [7], and then combined with the minimization Eq. (2), the Euler equation of the CP-GVF field can be calculated, as shown in Eqs. (3) and (4).

$$\mu = \Delta u(x, y) - [u(x, y) - W(x, y) \bullet f_x(x, y)] \tag{2}$$

$$\nabla f = [W(x, y) \bullet f_y(x, y)]^2 \tag{3}$$

In the formula, ∇ stands for Laplace operator, and partial differentiation is expressed by f_x and f_y. The function of removing noise changes is as follows:

$$u = u(x, y, t) \tag{4}$$

$$v = v(x, y, t) \tag{5}$$

Because $W(x, y) \bullet f_x(x, y)$ and $W(x, y) \bullet f_y(x, y)$ are both scalar products, the solution to the above equation is similar to the GVF model, and $f_x(x, y)$ and $f_y(x, y)$ need to be multiplied by $W(x, y)$, respectively.

Median filtering was proposed by Turky in 1971. It was originally used for time series analysis of one-dimensional signal processing techniques, and was later used in two-dimensional image processing, and achieved good results in denoising restoration. Median filtering is a nonlinear signal processing technique based on the theory of sorting statistics that can effectively suppress noise. The basic principle of median filtering is to use the value of a point in a digital image or a sequence of numbers to use a point value in a neighborhood of the point. The median value is substituted, so that the surrounding pixel values are close to the true value, thus eliminating isolated noise points [8]. The median filter replaces the pixel value of each point in the image by the median of the pixel values in the corresponding filtered region R. If the number of pixels in the filter area of the median filter is even, then it usually introduces new pixel values. Since the median is a permutation order statistic, it can be understood in a sense that the result is determined by the "majority" of the pixels involved, and the individual particularly high or very low pixel values will not produce too much result. For large influences, they simply push the result forward or backward by a value, so the median is considered a robust quantity.

The basic operations include: decaying candle, expansion, opening and closing operations. It is worth noting that when using morphological filters, structural elements with different shapes, sizes, and directional characteristics should be selected for different purposes. In addition, morphological open and closed operations are idempotent, which means that one filter filters out all structural element-specific noises, and repeating them will not produce new results. Morphological filtering can change the local structure in a predictable way, a property not available in classical methods. Since the morphological operation is image processing from the geometrical point of view of the image, this excellent nonlinear filter can keep the image structure from being ironed while filtering.

2.2   Spatial Scene Image Transcoding Preprocessing

The original image of the spatial scene is usually in YCbCr format, because this method is aimed at RGB format images in the image processing, so it is necessary to use the big data analysis technology to convert the image format. The YCbCr format is one of many color spaces, usually used for continuous image processing of movies, or used in digital photography systems. The YCbCr format is one of many color spaces, usually used for continuous image processing of movies, or used in digital photography systems. In the IMGLIB function library, a function is provided specifically for image format transcoding, which converts YCbCr components into RGB components. The output image occupies 2 bytes per pixel, where R component is 5 bits, G component is 6 bits, B component is 5 bits, and a total of 16 bits, which is exactly 2 bytes. Therefore, before using the image, you need to extract the corresponding RGB component from the 2 bytes of space corresponding to each pixel and convert it to 8bit standard RGB. After the DSP obtains the RGB image to be processed, the image is first intercepted. Because the original image size is 576×720, a total of more than 410,000 pixels, preprocessing all these pixels is not only time-consuming but also unnecessary. In fact, the material is fixed at the position of the image each time, roughly in the photoelectric Near the switch. Quality, which in turn affects the final discriminant result, so this step is particularly important. The method first performs noise removal processing in the preprocessing step. The IMGLIB function library has a median filter function IMG_-median_3 $\times$ 3, which takes a 3×3 pixel matrix and selects the intermediate value as the gray value of the current pixel to perform median filtering.

2.3   Space Scene Image Distortion Correction

In the process of image acquisition of space scenes, due to camera process problems or the wide-angle imaging system itself, the captured images tend to be distorted, causing the captured images to be severely distorted and cannot be directly sorted [9]. Therefore, before performing the sorting process, the source image is corrected for distortion and restored to an ideal state. In order to facilitate the acquisition of real source images, the method adopts a distortion correction method that is simple, efficient and meets real-time requirements. Firstly, the distortion point and the corresponding ideal point are extracted by using the drawn template, and the transformation model is established according to the corresponding coordinate position relationship [10]. When the correction is performed, each pixel point in the distortion image is reduced to an ideal point by the model, and finally passes through the bilinearity. The interpolation calculates the actual gray value of each pixel, and then obtains the complete ideal image. Distortion correction can be roughly divided into three parts, i.e., obtaining control points, obtaining distortion map and ideal point mapping relationship, and bilinear interpolation method for calculating gray value. The first two parts are calculated in advance on the PC. The third part is first simulated on the PC to detect whether the corrected image meets the requirements. Then the distortion corrected code is transplanted into the DSP according to the weights obtained in the first two parts. Realize the real-time operation in the sorting process and complete the correction of the distorted image in the DSP.

In the correction of the distorted image, the control point is manually set in advance. The more the number of control points, the smaller the average error of the calculated result. Therefore, the complete image is used when acquiring the control point, and more control points are set as much as possible [11]. The method adopts a chess board template similar to a chess board. The black and white grid is drawn by equally spaced horizontal lines and vertical lines, wherein the intersection of the horizontal line and the vertical line is set as a control point. When the grid template is placed, the center of the camera's optical axis coincides with the center of the grid. This method first uses the open source Open CV computer vision library on the PC side, and uses the Harris corner detection method to perform corner detection on the checkerboard template image captured by the camera. The actual position of each distortion control point can be obtained very quickly. Because the sides of the mesh are consistent, the ideal coordinates of other control points can be derived from the coordinates of the center point of the image, and the corresponding distortion coordinates can be calculated from the extracted feature point images. The mapping relationship between the distortion point and the ideal point can be established from different aspects. The more common conversion relationship between the corresponding x-axis and y-axis coordinates. Because the radial distortion is mainly considered in the image distortion, the pixel is only distorted in the direction from the center of the optical axis to the pixel, and the angle does not change. The degree of distortion is only related to the distance between the pixel point and the center of the optical axis. Therefore, the method uses the mapping relationship to the center distance of the optical axis to simultaneously measure the values of different pixel points to obtain the mapping relationship between the distortion point and the ideal point, and then use the distortion point and the ideal point. The mapping relationship corrects the distorted image.

2.4   Spatial Scene Image Smoothing

The RGB color scheme encodes colors into a mixture of three primary colors. This scheme is widely used in the transmission, presentation and storage of color images, not only in analog devices such as televisions, but also in digital devices. Under the application of big data analysis technology, many spatial scene image processing and spatial scene image programs use RGB scheme as the internal representation of spatial scene color image [12], and many language libraries use it as a standard spatial scene image representation scheme. The spatial scene color image and the spatial scene gray image are represented in the same way, and are expressed by a series of pixels, but the models used to express the order of the respective color components are different. Each color component in a color image of a spatial scene can be divided into a component arrangement and a packed arrangement method. In the component arrangement, each color component is assigned to a different array of the same size. At this time, the color image of the space scene can be regarded as a set of related intensity images I_R, I_G, I_B, so $I = <I_R, I_G, I_B>$; The RGB component values of the spatial scene color image I at the point (u, v) can be obtained by obtaining all three intensity images as follows:

$$\begin{bmatrix} R \\ G \\ B \end{bmatrix} = \begin{bmatrix} I_R(u, v) \\ I_G(u, v) \\ I_B(u, v) \end{bmatrix} \tag{6}$$

A packed array of component values representing a particular pixel color is packaged in a single primitive representing the array of images, as shown in Fig. 1. So $I(u, v) = <R, G, B>$, the RGB values of a packed image at points (u, v) can be obtained by accessing the components of the color separately as follows:

$$\begin{bmatrix} R \\ G \\ B \end{bmatrix} = \begin{bmatrix} \text{Red}(I(u, v)) \\ \text{Green}(I(u, v)) \\ \text{Blue}(I(u, v)) \end{bmatrix} \tag{7}$$

In the Java Image API, RGB c olor images are applied in a packed arrangement. Each color pixel is represented by an integer value of one bit. There are 8 bits of transparency, the alpha component. The access functions Red(), Green(), and Blue() depend on the specific implementation of encoding the color pixels.

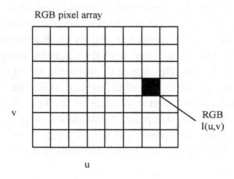

Fig. 1. Schematic diagram of RGB color image pixel storage in Java

In order to decompose the three component values from the packed integer values, appropriate shift and mask operations need to be applied. Since the R, G, and B component values are constructed in the opposite direction, the bit or operation and the left shift operation are applied, and the Oxff is used for the mask operation. This method directly accesses the pixel array, which can improve the program running efficiency. It saves time compared to using Java's image API functions.

The description of the grayscale image of the spatial scene, like the color image of the spatial scene, still reflects the distribution and characteristics of the overall and local chromaticity and brightness levels of the entire image. In the RGB model, if $R = G = B$, the color represents a grayscale color, and the value of $R = G = B$ is called a gray value. Therefore, the spatial scene gray image requires only one byte per pixel to store the gray value, and the gray scale ranges from 0 to 255. The grayscale processing of spatial scene images generally has the following three design schemes:

Weighted average method. According to the importance and other indicators, the three components R, G, and B are weighted and averaged with different weights. Since the human eye is most sensitive to green and the least sensitive to blue, the weighted average of the three components of R, G, and B using Eq. (8) can obtain a more reasonable spatial scene grayscale image;

$$f(i, j) = 0.3R(i, j) + 0.59G(i, j) + 0.11B(i, j) \qquad (8)$$

(1). Average method. The luminances of the three components R, G, and B in the color image of the space scene are simply averaged, and the average value is output as a gray value to obtain a grayscale image.
(2). Maximum method. The maximum value of the luminance in the three components R, G, and B in the color image of the space scene is used as the gray value of the grayscale image.

2.5 Spatial Scene Image Segmentation

Image segmentation of spatial scenes is an important step in the accurate transmission of visual information. Therefore, spatial image segmentation based on big data analysis technology is an important image processing link. It can not only compress data in large quantities, but also reduce storage capacity. Simplify later analysis and processing steps to provide a basis for subsequent classification, identification, and retrieval. The image threshold segmentation uses the difference between the target and the background in the image, such as color, geometric shape, spatial texture, etc., and selects one or several suitable thresholds to divide the original image into several disjoint regions to determine the image. Each pixel in the image should belong to the target or the background area, that is, the target is separated from the background image, and then according to the need, whether the corresponding binary image is further generated, and then the segmented image is analyzed. To put it simply, the analysis target is extracted from the background to facilitate subsequent processing of the image.

Pre-processing of image segmentation includes image input, denoising, smoothing, enhancement processing, color space conversion, and the like. Images can be read from a local storage device or read from a remote device via the network and file system. In order to achieve a good segmentation effect and reduce the influence of noise, the general image needs to be smoothed before segmentation, such as using a Gaussian filter for denoising. In some edge-based algorithms, it is necessary to perform Laplacian variation on the edge and enhance the edge information to complete the image segmentation. In addition, different color spaces represent different information of the image. The specific segmentation algorithm always depends on the specific color space, while the general image is stored as (R, G, B), three-channel color information, so it needs to be converted to specific The color space, such as (R, G, B) → (H, S, I), (R, G, B) → (H, L, S). The core segmentation algorithm is the soul of the image segmentation system and includes all of the various algorithms listed above. The efficiency of the core algorithm determines the efficiency of the entire segmentation system, so it is necessary to ensure the stability of the core algorithm in the system. In practical applications, we need to analyze the running complexity of the core algorithm in time and space, and optimize the algorithm and program to the maximum extent to achieve the optimal use of computing resources. After the image is segmented, there may be many small areas, or the edges of the area are not closed, so it needs to be refined by a specific post-processing algorithm.

2.6   Realization of Accurate Information Transmission

After the vision system has processed the spatial scene image and obtained the sorting result, the result needs to be sent to the control system to realize the accurate communication of the spatial scene visual information. Therefore, it is necessary to select a communication method suitable for the method. Currently commonly used communication methods are: serial communication, parallel communication, wireless communication, and the like. The serial communication only needs a pair of transmission lines to realize two-way communication, which is relatively simple and difficult to be interfered with; the parallel port requires more transmission lines, and the implementation is more complicated, and the 8 channels are susceptible to interference; Wireless communication mainly uses electromagnetic wave signals to propagate information in free space, which is more complicated and can get rid of the dependence on cables in traditional communication. There is no wireless communication requirement in the method, and serial communication is selected as the communication mode between the vision system and the control system for the sake of easy maintenance and low cost.

3   Experimental Results and Analysis

In order to ensure the effectiveness of the accurate communication method of spatial scene visual information based on big data analysis, a simulation experiment is designed. During the experiment, the visual information of a certain spatial scene is taken as the experimental object, and the visual information of the spatial scene is accurately transmitted. In order to ensure the validity of the experiment, the traditional spatial scene visual information communication method is compared with the spatial scene visual information accurate communication method based on big data analysis, and the test results are observed. Spatial scene visual information communication requires noise iteration of spatial scene images to obtain high-definition images to realize visual information transmission of spatial scenes. The basic requirement for noise iterative performance of spatial scene visual information communication methods is fast adaptive convergence.

Figure 2 is the noise average curve of the visual information image of a space scene,As shown in Fig. 2, the noise of the visual information image of the spatial scene is collected. The curve shows that as the number of frames increases, the image noise also increases. When the number of frames is 12, the image noise is 0.54 dB.

Figure 3 is a comparison of the iterative times and the noise average curve of the traditional spatial scene visual information transmission method and the spatial scene visual information accurate communication method based on big data analysis. In Fig. 2,

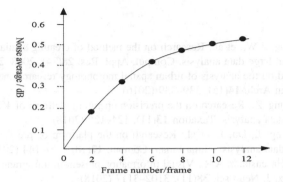

Fig. 2. Noise average curve of visual information image of a spatial scene

the maximum noise average is around 0.54 dB, and the overall upward trend of noise is relatively flat. In Fig. 3, after the processing of the spatial scene visual information transmission method, the image noise is reduced, and the spatial scene visual information accurate communication method based on the big data analysis can perform fast adaptive convergence under the same number of iterations, and iteratively 6 times It can reduce the image noise to 50%, and has better noise iteration performance than the traditional spatial scene visual information transmission method.

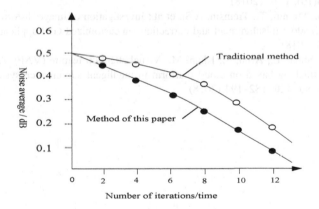

Fig. 3. Comparison of the number of iterations and the noise mean

4   Conclusion

The accurate communication method of spatial scene visual information based on big data analysis can realize the precise processing and accurate communication of spatial scene visual information images. The application of this method will help to improve the communication level of visual information in space scenes.

References

1. Zhao, M., Tang, Z.W., et al.: Research on the method of creating spatial weights based on Map Reduce in large data analysis. Comput. Appl. Res. **28**(24), 2068–2070 (2016)
2. Wan, J.: Based on the analysis of urban spatial morphology research methods in the era of big data. Urban Arch. **14**(15), 339–339 (2016)
3. Fang, M., Yating, Z.: Research on the precision operating platform of 4G telecom operators based on big data analysis. Taxation **13**(11), 124–125 (2018)
4. Wang, H., Deng, J., Du, J., et al.: Research on the platform of precise marketing service based on big data analysis. China Finan. Comput. **17**(18), 140–144 (2014)
5. Morrill, R.J., Hasenstaub, A.R.: Visual information present in infragranular layers of mouse auditory cortex. J. Neurosci. **38**(11), 3102–3117 (2018)
6. Liu, S., Lu, M., Li, H., et al.: Prediction of gene expression patterns with generalized linear regression model. Front. Genet. **10**, 120 (2019)
7. Schubert, T., Reilhac, C., McCloskey, M.: Knowledge about writing influences reading: dynamic visual information about letter production facilitates letter identification. Cortex **103**, 302–315 (2018)
8. Chelazzi, L., Santandrea, E.: High-Acuity information is retained through the cortical visual hierarchy of primates. Neuron **98**(2), 240–242 (2018)
9. Shuai, L., Gelan, Y.: Advanced hybrid information processing. Springer International Publishing, USA (2019)
10. Lee, D.H., Lee, D.W., Henry, D., et al.: Minimisation of signal intensity differences in distortion correction approaches of brain magnetic resonance diffusion tensor imaging. Euro. Radiol. **28**(10), 1–10 (2018)
11. Vitucci, G., Minniti, T., Tremsin, A.S., et al.: Investigation of image distortion due to MCP electronic readout misalignment and correction via customized GUI application. J. Instr. **13**(04), 4028 (2018)
12. Zheng, P., Shuai, L., Arun, S., Khan, M.: Visual attention feature (VAF): A novel strategy for visual tracking based on cloud platform in intelligent surveillance systems. J. Parallel Distr. Comput. **120**, 182–194 (2018)

Fast Detection Method for Local Search Target of Community Structure Under Big Data

Wang Jing-hua[✉] and Zhou Jing-quan

College of Electronic and Optical Engineering and College of Microelectronics,
Nanjing University of Posts and Telecommunications, Nanjing 210003, China
xiewenda4376@163.com

Abstract. The traditional detection method has the problems of complicated operation and slow search speed, which brings great impact to the efficient operation of the local search system of community structure. To this end, it studies the rapid detection method of local search target of community structure under big data. Analyze the key technologies for constructing detection methods, use quantitative algorithms to achieve rapid target location, perform resource entry on targets, and calculate data convolution kernel size. The convolution data is statistically analyzed, and the detection result is subjected to parsing and storage, thereby realizing the extraction of the target and completing the rapid detection of the local search target of the community structure. It is proved by experiments that the fast detection method of local search target of community structure has obvious advantages in search time consumption and has a good development prospect.

Keywords: Big data · Community structure · Local search · Target · Rapid detection method

1 Introduction

In the era of big data, the rapid detection method of local search targets in community structure is very important. The reason is that the reconnaissance system can automatically switch the data model to achieve further accurate segmentation of the target, extract the essential features of the target of interest, and lay the foundation for rapid feature matching and automatic determination of the target type. The current local search target detection methods mainly include: a template matching based detection method, which is simple and mature, but requires a target template and is only suitable for target instance detection. The second is based on the detection method of key points, which is invariant to image noise, rotation, scale and illumination changes, but requires a target template and cannot obtain the target area. The third method is based on the segmentation detection method. The method is less affected by noise and the segmentation region is accurate, which is beneficial to the intelligent target recognition, but the segmentation result is unreliable and the calculation amount is large. The fourth method is based on the sliding window detection method. This method is simple and real-time, but the target area cannot be obtained. The classifier needs more supervision information when training. The fifth type is based on the partial detection method. This

S. Liu and L. Xia (Eds.): ADHIP 2020, LNICST 348, pp. 355–365, 2021.
https://doi.org/10.1007/978-3-030-67874-6_33

type has good detection effect on the deformed target and the occluded target, but the target representation is complex and computationally intensive, and high-resolution images are needed, which is not suitable for detecting small targets. In this chapter, aiming at the detection speed of traditional detection models, a fast detection method for local search targets of social structures is proposed. By redesigning the social structure, the network depth and network performance are weighed. This paper first introduces the design principle of the target fast detection method, and proposes the design form of various methods according to the principle, and gives the design scheme of the final detection method of the local social structure local search target. Then quantitative analysis of each structure is carried out to judge the rationality of the detection method in advance. Then the local search target fast detection method is merged into the Faster R-CNN model to obtain the target fast detection network, and the self-built data set is used for parameter training to obtain the final detection model. Finally, the validity of the network is verified by experiments.

2 Analysis of Rapid Detection Method for Local Search Target of Community Structure Under Big Data

2.1 Key Technology Analysis of Target Rapid Detection Method

Community structure local search target rapid detection method construction involves many aspects, and each content is closely related. For example, the ontology classification and the attribute description of the ontology class determine the construction of the judgment matrix between the organizations, and also affect the design of the storage table of the device component instance. The judgment matrix affects the selection of the device component type. The design of the storage table of the device component instance will affect the efficiency and quality of the instance query [1]. Therefore, based on the analysis of relevant research, combined with the current status of detection methods, the key technical analysis of the rapid detection method of local search target of community structure is as follows: The first detection equipment component modular analysis and modeling technology, the detection equipment can be regarded as a mechatronics equipment. Different from general mechanical equipment, electromechanical equipment has higher accuracy requirements, faster response speed, better stability and better rigidity. It also has to have good reliability, light weight, small volume, long life and other requirements. In order to meet these requirements, the testing equipment shall classify the components that constitute the non-standard testing equipment in detail, and describe the performance attributes of each component in detail. How to use effective modeling methods to divide non-standard components and describe their attributes will be the focus of this paper. The second detection device component instance storage technology, and the instance storage of the detection device component is the basis and premise for implementing the selection query. It is also one of the research points to effectively store non-standard equipment components with strong heterogeneity to ensure the efficiency of query matching. The third device component selection and instance detection technology, after the device resource library is constructed, realizes the classification and storage of the device resource library, and constructs the premise of device component selection and instance retrieval. How to select the fast equipment components when the new testing equipment requirements come, and to find and match the existing resources in the resource library is also one of the focuses of this paper.

2.2 Overall Model Framework

The overall architecture of the local search target fast detection method is shown in Fig. 1.

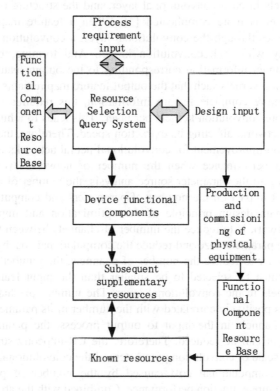

Fig. 1. Overall architecture of the rapid detection method

First, under the guidance of industry experts, build a library of device feature resources that contain existing knowledge and open interfaces for future entry of new resources. The resource in the resource library contains two parts, one of which is the description of the device component, including the device component class description, the device component attribute description, and the judgment matrix between the attributes; the second is the instance storage of the device component [2]. After the resource library is built, when the detection device requires input, the query matching is performed through the selection query system, the selection of the device components is completed, and the available device functional components are output. The designer uses the functional component as a guide to design the equipment. After the design is completed, the virtual prototype model of the non-standard equipment key components is constructed. The simulation software is used to simulate the key components, and the software design is used to analyze the existing design scheme. If the design scheme is satisfied, the production and debugging of the physical equipment can be performed. If it is not satisfied, the design will be modified according to the generated problem [3].

2.3   Web Search Model

The parameters in the convolutional neural network are derived from the number of convolution kernels in each convolutional layer, and the structure of the single-layer convolutional layer is more complicated [4]. When a feature map with a channel number of M passes through the convolutional layer, a convolution operation is performed separately with each convolution kernel. And forming, by the activation function, one channel information corresponding to the output feature map, the plurality of convolution kernels such that the output feature map remains multi-channel. In order to obtain more complete feature information, this requires a deeper level of network, with more convolution kernels per layer. This leads to a huge computational overhead of the network, affecting its extraction speed. Therefore, the design direction of optimization and improvement for convolutional neural networks is to achieve good feature extraction performance when the number of network layers is shallow. In addition, according to the parameter source analysis, the number of channels between layers has a direct impact on the number of parameters and computational overhead. Therefore, the main design principle of the optimization and improvement of the convolutional network is to reduce the number of channels between the convolutional layers, reduce the parameter size and reduce the computational cost by the low channel number [5]. In order to reduce the number of channels, the number of points convolution channels must be selected to be smaller than the input feature map and the number of channels in the convolutional layer. The number of channels in the convolutional layer is positively correlated with the number of its parameters. By reducing the number of channels in the input to output process, the parameter size of the convolutional layer can be reduced. Therefore, the CR-mlpconv structure can effectively reduce the scale of convolution parameters of the convolutional layer and reduce the complicated computational cost caused by the number of parameters, while maintaining its feature extraction performance. Combined with the above structure, this paper proposes a hybrid structure CNN model [6]. The model adopts a full convolution structure design. Only 6 layers of convolutional layers are used in the network for concatenation, which reduces the information loss caused by the pooling layer and the excessive parameter size brought by the full connection layer [7]. In addition, in order to further reduce the parameter size of the network, according to the main design principles of structural optimization, the CR-mlpconv convolutional layer structure proposed in this paper replaces the original convolutional layer and uses the C.ReLU strategy for synergy. Therefore, in the hybrid structure convolutional neural network structure designed in this chapter, the first layer is still the standard convolutional layer, and the other five layers are all CR-mlpconv convolutional layer structure. At the same time, in the first three layers of the network, the C.ReLU strategy is adopted, that is, the second and third layers are further mixed of the CR-mlpconv structure and the C.ReLU policy.

2.4 Convolution Nuclear Quantitative Analysis

In a typical convolutional neural network, assume that the k-th layer convolutional layer input feature map size is $D_M \times D_M \times M$, where M is the number of input image channels. After the convolution operation, the output feature map size is $D_M \times D_M \times N$, where N is the number of output channels, it can be clarified that the convolution layer is composed of N convolution kernels. Assuming the size of each convolution kernel is, there are:

$$D_N = \frac{D_M - D_K + 2P}{S} + 1 \tag{1}$$

Where P represents the width of the edge fill and S represents the sliding step size of the convolution kernel [8]. This paper assumes that the input length and width of a typical convolutional neural network are equal and there is no bias term. The parameter number Pk and computational cost Ck of the k-th layer convolutional layer are as follows:

$$P_K = M \times N \times D_K \times D_K \tag{2}$$

The above formula can be seen that the number of channels is positively correlated with the number of parameters and the computational cost, which indicates that reducing the number of channels can reduce the parameter size of the network model [9]. For the CR-mlpconv structure, the selection criterion of the number of channels N' of the dot convolution is that the number of input channels M and the number of output channels N are about three times N'. Then the number of parameters of the kth layer convolution layer becomes $p'_k = M \times N' \times 1 \times 1 + N' \times N \times D_K \times D_K$. The computational overhead becomes $C'_K = p'_k \times D_N \times D_N$, compared to the original convolution structure:

$$\frac{p'_k}{P_K} = \frac{C'_K}{C_K} = \frac{N'}{N \times D_K \times D_K} + \frac{N'}{M} \tag{3}$$

According to the appropriate choice of N', the CR-mlpconv layer structure can reduce the number of parameters and computational cost of about 45% to 80%.

3 Realize the Rapid Detection of Local Search Target of Community Structure Under Big Data

3.1 Extraction of Local Search Targets

The user can use the Client to perform resource entry and process requirement submission of the resource library, and present the selection result to the designer for guiding design after Sever performs device component selection feedback. So the Client operation is simple and the results are intuitive. C#.Net is an object-oriented programming language released by Microsoft [10]. Its Windows Forms program development module is suitable for writing interface software under Windows platform. Therefore, based on the VS2013 development platform, this article uses the Windows Forms application module under C#.Net to write the client software. The Client may be logged in by multiple people at the same time, so the user name and password are required for user differentiation. After logging in to the page, first confirm the IP and port number of the data center Sever, which is used to establish Socket communication with Sever. The IP and port numbers are the default under normal circumstances, and can be modified without modification. Only after the Sever address changes will be modified. Enter the username and password. Different user name suffixes represent different permissions. "username_Add" means the user who adds data to the server. "username_Submit" indicates the submitting user of the process requirement [11]. There is a click interface "Sever connection and user login test button", the verification pass shows "Severe connection Sever and verify user information through", if there is no pass, there will be corresponding prompt information. For example, "User name and password are incorrect", "Server connection error", etc. After the verification is passed, click OK to display the next level application interface [12, 13].

Simulating the community structure is the basis and premise of the local search target detection method. Simulating the community structure can reduce the delay of local search target detection. At present, there are two community structure simulation methods, one is based on the physical model, and the other is based on the mathematical model. Firstly, the local search target status of clock synchronization is analyzed through the simulation calculation of the community structure, and then the local search target detection node is set on the community structure, and the local search target detection method is analyzed through calculation.

The relationship between the background position q_{i-leak} of the community structure node and the local search target position is as follows:

$$q_{i-leak} = aQ_i^{red}H^{avl} \tag{4}$$

In the formula, Q_i^{red} represents the location of the local search target node i of the community structure; a represents the local search target node simulation parameter; H^{avl} is the node background location coefficient.

The relationship between the position of the local search target node and the positioning coefficient is shown in Eq. (5):

$$Q^{val} = \begin{cases} Q^{red}, H^{red} \geq H^{des} \\ Q^{red}\left(\frac{H^{avl}-H^{min}}{H^{des}-H^{min}}\right)^{\frac{1}{a}}, H^{min} \leq H^{avl} \leq H^{des} \\ q_{i-leak}, H^{avl} \leq H^{min} \end{cases} \quad (5)$$

In the formula, Q^{val} represents the actual position of the local search target node; Q^{red} represents the detection position of the local search target node; H^{des} represents the critical positioning coefficient of the local search target node; H^{min} represents the minimum positioning coefficient of the local search target node.

The addition of instance data is divided into transmission mechanism attribute entry, detection sensor mechanism attribute entry, guidance mechanism attribute entry, and actuator attribute entry. It also adds an interface to quickly check that the matrix is consistent, as shown in Fig. 2.

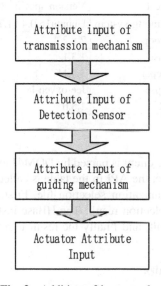

Fig. 2. Addition of instance data

The addition of each instance data includes four parts, one is the addition of public attributes, that is, the information attributes mentioned in Chapter 3, and the second is the attribute addition of the instance type. Three and four are manually added for the public attribute manual addition and the transmission attribute, respectively, to add unique attributes that are not intersected between the same type of equipment components. After the instance data is added, click the OK button. The Client encapsulates the attribute into an XML file and transfers it to Sever for data storage via Socket. After the data is stored, in order to facilitate the information exchange between the client and the server in the programming, and in order to facilitate the subsequent query, the data needs to be encoded first. The class hierarchy is coded according to the class structure diagram of the ontology design, and each level code is divided into two parts. The first part is the first two acronyms and numbers of the class name, and the class level code 0 number can be in the order of modeling in the ontology. The other class hierarchy codes are below the level code 0, the order in the modeling tool is arranged in lexicographic order, and the layer codes outside the 0 level are numbered according to the acronym and the lexicographic order. The existence of the lexicographic number can make it impossible for the class hierarchy to be the same even if the acronyms are the same.

3.2   Local Search Target Rapid Detection Process

The server is running the Hadoop platform. The development environment is shown in Table 1.

Table 1. Sever development environment

Project	Version specification
Operating system	Ubuntu16.0.43 LTS
Hadoop	2.8.2
HBase	1.3.1
Zkeeper	3.4.11
Eclipse	Jee-oxyen-la
JDK	1.8.0

The task submitted by the client is parsed by DOM4J to determine the task file type. If it is an instance storage file, the API in the table is called for instance storage. If not, the task file type is the organization selection file. First, the hierarchical analysis is performed to generate the selection result. The HBase instance query is performed by the hierarchical analysis result, and finally the result is returned to the package.

4   Experimental Results

The experiment performs target detection tasks for common community structure goals, and the pre-training network is trained by the classification task of 24 categories of targets. The pre-training network training dataset is derived from the SubImageNet dataset, and both the training set and the test set data of the target detection network are derived from the SubVOC dataset. In the experiment, the target rapid detection network is trained. At the same time, the two common network models are trained by using the same data set and the same training mode. The obtained network model is compared with the target rapid detection model of this paper. The experimental results are shown in Table 2.

Table 2. Experimental results

Model	Parameter quantity	Computational overhead	Running memory(MB)	Running time(ms/fps)	MAP
Fast target detection network	1187 K	451.4 M	895	32/31	44.5
Fastr R-CNN (VGG-16)	34 M	13.5G	5525	128/7	56.2
Faster R-CNN (ZFNet)	3725 K	1.2G	1227	47/21	43.5

It can be seen from Table 2 that the detection performance of the target rapid detection method proposed in this paper is 44.5%, which is −11.7% and +1.0% compared with the detection performance of the other two models. However, in terms of parameter quantity and computational cost, the advantage of the target fast detection data presented in this paper is obvious. The model can reduce the number of parameters by 69% and 95%, respectively, and reduce the computational cost by 69% and 95% respectively.

In order to ensure the effectiveness of the rapid detection method of local search target of community structure under the big data proposed in this paper, experimental demonstration is carried out. Compared with the traditional detection method, the search target consumption time is compared, and the experimental results are shown in Table 3.

Table 3. Search target consumption time comparison table

Number of experiments/time	Traditional method/s	Improved method/s	Time ratio
1	2.565	0.621	4.13:1
2	3.192	0.782	4.07:1
3	2.506	0.930	4.03:1
4	3.441	0.622	3.71:1
5	3.494	1.0922	3.02:1

It can be seen from Table 3 that under the same conditions, the traditional method takes about 3 s, while the proposed method takes about 0.9 s; the rapid detection method of community structure local search target has obvious advantages, and this method can greatly improve The operating speed is of great significance.

5 Conclusion

The so-called target detection is to simulate human visual organs and brain systems by combining techniques such as image processing and machine learning algorithms, and to accurately locate and accurately represent the target in an unknown image. It usually includes two subtasks: target location and target recognition. Target location is to search for the target in the image and frame the location of the target. Target recognition is to classify and identify the target of the search location. In the context of the era of big data, the rapid detection method of local search target of community structure is one of the important basic research topics at present, and also the cornerstone of many related research questions. The main deep search model used is convolutional neural network, which is also the research focus of this topic. Through the analysis, optimization and reconstruction of the CNN model, combined with the existing target detection methods, a target rapid detection network is established. Then, according to the impact of purely improving the detection speed on the network detection accuracy, the corresponding optimization strategy is adopted to further weigh the detection speed and detection accuracy of the CNN-based target detection model, which has important research value and significance in application engineering.

Acknowledgments. 1. National natural science foundation of China (61003237, 61401225); 2. Innovation program for postgraduates of professional degrees in ordinary universities and colleges of jiangsu province (SJLX15_0377).

References

1. Yan, Y.: Improvement of massive multimedia information filtering technology in big data environment. J. Xi'an Polytech. Univ. **04**, 569–575 (2017)
2. Huantong, G., Zhengpeng, C., Zhe, C., et al.: Multiobjective particle swarm optimization based on balanced search strategy. Pattern Recogn. Artif. Intell. **30**(3), 224–234 (2017)
3. Sanyi, L., Wenjing, L., Junfei, Q.: Density-based local search NSGA2 algorithm. Control Decis. Making **16**(1), 60–66 (2018)
4. Yulian, C., Wenfeng, L., Zhang, Yu.: Hybrid particle swarm optimization algorithm based on Quasi-entropy adaptive start-up local search strategy. J. Electron. **46**(1), 110–117 (2018)
5. Jinyi, Z., Bin, L., Dikai, T., et al.: Fast ICP-SLAM under coarse matching and local scale compression search. J. Intell. Syst. **12**(3), 413–421 (2017)
6. Flanagan, R., Lacasa, L., Towlson, E.K., et al.: Effect of antipsychotics on community structure in functional brain networks. J. Complex Networks **7**(6), 932–960 (2019)
7. Sun, Y., Zhao, J., Chen, L., et al.: Methanogenic community structure in simultaneous methanogenesis and denitrification granular sludge. Front. Environ. Sci. Eng. **12**(4), 10 (2018)
8. Ahlberg-Eliasson, K., Liu, T., Nadeau, E., et al.: Forage types and origin of manure in codigestion affect methane yield and microbial community structure. Grass Forage Sci. **73**(3), 740–757 (2018)
9. Shuai, L., Weiling, B., Nianyin, Z., et al.: A fast fractal based compression for MRI images. IEEE Access **7**, 62412–62420 (2019)

10. He, R., Zhao, D., Huimin, X., et al.: Abundance and community structure of ammonia-oxidizing bacteria in activated sludge from different geographic regions in China. Water Sci. Technol. J. Int. Assoc. Water Pollut. Res. **77**(6), 56 (2018)
11. Fu, W., Liu, S., Srivastava, G.: Optimization of big data scheduling in social networks. Entropy **21**(9), 902 (2019)
12. Bates, K.A, Clare, F.C., O'Hanlon, S. et al.: Amphibian chytridiomycosis outbreak dynamics are linked with host skin bacterial community structure. Nature Commun. **9**(1), 693 (2018)
13. Liu, S., Liu, G., Zhou, H.: A Robust Parallel Object Tracking Method for Illumination Variations. Mobile Networks Appl. **24**(1), 5–17 (2019)

Research on Adaptive Segmentation Algorithm of Image Weak Target Based on Pattern Recognition

Tao Lei[1] and Xiao-gang Zhu[2]([✉])

[1] South China Normal University, Guangzhou 510006, China
hcy211727@163.com
[2] Nanchang University School of Software, Nanchang 330047, China
csy22215@163.com

Abstract. Based on the comprehensive research of image segmentation technology, an adaptive segmentation algorithm based on pattern recognition for image weak targets is proposed. By systematically designing the image segmentation algorithm by analyzing the algorithm requirements and principles, the modules such as image preprocessing, weak target detection, image feature extraction and adaptive threshold selection are designed and implemented according to the algorithm implementation flow. In order to verify the experimental performance of the algorithm, experimental analysis shows that the adaptive image segmentation algorithm can be used to preserve image details, improve the quality of the segmented image, and shorten the image segmentation time.

Keywords: Pattern recognition · Image adaptation · Segmentation algorithm

1 Introduction

The transmission of pictures and the processing of images are one of the important symbols of new media [1]. Some special operations on images, such as recognition, restoration, compression and segmentation, can be realized by using new media technologies. Image segmentation is the technique and process of dividing an image into specific regions of unique nature and proposing objects of interest. It is a key step from image processing to image analysis. The process of image segmentation is a marking process, image segmentation is a critical preprocessing for image recognition and computer vision. Without proper segmentation, it is impossible to have a correct identification. However, the only basis for segmentation is the brightness and color of the pixels in the image. When the computer automatically processes the segmentation, various difficulties will be encountered. Therefore, image segmentation is a technology that needs further research. Especially for some weak image targets in pattern recognition, it is necessary to design an adaptive segmentation algorithm to achieve image segmentation. Pattern recognition is the use of computer mathematical methods to study the automatic processing and interpretation of patterns. An important form of the information processing process is the identification of the environment and the object by the living body. This recognition method is applied to image segmentation to obtain higher quality segmentation results. Nowadays, there are many image segmentation

S. Liu and L. Xia (Eds.): ADHIP 2020, LNICST 348, pp. 366–378, 2021.
https://doi.org/10.1007/978-3-030-67874-6_34

algorithms. There is no uniform solution to the image segmentation problem. This technology is usually combined with the knowledge of related fields, so that the image segmentation problem in this field can be more effectively solved.

2 Principle of Image Segmentation

The principle of image segmentation-based image weak target adaptive segmentation algorithm is to divide the image into regions with specific characteristics and extract the target of interest. The characteristics include grayscale, color, texture, etc., and the selection of the target may be a corresponding single area, or may correspond to multiple areas. Image segmentation is a key step from image processing to image analysis and a basic computer vision technology. With the concept of collection, image segmentation can be given the following more formal definitions: Let the set R represent the entire image area, and the segmentation of R can be regarded as dividing R into a number of non-empty subsets R_1, R_2, ..., R_n satisfying the following conditions: The sum (set) of all sub-regions obtained by segmentation should be able to include all pixels in the image, or segmentation should divide each pixel in the image into a sub-region. Each sub-region does not overlap with each other, or one pixel cannot belong to two regions at the same time; pixels belonging to different regions obtained after segmentation should have such different characteristics. Pixels within the same sub-area should be connected. The segmentation of the image is always based on the criteria of a segmentation [2]. There are many types of segmentation algorithms, among which the main application is the combination of the threshold segmentation algorithm and the region segmentation algorithm. The principle of the threshold segmentation algorithm is: Assuming that the object and the background are at different gray levels, the image is polluted by zero-mean noise. The gray distribution curve of the image approximately represents two normal distribution probability density functions representing the objective function and the background histogram respectively. Using the composite curve of these two functions to fit the histogram of the overall image, the histogram of the image will have two separate peaks. Then, according to the minimum error theory, the threshold value of the segmentation is obtained for the gray value corresponding to the valley between the two peaks of the histogram. After determining the appropriate threshold, the threshold is compared with the gray value of the pixel one by one, and pixel segmentation can be performed for each pixel in parallel, and the result of the segmentation is directly given to the image area.

3 Weak Target Adaptive Segmentation Algorithm

3.1 Image Preprocessing

Image preprocessing for pattern recognition includes image filtering, contrast enhancement and histogram enhancement, and effective preprocessing for infrared image features. Image filtering includes neighborhood averaging and median filtering. The neighborhood averaging method takes the average gray value of all the pixels in the neighborhood of the pixel on the input image as the output value of the pixel, so that the noise of the original image can be reduced. After smoothing the image with this filtering method, it can be reflected from the visual effect that the image has become softer than the original image, the noise is reduced, and the gray level changes more smoothly. However, this method does not carefully consider the actual difference between edge jump and noise, so the filtering effect is general. The median filtering method is a nonlinear image enhancement technique that has a good suppression effect on the interference pulse and the point noise, and can better maintain the edge of the image. Its working steps include roaming the template in the image and aligning the center of the template with a pixel location in the image; the gray value of each corresponding pixel in the template is read; the gray values are arranged in a row from small to large; one of the values is found in the middle; and the intermediate value is assigned to the pixel corresponding to the center position of the template. Different shapes of windows produce different filtering effects [3], which must be selected according to the content of the image and different requirements. One way is to use a small-scale window first, then gradually increase the window size until the median filter has more disadvantages than the benefit. Another method is to use a one-dimensional filter and a two-dimensional filter alternately. There is also an iterative operation that performs the same median filtering of the input image until the output no longer changes. Contrast enhancement can be performed after the image filtering process is completed. Contrast enhancement is a relatively simple but important method in image enhancement technology. The method is to modify the gray level of each pixel of the input image according to certain rules, thereby changing the dynamic range of the image gray level. It can expand the gray dynamic range, or compress it, or segment the grayscale, and compress it in a certain interval according to the characteristics and requirements of the image to expand in another interval. In practice, due to insufficient exposure or nonlinearity of the imaging system, the contrast of the image is not high, and the contrast enhancement can effectively improve the image quality. Contrast enhancement can take either a grayscale linear transformation or a grayscale nonlinear adjustment. Such as logarithmic transformation, exponential transformation. Logarithmic transformations are commonly used to extend low gray values and compress high gray values, which makes low grayscale image details easier to see. In addition to contrast enhancement, the enhancement method also has histogram enhancement. The histogram enhancement is based on probability theory, and the gray point operation is used to realize the transformation of the histogram. Histogram equalization is the transformation of an image of a known gray probability distribution into a new image with a uniform gray probability distribution. Since the new grayscale has a uniform probability distribution, the image looks very clear. Histogram equalization is a form of transformation, which is a transformation algorithm that makes the output image histogram become approximately uniform distribution. The calculation process is: List the original image gray level f_j, $j = 0,1,\ldots, k,\ldots, L -1$, where L is the number of gray levels; then count the number of pixels n_j of each gray level; Calculate the original image histogram and cumulative distribution function using Eq. 1;

$$P_f(f_f) = \frac{n_j}{n}, \; j = 0, 1, \ldots k, \ldots L - 1$$

$$c(f) = \sum_{j=0}^{k} P_f(f), \; j = 0, 1, \ldots k, \ldots L - 1 \qquad (1)$$

The calculated result transfer function is used to calculate the gray level g_j of the output after the mapping, and the calculated value of g_j is rounded. The number of pixels of each gray level after the mapping is counted and the output image histogram is calculated. The mapping relationship between f_j and g_j is used to modify the gray level of the original image to obtain an output image with a nearly uniform distribution of the histogram. Histogram equalization belongs to a gray-scale nonlinear transformation. In the equalization process, the gray level corresponding to the pixel transformation depends on the gray probability distribution of the entire image and its own occurrence probability. This is different from the grayscale contrast broadening. In the gray-scale contrast broadening, after determining the parameters, the gray-scale after pixel transformation usually only depends on its own gray-scale level, and has nothing to do with the entire gray-scale distribution. The histogram equalization causes the gray level interval with fewer pixels to narrow in the gray level interval corresponding to the transformed result image. Due to the proper quantization, the gray layer with lower probability of occurrence is integrated into other gray layers after the transformation. Therefore, the histogram can be balanced to overcome the problem that the smaller pixels existing in the linear stretching process occupy a large gray interval.

3.2 Weak Target Detection

The weak targets in the image mainly include noise and clutter [11–13]. The noise source comes from the structural noise generated by the imaging system during the photoelectric conversion process. In the imaging work, it will be affected by many factors, such as external environmental conditions during image shooting, such as temperature, humidity, etc., as well as the performance of the sensor itself will decrease with the increase of use time. At the same time, when the signal is digitally extracted after the image is taken, noise is introduced due to the error. There are two methods for detecting weak targets in the image: normality test and whiteness test. The normal test is to subtract the original infrared image from the estimated background image after background estimation of the infrared image. In the ideal case, the suspicious target point and the residual Gaussian white noise in the image can be obtained from the image. But we hope to have a more intuitive way to test the normality of residual noise. Since we can't simply plot the probability density function from the gray histogram of the image, we can reflect the distribution of gray values at different frequencies. After the noise is detected, the clutter is detected and suppressed. The detection and suppression of image background clutter is the premise of target tracking and detection. The main methods of background clutter suppression are divided into linear and nonlinear methods [4]. Linear methods such as low-pass filtering and adaptive enthalpy methods have the characteristics of simple structure, high computational efficiency, and high target-target signal clutter ratio gain output. The disadvantage is that the filtering performance of the non-stationary image background is poor. The nonlinear method can effectively deal with the non-stationary image background. Since this paper assumes that each image sub-block is short-term stationary, a linear filtering method is used to perform background suppression of image clutter. The linear background clutter suppression method is a spatial filtering method based on nonparametric regression estimation, that is, a spatial domain adaptive filtering method.

3.3   Feature Extraction

In the segmentation image, the feature needs to be extracted to facilitate the delineation of the boundary. The feature extraction includes color feature extraction, texture feature extraction, boundary feature detection and region feature tracking. Color is the most important feature of color image. Firstly, the image is converted from RGB color space to X*Y*Z* space, and the average color of X, Y and Z channels is extracted as the 3-dimensional color feature of each region; The shape feature is the most powerful tool for describing the contour of the region, the 1-dimensional density ratio of the extraction region, the 2-dimensional centroid, the 4-dimensional rectangular box, and the 7-dimensional invariant moment as the 14-dimensional shape feature; the texture feature describes the texture characteristics of the image, calculates the co-occurrence matrix of the region, and extracts four statistical properties of energy, inertia, enthalpy and uniformity as 16-dimensional texture features. Thus each segmentation region is represented by a 33D feature vector. The texture features are described by means of a co-occurrence matrix. The co-occurrence matrix is calculated by the spatial dependence of the pixels of the gray image. The statistical methods are used to calculate 14 second-order statistics, and these second-order statistics are used as texture features. However, among the 14 texture features, 4 features are generally used to extract the texture features of the image, namely energy, inertia, enthalpy, and uniformity [5–7]. Energy is the sum of the squares of the values of the symbiotic matrix elements, reflecting the degree of uniformity of the gray distribution of the image and the degree of texture thickness. The energy of the fine texture is relatively small, and the energy of the coarse texture is relatively large; Inertia can reflect the complexity of image grayscale, the inertia of a simple grayscale image is small, and the inertia of a complex image is large; Di is a measure of the amount of information contained in an image [8–10]. The image with a fine texture is larger, and the value of the image with less texture is smaller. When the image does not contain any texture, the value of Di is close to zero. The uniformity reflects the local homogeneity of the image. When the symbiotic matrix is concentrated along the diagonal, the uniformity value is relatively large. After extracting the color and texture features in the image, the edge of the image needs to be detected. The edge of the image is the most basic feature of the image. The so-called "edge" refers to a collection of those pixels whose pixel gray level has a step change or a roof change, and the specific form is as shown in Fig. 1.

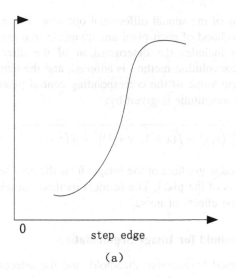

step edge

(a)

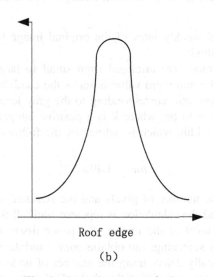

Roof edge

(b)

Fig. 1. Step edge and roof edge

The edge is widely present between the object and the background, between the object and the object. Therefore, it is an important feature that image segmentation relies on. The edge detection segmentation method detects the edge points in the image first, and then joins the contours according to a certain strategy to form a segmentation region. Edges are the result of discontinuities in gray values, which can often be easily detected using first and second derivatives. In fact, the derivative in the digital image is performed by differential approximation, so the detection of the edge is often done by convolution by means of the spatial differential operator. The edge detection operator examines the neighborhood of each pixel and quantifies the grayscale rate of change, which typically also includes the determination of the direction. In the operator operation, a similar convolution method is adopted, and the template is moved on the image and the gradient value of the corresponding central pixel is calculated at each position. Its gradient magnitude is given by:

$$G\,\mathrm{rad} = \sqrt{[f(x,y) - f(x+1,\ y+1)]^2 + [f(x+1,\ y) - y(x,\ y+1)]^2} \qquad (2)$$

In Eq. 2, Grad is the edge gradient of the image, f() is the gray level of the pixel, and x and y are the positions of the pixel. The formula method not only detects edge points but also suppresses the effects of noise.

3.4   Adaptive Threshold for Image Segmentation

The image is segmented by adaptive threshold, and the selection of the threshold is based on the optimal group pitch iteration strategy. The implementation steps are as follows:

(1). Obtaining all equal weekly rates of the original image by improving the equal-peripheral rate method;
(2). The equal-period ratios are arranged from small to large, and the gray level t corresponding to the minimum value is set as the candidate threshold value;
(3). The equal-peripheral ratio corresponding to the gray level in the range of [t–k * bwt + k * bw] is set to be, where k is a positive integer and bw is the optimal histogram interval width, which is defined by the following formula:

$$bw = 3.49\,\alpha\,N^{-\frac{1}{3}} \qquad (3)$$

Where N and α are the number of pixels and the standard deviation of the original image, respectively, and the calculation is repeated until all the equal-period rates are ∞. For the full gray level of the image, this paper firstly uses the above iterative strategy for preliminary screening, and obtains some candidate threshold values. Then, the method of dynamically determining the number of node clusters is introduced to automatically determine the number of thresholds D. Finally, the previous D candidate thresholds perform multi-level segmentation on the image. In order to realize the adaptive selection function of the threshold, the automatic determination of the threshold and the number of thresholds is set. Dynamically determining the number of enthalpy values has always been a difficult problem in the multi-level enthalpy segmentation algorithm. It is solved by introducing the criterion Q for determining the number of node clusters. The definition criterion Q is as follows:

$$Q(P_k) = \sum_{c=1}^{K} \left[\frac{A(V_c, V_c)}{A(V, V)} - \left(\frac{A(V_c, V)}{A(V, V)} \right)^2 \right] \tag{4}$$

Where: P_K is the K division of the image G, $\frac{A(V_c, V_c)}{A(V, V)}$ and $\frac{A(V_c, V)}{A(V, V)}$ respectively indicate the experimental probability p of the two end nodes of any one of the graphs G in the class C and the experimental probability p of the at least one end node in the cth class. Then the criterion Q can be regarded as the degree of deviation of p. In the formula, A(V, V) is a constant, and the number D of thresholds can be determined. The function of adaptive threshold determination can be realized by algorithm formula, and the threshold is optimally selected in order to improve the effect of segmentation image. Draw a Euler curve formed by a Euler number and a corresponding closed value, and define the Euler angle point on the Euler curve to the point on the straight line passing through the start point and the end point. The threshold corresponding to the point is the corresponding threshold, that is, the optimal threshold. After the threshold segmentation process is completed, the foreground and background regions of the image are basically separated. Since some isolated heterogeneous regions may appear inside the foreground or background region during processing, which will directly affect the subsequent image feature point extraction, these isolated regions should be post processed. Using the morphological method, the noise region is eliminated by the open operation and the closed operation to obtain a more connected cluster cluster. Generally, the open operation can remove the false region in the foreground, and the closed operation can remove the error region in the background. It is considered that in the background of the image, some noise similar to the image texture sometimes appears, such as a straight line, a curve, etc., and a circle having a radius r is used as a structural element for the opening and closing operation. The use of circular structural elements to treat strip noise is good, while smoothing the edges of the fingerprint. To get a finer edge, you can use the Gaussian template to smooth out the final result. Because the running process is more complicated, the hardware implementation uses time-sharing operation, that is, the real-time processing part can be divided into odd field and even field to process, so that the time for processing one frame can be doubled. The peripheral circuits for the memory and DSP processor sections can be implemented with an FPGA to save area and achieve non-uniformity correction of the image from the detector in the FPGA.

In summary, the specific process of the image weak target adaptive segmentation algorithm based on pattern recognition is shown below (Fig. 2).

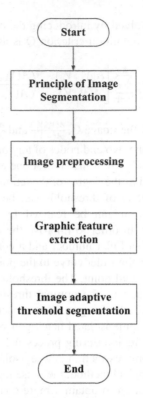

Fig. 2. Specific flow chart

4   Simulation Experiment

In order to ensure the validity and feasibility of the image segmentation-based image weak target adaptive segmentation algorithm proposed in this paper, experimental demonstration is carried out. The experimental demonstration uses all the image resources in the same image database, and has the same resolution and image file parameters for image segmentation experiments. In order to ensure the rigor of the experiment, the traditional image segmentation method is used as the experimental argumentation comparison, and the precision and the quality of the segmented image are counted. The precision is calculated as: precision = e/(e + s), where e is the number of similar images retrieved, s is the number of dissimilar images retrieved, and the value of the precision is expressed as a percentage. The average of all image precisions in the image library is defined as the average precision. Based on image pixel statistics, peak signal-to-noise ratio and mean square error are two common quality evaluation methods. They measure the quality of the image to be evaluated from a statistical point of view by calculating the difference between the gray value of the pixel corresponding to the image to be evaluated and the reference image. The experimental demonstration result curve is shown in Fig. 3 Fig. 4.

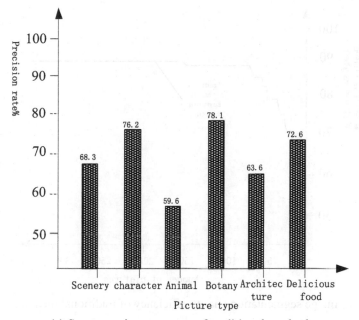

(a) Segmentation accuracy of traditional methods

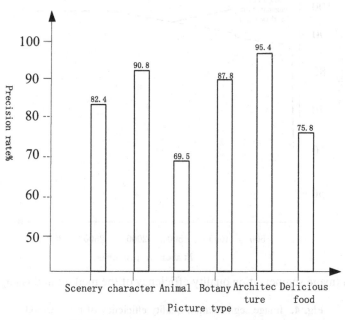

(b) The segmentation accuracy of this method

Fig. 3. segmentation accuracy of the two methods

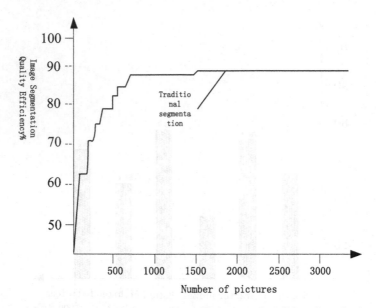

(a) Image segmentation quality efficiency of traditional methods

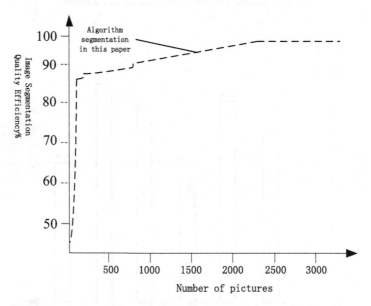

(b) Image segmentation quality efficiency of the method in this paper

Fig. 4. Image segmentation quality efficiency of two methods

It can be seen from the experimental results that compared with the traditional image segmentation method, the proposed image segmentation of the image weak target adaptive segmentation algorithm has a relatively high precision in segmenting any type of image. It can be seen from Fig. 3 that it is particularly evident in the pictures of buildings; In terms of image segmentation quality, after calculation and comparison analysis, as the number of segments increases, the segmentation quality will gradually increase, and the algorithm segmentation method will always rise. According to the curve (b) in Fig. 4, it can be concluded that the image quality of the algorithm segmentation is always higher than that of the traditional method segmentation. It fully reflects the feasibility and use value of the algorithm, and provides a more comprehensive segmentation processing method for China's image processing business.

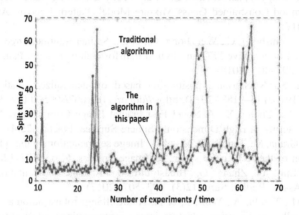

Fig. 5. Comparison of segmentation time between two algorithms

In order to verify the effectiveness of the algorithm in this paper, the adaptive segmentation time of the weak target image of the algorithm in this paper and the traditional algorithm is compared and analyzed. The comparison result is shown in Fig. 5.

According to Fig. 5, the image weak target adaptive segmentation time of the algorithm in this paper is within 20 s, while the image weak target adaptive segmentation time of the traditional algorithm is within 68 s, which shows that the image weak target adaptive segmentation time of this algorithm is longer than the image weak target of this algorithm. The adaptive segmentation time is short.

5   Conclusion

Image segmentation processing is an important processing link in image processing. However, in the process of image acquisition and imaging, there are inevitable degradation and degradation processes such as blur, motion deformation and noise, which not only affect people's visual perception of images, but also greatly reduce the use of effective information in images. The image segmentation based image weak target adaptive segmentation algorithm is used to segment the degraded image to improve the segmentation accuracy and improve the image quality.

References

1. Qin, L., Xiong, Y., Ke, L., et al.: Similarity calculation method of scene image based on adaptive weighting of similarity matrix. Pattern Recogn. Artif. Intell. **30**(11), 1003–1011 (2017)
2. Wang, J., Hong, Z.: Target tracking algorithm with saliency target indication and background adaptive constraints. Pattern Recogn. Artif. Intell. **30**(10), 875–884 (2017)
3. Quanhua, Z., Hongyun, Z., Xuemei, Z., et al.: Fuzzy Clustering Image Segmentation Based on Neighborhood Constrained Gauss Mixture Model. Pattern Recogn. Artif. Intell. **30**(3), 214–224 (2017)
4. Genggeng, L., Zhisheng, C., Wenzhong, G., et al.: Steiner minimum tree algorithm with X structure based on adaptive PSO and hybrid transformation strategy. Pattern Recogn. Artif. Intell. **31**(5), 398–408 (2018)
5. Lu, M., Liu, S.: Nucleosome positioning based on generalized relative entropy. Soft. Comput. **23**(19), 9175–9188 (2018). https://doi.org/10.1007/s00500-018-3602-2
6. Liu, S., Pan, Z., Cheng, X.: A Novel Fast Fractal Image Compression Method based on Distance Clustering in High Dimensional Sphere Surface. Fractals **25**(4), 1740004 (2017)
7. Moreno, R., Graña, M., Ramik, D. M., et al.: Image segmentation on the spherical coordinate representation of the RGB color space. Iet Image Process. **6**(9), 1275–1283 (2018)
8. Huang, Q., Luo, Y., Zhang, Q.: Breast ultrasound image segmentation: a survey. Int. J. Comput. Assist. Radiol. Surg. **12**(3), 493–507 (2017)
9. Khan, A.U.M., Torelli, A., Wolf, I., et al. AutoCellSeg: robust automatic colony forming unit (CFU)/cell analysis using adaptive image segmentation and easy-to-use post-editing techniques. Sci. Rep. **8**(1), 7302–7310 (2018)
10. Liu, S., Glowatz, M., Zappatore, M., Gao, H., Gao, B., Bucciero, A.: E-Learning, E-Education, and Online Training, pp. 1–374, Springer, Bethesda
11. Mariano, R., Oscar, D., Washington, M., et al.: Spatial sampling for image segmentation. Comput. J. **3**, 313–324 (2018)
12. Chen, L.C., Papandreou, G., Kokkinos, I., et al.: Deeplab: semantic image segmentation with deep convolutional nets, atrous convolution, and fully connected CRFs. IEEE Trans. Pattern Anal. Mach. Intell. **40**(4), 834–848 (2018)
13. Costa, H., Foody, G.M., Boyd, D.S.: Supervised methods of image segmentation accuracy assessment in land cover mapping. Remote Sens. Environ. **205**, 338–351 (2018)

Target Tracking Algorithm for Multi-channel Information Transmission in Large Data Environment

Zhu Xiao-gang[1(✉)], Yu Zhi-wei[2], and Lei Tao[3]

[1] Nanchang University School of Software, Nanchang 330047, China
csy22215@163.com
[2] State Grid Jiangxi Electric Power Co., Ltd., Nanchang 330047, China
[3] South China Normal University, Guangzhou 510006, China

Abstract. Because the traditional single-channel information transmission algorithm ignores the real-time control of the transmission data, the signal transmission tracking accuracy is low. For this reason, a target tracking algorithm for multi-channel information transmission in a big data environment is proposed. The algorithm solves the echo signal of each point, determines the transmission range of the multi-channel information, uses the interrupt mechanism to optimize the decoding algorithm, and obtains the position of the data of each point through the classification of the classifier, so as to realize the transmission target tracking of the multi-channel information. The traditional single-channel information transmission algorithm and the target tracking algorithm of multi-channel information transmission are compared and analyzed. The experimental results show that the information transmission and tracking accuracy of the multi-channel information transmission target tracking algorithm in the big data environment is better than that of the traditional single-channel information transmission algorithm The information transmission tracking accuracy is high, and it has a better information transmission tracking effect.

Keywords: Digital signal · Multi-channel · Synchronous transmission · Transmission rate

1 Introduction

Multichannel information transmission is a kind of long-distance integrated detection technology. Through network remote instruments, long-distance resources can be telemetered and controlled under non-contact conditions. Since the first laser came out in the 1960s, laser technology has been widely used in many fields, such as long-distance ranging, space military, atmospheric research, etc. because of its monochrome, good directivity and coherence. With the continuous progress and breakthrough of laser technology, laser is more and more used in the field of signal transmission in space network. The traditional optical passive network signal is transformed into an active network signal transmission, and the physical and spatial characteristics of the surface can be acquired more accurately and efficiently. Efficient transmission of network

S. Liu and L. Xia (Eds.): ADHIP 2020, LNICST 348, pp. 379–388, 2021.
https://doi.org/10.1007/978-3-030-67874-6_35

random digital signals is one of the main methods to obtain high-resolution three-dimensional images. Its measurement method is to transmit a set of long-distance, high-brightness laser beam pulse signals to the target object, and then detect the height information carried by the reflected pulse echo [1].

The traditional single-channel information transmission algorithm cannot process massive network random digital signals, and the information transmission efficiency is low. At the same time, the stability and accuracy of information transmission can not be guaranteed. A target tracking algorithm for multi-channel information transmission in large data environment is proposed. The target tracking algorithm for multi-channel information transmission is realized by determining the transmission range of multi-channel information and complementary gain power transmission mode. The algorithm can real-time control the sampled data, ensure the integrity and non-distortion of the data, and ultimately achieve efficient transmission of network random digital signals. Experimental results show that the proposed multi-channel information transmission target tracking algorithm has high information transmission efficiency, and has good performance in data transmission accuracy, method stability and comprehensive error control [2].

2   Target Tracking Algorithm for Multi-channel Information Transmission

2.1   Determine the Range of Multi-channel Information Transmission

After the user interface represented by Windows enters the host computer, it requires the ability of high-speed elevation detection and I/O processing. This requires different measurement methods to improve its performance, but puts forward new requirements for the range of multi-channel information transmission. In fact, the peripheral speed of laser beam pulse signal has been greatly improved, such as the data transmission rate between hard disk and controller has reached more than 10 MB/s, and the data transmission rate between controller and display has also reached 69 MB/s. It is generally believed that the speed of laser pulse signal should be 3–5 times that of peripheral device. Therefore, the range set in the past is far from meeting the requirements, and has become the main bottleneck of the whole system. Therefore, a higher performance requirement is put forward for laser technology. An advanced local bus is used to deal with a large number of network random digital signals. The problems of low efficiency, instability and poor accuracy of information transmission [3]. A local bus is a local bus that is not attached to a specific processor. Structurally, the local bus is the first-level bus inserted between the original system bus, which is managed by a bridge circuit, and the interface between the upper and lower is realized to coordinate data transmission. The manager provides signal buffering to support 10 peripherals and maintain high performance at high clock frequencies. Local bus also supports bus master control technology, allowing intelligent devices to acquire bus control when needed to speed up data transmission [4]. The main performance and characteristics of local bus are as follows.

The main performance of local bus: support 10 peripherals; bus clock frequency 33.3 MHz 166 MHz; maximum data transmission rate 133 MB/s; clock synchronization mode; independent of CPU and clock frequency; bus width 32 bits (5 V) 164 bits (3.3 V); can automatically identify peripherals; especially suitable for working with Intel CPU.

The characteristics of local bus are as follows: it has the ability to operate in full parallel with processor and memory subsystems; it has implicit central arbitration system; it reduces pin number by multiplexing (address line and data line); it supports 64-bit addressing and full multi-bus master control capability; it provides parity check of address and data; it can convert SV and 3.3 V signal environment. It can be seen that the local bus provides a wider and faster access to the microprocessor for peripheral devices, and effectively overcomes the bottleneck of data transmission. At present, local bus interface is the preferred interface for many adapters, such as network adapter, built-in modem card, voice adapter, etc. At present, most motherboards have PCI slots [5].

In summary, multi-channel recording system needs to transmit multi-media information at the same time, and it also requires high transmission capacity: large capacity, high bandwidth, low latency transmission; in order to meet the needs of system synchronization, it needs to support resource management requirements including acceptability testing and resource scheduling functions: to facilitate the system to achieve distributed voice acquisition through multiple channels. The local bus can deal with the above problems well, but it needs to determine the transmission range of multi-channel information according to the data of each point, and adjust the frequency of network random digital signal according to the change of curve slope [6]. If the sampling regression curve of network random digital signal is $y = f(x)$, the sampling set t can represent $t = [(x_1, y_1), (x_2, y_2), \ldots (x_n, y_n)]$. The sampling time of each sampling point in the above formula is $t_1, t_2, \ldots, t_n$, and the relationship between sampling time of each point is $t_1 < t_2 < \ldots < t_n$. Using the nearest n information sampling points, a regression curve is fitted, and the regression equation of one-dimensional line is applied. The formula is as follows:

$$\left\{ \chi = \frac{\sum\limits_{k=i-1} i = n(t_n - t)}{\sum (t_n - t)} \right\} \tag{1}$$

In formula (1), χ represents the slope coefficient. n is the sampling value of sampling point. If χ is larger, the sampling time interval is smaller. If χ is smaller, the sampling time interval is larger. In the process of controlling each sampling point, in order to improve the transmission accuracy of the network random digital signal, it is necessary to consider the specific expression of the echo signal when discretizing the network signal. The corresponding series values of the network signal obtained after sampling are formula (2), as shown below:

$$y = f(x) = \sum_{i=1} \beta \left(\frac{n(t_n - t)}{t^2} \right) \tag{2}$$

Formula (2) denotes the corresponding series values of network signal echo signals. To achieve a precision level for information transmission, the echo signals of each point can be obtained by solving them. According to the above steps, the multi-channel synchronous and accurate transmission of network random digital signals can be completed, and the range of multi-channel information transmission can be determined [7].

2.2   Optimized Decoding Algorithms

Multichannel inevitably requires that the system has a higher processing efficiency, requiring that the system can process up to dozens of channels of media information in a unit time. On the one hand, the processor of the system should have a higher computing speed, or the system should choose a more efficient media decoding algorithm. On the other hand, we can decompose the processing operations into the main part of the system by constructing a distributed architecture. In order to reduce the operation pressure of the main processor, the processing efficiency can be improved in other parts besides the processor. In multimedia recording system, multimedia I/O and multimedia encoding and decoding process are responsible for processing media information. To improve the processing efficiency of the system, we must start from these two layers and analyze their functions, structures, and applicable methods. Similarly, the improvement of processing efficiency of the system will mainly be reflected in these two layers [8]. The multi-channel information transmission structure is as follows.

Figure 1 is a multi-channel information transmission structure. The "transport layer" includes multimedia real-time transmission and control protocols. It provides packages, sending and receiving functions, data transmission scheduling for compressed media streams, and abstracts application-level parameters such as data burst and average data rate, so as to make them correspond to communication-level parameters and maximize transmission capacity. This layer shields network resources downward, provides media transmission interface upward, and provides the function of system resource management and use. The transport layer can be divided into three parts: one is the interface and protocol with the multimedia processing layer, the other is the interface and protocol with the storage management layer, and the other is the data transmission scheduling and control. "Storage Management Layer" is responsible for storing, querying and managing all kinds of multimedia records. It is the core of multimedia recording system [9].

The traditional single-channel information transmission method cannot cope with the massive network random digital signals, and the information transmission efficiency is low. At the same time, the stability and accuracy of information transmission can not be effectively guaranteed. Furthermore, only one 12 ADCI converter is introduced into the multimedia processing layer. There are 18 input channels in the converter, including 16 external input channels and 2 internal input channels. Four external input channels are used to input four sensor acquisition signals and one internal input channel is used to input internal reference voltage signals. Before using ADC 1 conversion, configure ADC 1. ADC 1 uses regular channel conversion sequence and converts the mode to multi-channel single conversion mode. And enable DMA mode. The sampling time from channel 0 to channel 7 is 28.5 cycles. The shorter

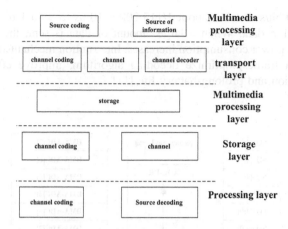

Fig. 1. Information transmission structure

the sampling time, the faster the sampling channel can be closed, the system can enter the sleep mode, reduce energy consumption, and prolong the battery life. However, the sampling time is too short, which may reduce the accuracy of sampling data. The final compromise is 28.5 cycles. Initialize the sampling value of 8 channels to ADC1 IN17, open 8 regular channels to SQ1–SQ8. The operation formula is as follows.

$$v_{ref\ln t} = \frac{v_{ref\ln t}}{4096} \times vdd \qquad (3)$$

In formula (3), $ref \ln t$ denotes the reference value, $v_{ref\ln t}$ denotes the theoretical value, vdd denotes the theoretical value. According to formula (3), it can be calculated that $ref \ln t$ enlarges 100 times and retains decimal places. Compensation factor is the ratio of theoretical sampling value to actual sampling value. When the sensor data is sampled, the compensation factor is calculated firstly, then multiplied by the actual sampling value of the sensor, divided by 100, the theoretical sampling value of the sensor after compensation can be obtained. The value of rule channel conversion is stored in a single data storage management layer. Once the conversion is started, the converted data will be stored in the storage management layer. For eight consecutive conversions, set the queues of the eight regular conversions to the same value, and then average the results of the eight conversions to obtain the final result [10]. The structure of multi-channel internal analog-to-digital conversion is shown below.

Figure 2 is a multi-channel internal analog-to-digital conversion structure. When the lower computer receives the data sent by the coordinator, it first checks the data frames to prevent errors caused by data transmission. If the check fails, the frame command is dropped directly. If the check passes, the target address of the frame command is judged to be itself or not, and if not, it is returned directly. If the target address is itself, the command is executed and the data collected by the corresponding sensor is returned. The timeout count is cleared and data is sent. Channel 1 sends the first frame of data, and activates the sending interrupt. The interrupt function completes the sending of the remaining data frames. The serial port baud rate designed in this paper is 115 200 bit/s. It takes about 1/115 200 = 10 us to send 1 bit data, and about 100 us to send 1 b data. When a large amount of data is sent, the time consumed increases and the power consumption increases. Interruption mechanism is used to send data, check data frames, optimize decoding algorithm, improve efficiency, reduce power consumption and prolong battery life [11].

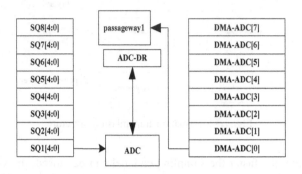

Fig. 2. Multichannel internal A/D conversion architecture

2.3   Implementation of Multi-channel Information Transmission Target Tracking

In large data environment, multi-channel information transmission target tracking algorithm finally transforms the tracking problem into the binary classification of samples. Through classifier classification, the sample location with the greatest similarity is the approximate location of each point data. Then, the location of each point data is optimized and the location of each point data is finally obtained. Sample collection in the algorithm is divided into two parts. In the first part, positive and negative samples for classifier updating are collected, which are carried out in the nth data frame of the known target location; in the second part, samples that need to be classified by the classifier are collected, which is carried out in the nth data frame of the tracking task. The two parts of sampling are carried out in different data frames, but the sampling process is consistent. In the process of sampling, the first step is to determine the location of the sampling area. In the algorithm, the sampling area is a square area centered on vertex (x, y) in the upper left corner of the target position, and the edge length of the square is twice the radius of the outer boundary of the sampling area. Sampling grids are generated in the region, and then appropriate samples are selected according to the sampling conditions. Sampling conditions can be expressed as $ind = (rd < prod) \cap (dist < inrad)$, which are mainly determined by two conditions in the process of sampling. First, samples must be selected in the sampling area. When the positive sample is taken, the radius $inrad$ of the outer boundary of the sample area is smaller.

In order to collect positive samples near the target location. When negative samples are sampled, the samples will be sampled in the annular region where the radius of the inner boundary is larger than that of the outer boundary, and the radius of the inner boundary is larger than that of the outer boundary when positive samples are sampled. When acquiring the samples to be classified, the samples are sampled near the same position in the current frame according to the target position obtained in the previous frame. The radius of the outer boundary of the sampling area is between the range of the radius of the outer boundary when the positive and negative samples are collected.

Since the classifier is updated frame by frame according to the position of the target, and does not retain any prior knowledge, if similar objects suddenly appear in the data sequence, it may lead to the classifier can not classify the samples accurately, which may lead to the failure of tracking. Once the target is lost in the tracking process, the classifier will update according to the wrong location, which will cause the subsequent tracking process to fail. Aiming at the problem of target tracking, the proposed algorithm is improved on the basis of the traditional target tracking algorithm, which adds the process of target location optimization and achieves multi-channel information transmission. At the same time, the weighted sequence tracking window is used to solve the target problem, correct the error in the centroid recalculation process, and improve the tracking efficiency and robustness of the algorithm. The formula for calculating the centroid weight is as follows.

$$\delta_{rd} = \sum_{i=1} w \left(\frac{n\left(t_n - \overline{t}\right)}{t^2} \right) \tag{4}$$

In formula (4), the discriminant weight of a data frame on w can be adaptively adjusted by the learning of discriminant weight, and the two features can be organically combined to improve the discriminant ability of the tracking algorithm. When the target information is similar, the discriminant ability of the internal analog-to-digital conversion structure becomes weak, and the corresponding weight decreases accordingly. When the sparse feature of the target is similar, the discriminant ability of the sparse representation coefficient feature becomes weak, and the weight of the discriminant weight updating strategy reduces its weight. It can be seen that the update strategy increases the accuracy and robustness of the tracking algorithm and reduces the occlusion of the target. So far, target tracking algorithm for multi-channel information transmission is realized [12–15].

3   Experimental Conclusion

In order to verify the effectiveness of the proposed multi-channel information transmission target tracking algorithm in large data environment, 14 common data sequences are selected from a large number of data sets provided in the database for estimation. The attributes of these 16 common data sequences are as follows.

Table 1 is the attributes of 16 general data sequences. According to the attributes of 16 general data sequences, the traditional single-channel information transmission algorithm and the multi-channel information transmission target tracking algorithm proposed in this paper are used to carry out information transmission tracking on the data. The transmission tracking accuracy of the two algorithms is compared. The comparison results are shown in Fig. 3 Show.

Table 1. Attributes for 14 common data sequences

Sequence name	Frame number	Challenge
Basketball	#725	IV. OCC. DEF. OPR. BC
Bolt	#350	OCC. DEF. IPR. OPR
Cardark	#393	IV, BC
David	#462	IV, SV, OCC, DEF, MB, IPR, OPR—
David3	#252	OCC. DEF. OPR. BC
Deer	#71	MB, FM. IPR. BC, LR
Doll	#74	IV. SV. OCC, IPR. OPR
Faceoccl	#128	OCC
Footballl	#900	IPR. OPR. BC
Human8	#1336	IV. SV. DEF
Lemming	#1000	IV. SV. OCC, FM. OPR. 0 V
Panda	#351	S V. OCC, DEF. IPR. OPR. 0 V. LR
Singed	#200	IV. OPR. SV. OCC
Skating 1	#500	IV. OCC. DEF. MB, FM. IPR. OPR. O V

Analysis of the data in Fig. 3 shows that the target tracking accuracy of the multi-channel information transmission proposed in this paper increases with the number of experiments, up to 98%, while the target tracking accuracy of the traditional single-channel information transmission algorithm The number of experiments continues to

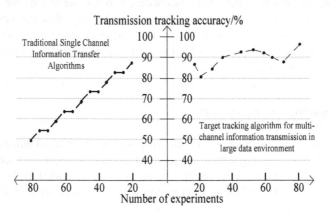

Fig. 3. Comparing the experimental results

increase, showing a downward trend, and the highest is only 88%. The target tracking accuracy of the multi-channel information transmission proposed in this paper is higher than that of the traditional single-channel information transmission algorithm, which shows that the target tracking of the multi-channel information transmission proposed in this paper The algorithm can track the target more robustly and accurately.

4 Concluding Remarks

There is a huge amount of information in the field of signal transmission in space network. The information extracted by traditional transmission methods is limited, and the accuracy can not be effectively guaranteed. A target tracking algorithm for multi-channel information transmission in large data environment is proposed, which can improve the efficiency of information transmission and ensure the accuracy of information acquisition. The experimental data prove the effectiveness of the multi-channel information transmission target tracking algorithm in large data environment.

References

1. Weibo, X., Yuanxiang, X., Wen, L.: Target tracking algorithm using complementary features of kernel correlation filtering. J. Overseas Chin. Univ. Natural Sci. Edition **3**, 429–434 (2018)
2. Ridong, Z., Xiaoyuan, Y., Jingkai, W.: Target tracking algorithm based on Fourier domain convolution representation. J. Beijing Univ. Aeronaut. Astronaut. **44**(1), 151–159 (2018)
3. Jizhou, W., Changhua, L., Weiwei, J.: A multi-moving target tracking algorithm based on random field. J. Electron. Meas. Instrum. **31**(6), 909–913 (2017)
4. Jizhou, W., Changhua, L., Weiwei, J.: A multi-moving target tracking algorithm based on random field. J. Electron. Measur. Instrum. **31**(6), 909–913 (2017)
5. Xiaoshu, C., Zexi, H., Yuefang, G. et al.: Real-time target tracking algorithm integrating color and space-time context information. Minicomputer Syst. **38**(3), 630–634 (2017)
6. Wu, Z., Dongliang, P., Gang, R., et al.: Multi-sensor Management Algorithms for Joint Information Increment and Covariance Control of R NYI. Fire Command Control **5**, 42–46 (2017)
7. Pei-man, Z., Jian, Z., Zebin, Z., et al.: Dynamic programming pre-detection tracking algorithm based on neural network. Modern Radar 39(11), 34–38 (2017)
8. Front, M., Chengwei, W., Huajie, C.: Multi-sensor multi-tracking task management method based on tracking continuity. Firepower Command Control **42**(9), 18–20 (2017)
9. Mengdi, Z., Hongxing, L., Yi, L., et al.: A non-orthogonal multi-carrier modulation technology in space-based Internet of Things. Radio Eng. **3**, 183–187 (2018)
10. Liu, S., Lu, M., Li, H., et al.: Prediction of Gene Expression Patterns with Generalized Linear Regression model. Frontiers Genet. **10**, 120 (2019)
11. Liu, G., Liu, S., Muhammad, K., et al.: Object tracking in vary lighting conditions for fog based intelligent surveillance of public spaces. IEEE Access **6**, 29283–29296 (2018)
12. Wei, G., Huqiang, L.: Cyclic linear 2-D table sorting algorithm for multi-channel video streaming transmission. Chin. J. Image Graph. **14**(10), 2149–2153 (2018)

13. Kar, A.K., Dhar, N.K., Mishra, P.K., et al.: Relative vehicle displacement approach for path tracking adaptive controller with multisampling data transmission. IEEE Trans. Emerg. Topics Comput. Intell. **3**(4), 322–336 (2019)
14. Liu, S., Glowatz, M., Zappatore, M., Gao, H., Gao, B., Bucciero, A.: E-Learning, E-Education, and Online Training, pp. 1–374, Springer, Bethesda (2014)
15. Huang, N.E., Qiao, F.: A data driven time-dependent transmission rate for tracking an epidemic: a case study of 2019-nCoV. Sci. Bull. **65**(6), 425–427 (2020)

Research on an Algorithm of Six Degrees of Freedom Manipulator Arm Moving with End Trajectory

Yun-sheng Chen[✉]

Guangzhou Huali Science and Technology Vocational College,
Guangzhou 511325, China
pofjha@sina.com

Abstract. Under manipulator trajectory motion algorithm, the time consumption of manipulator motion is long, and the stability coefficient is low, and the accuracy of obstacle avoidance is poor. An algorithm that six degree of freedom manipulator moves with end trajectory based on particle swarm optimization was proposed. By analyzing the positive solution of kinematics and inverse solution of kinematics of manipulator, the expected pose of manipulator was obtained to realize the analysis of manipulator trajectory. According to the constraint condition of the short time consumption of motion, the high accuracy of obstacle avoidance and the strong motion stability, the model that six degree of freedom manipulator ARM moved with end trajectory is built. The model was subdivided into particle swarm optimization algorithm to realize the solution of model. Then, some parameters such as initial position and speed of particle swarm were set, and particle swarm fitness function was calculated to get the optimal solution. Finally, we determined whether the current optimal solution was the global optimal solution. Thus, we obtained the optimal planning results that the manipulator moves with the end trajectory. Experiment shows that the time consumption of manipulator motion is short. The average accuracy of obstacle avoidance is 95%. The stability coefficient is high. This algorithm can effectively solve the problem of current algorithm, which has practicality.

Keywords: Six degrees of freedom · Manipulator arm · End trajectory motion · Particle swarm

1 Introduction

With the continuous development of mechanization, the form of robot becomes more advanced. Currently, the mechanical arm is widely used. There is still a large gap between China and foreign countries in developing and using mechanical arms. To research the whole operation of mechanical ARM is a very effective way. The trajectory tracking control of mechanical ARM is an important part of development of robot technology, which has high research value [1]. Due to the high research value of mechanical arm, many experts and scholars begin to analyze and study them in succession. Therefore, there have been some excellent research results.

S. Liu and L. Xia (Eds.): ADHIP 2020, LNICST 348, pp. 389–398, 2021.
https://doi.org/10.1007/978-3-030-67874-6_36

Reference [2] proposes a method for planning movement track of manipulator based on Cartesian space. The D-H matrix is used to realize the mathematical modeling of position and pose of manipulator. Then, circular interpolation method in Cartesian space is used to complete the path planning and obtain the smooth movement track. Finally, the artificial potential field is used to make the manipulator arm reach the termini without obstacle. Experiment proves that the proposed algorithm improves the accuracy of obstacle avoidance of manipulator, but it has a long time-consuming problem. Reference [3] proposes that the structural characteristic of picking manipulator is used to realize the modeling of structure parameters of mechanical arm through Denavit-Hartenberg and obtain the positive kinematics model. In solving inverse kinematics, the inverse kinematics equation is used to change the complex matrix into the easy algebraic equation, so as to complete the analysis of inverse kinematics. Thus, the optimal inverse solution is calculated based on the minimum energy principle. Experimental results show that the motion of manipulator arm is relatively stable with this algorithm, but the accuracy is poor in avoiding obstacles. Reference [4] proposes a robot trajectory method based on inverse solution of kinematics. On the basis of coordinated hoisting and transporting of multi robots for the same object structure, the kinematics model of coordinated operation system is built. The analytic formula of inverse kinematics of coordinated operation system is analyzed. Meanwhile, the analytic formula of inverse kinematics is given. The experimental platform is constructed through UG/ ADAMS/ MATLAB simulation system. Experiment shows that the operation process of this method is relatively simple, but the multi-robots coordination system cannot avoid the obstacle accurately by using this method. In reference [5], a tracking motion algorithm of manipulator arm based on predictive control is proposed. The analytical mechanics material is used to infer the difficulty formula under the load of mechanical arm, so as to obtain the kinematics model of manipulator system. On the basis of above, the simplified processing of kinematic model of manipulator system is realized, such as reduced order and linearization. The motion priority level is determined through the range of motion of each joint in model and the accuracy of positioning. Meanwhile, the real-time updates and the rolling optimization are used to ensure the timeliness of control. Experimental results show that this algorithm solves the motion redundancy, but the motion stability of manipulator ARM is poor.

At present, some related algorithms about the manipulator trajectory motion cannot realize the goal of high accuracy, high stability and low delay. Therefore, this article proposes an algorithm that six degrees of freedom manipulator tracks with the end trajectory based on particle swarm. The main structure is:

(1) According to the positive solution of kinematics and inverse solution of kinematics, we analyze that the six degrees of freedom manipulator tracks with the end trajectory, and build the model.
(2) Use particle swarm optimization algorithm to solve the model that six degrees of freedom manipulator tracks with the end trajectory, get the best planning way of manipulator arm to track the end trajectory.
(3) The feasibility of proposed algorithm is proved by the experiment.
(4) Summarize the full text and give the future research direction.

2 Algorithm of Six Degrees of Freedom Manipulator Arm for Following End Track

2.1 The Establishment of Trajectory Model and the Analysis of End Trajectory Motion of Manipulator Arm

To analyze that six degree of freedom manipulator moves with the end trajectory and build the trajectory kinematic model needs to construct the coordinate system of manipulator arm. Figure 1 is the coordinate system of manipulator arm.In Fig. 1, there are six joints in the six degree of freedom manipulator structure. For the purpose of describing the position and posture of end effect or of manipulator arm in space, a coordinate system is built on each joint. Thus, the position of end effect or is described through the relationship between coordinate systems.

(1). Positive solution of kinematics. The process of positive solution of kinematics of manipulator arm is the process using rotation angle of known joint variables θ_1, θ_2, θ_3, θ_4, θ_5 and θ_6 to get the pose of end gripper relative to reference coordinate system. The reference coordinate system is set on the base of six degree of freedom manipulator arm by the D-H method of standard upper joint [6]. It begins to change from base to the first joint, then to the second joint, and so on. Finally, it changes to the end gripper.

(2). Inverse solution of kinematics. The process of solving inverse solution of kinematics is the process of using pose of known end gripper relative to the reference coordinate system to get joint variable θ_1, θ_2, θ_3, θ_4, θ_5 and θ_6. It is the basis of the trajectory planning and trajectory control. It is also the most important part of kinematics [7]. According to the content above, the formula of the expected pose of manipulator arm is given.

$$TH = \begin{pmatrix} n_x & o_x & a_x & p_x \\ n_y & o_y & a_y & p_y \\ n_z & o_z & a_z & p_z \\ 0 & 0 & 0 & 1 \end{pmatrix} \tag{1}$$

Where, n denotes normal. o denotes the direction. a denotes the approaching vector. p denotes the origin of coordinate system of manipulator end gripper. Corresponding to the position vector of base coordinate system, x, y and z denote three coordinate directions of coordinate system. Generally, p can be given by using the position of work piece. n, o and o can be given by using rotation of rolling angle, pitching angle, and drift angle [8].

For the purpose of convenient calculation, we combine the joint 1 and joint 2 with the joint 5 and joint 6. Thus, the total transformation between the base and the hand of the six degree of freedom manipulator can be expressed as formula (1). According to the requirement of desired position in formula (1), manipulator may have multiple combinations of joint rotation in the same posture, so we must choose the optimal solution based on the actual structure of manipulator arm. In order to achieve the purpose of short time consumption, high accuracy of obstacle avoidance and strong

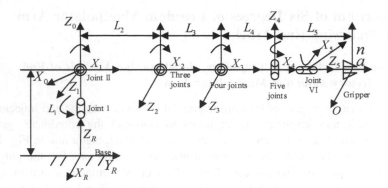

Fig. 1. Coordinate system of six degrees of freedom manipulator arm

motion stability, we take the three points as constraint conditions to build the model that a six degree of freedom manipulator moves with end trajectory.

According to the above content and the position of terminal actuator described by the six degree of freedom manipulator coordinate system, the positive solution of kinematics and inverse solution of kinematics are analyzed. The desired posture of manipulator ARM is calculated. According to the constraint condition of the short time consumption of manipulator motion, the high accuracy of obstacle avoidance and the strong motion stability, the objective function of model that six degree of freedom manipulator ARM moves with end trajectory is built.

2.2 To Solve Trajectory Model of Manipulator Arm Based on Particle Swarm

The particle swarm optimization algorithm is used to solve the model established in Sect. 2.1. The detailed process is as follows:

To randomly generate M particles to construct the initial particle swarm, and set the initial position and initial speed of particle, the fitness function is calculated according to parameters [9]. Supposing that any function of three objective functions in the model established by Sect. 2.1 is not consistent with the formula (3), then the fitness function of particle swarm can be expressed as $|U(J,I)|$, and the particle swarm optimization algorithm will decrease $|U(J,I)|$ in the process of loop iteration until it meets the constraint condition.

$$f(U) = \sum M \times |U(J,I)| \times A_{JI} \qquad (2)$$

Where, $f(U)$ denotes the constraint function of all joint motion in the six degree of freedom manipulator arm. A_{JI} denotes the influence coefficient selected by the fitness function of particle swarm. J denotes J th group in M groups of populations, and I denotes the number of populations which are not satisfied formula (2). Supposing that the three objective functions in model built by Sect. 2.1 are all conform to the formula (2), the fitness function can be expressed as $-|U(J,I)|$.

The optimal value after the k-th iteration and the optimal value of groups are selected by using the fitness function. The population is reintegrated and new particle swarm is formed. The formula of current optimal value is as follows:

$$Best(U) = \frac{f(U) \times (J,I) \times A_{JI}}{\sqrt{TH} \times k} \tag{3}$$

The calculation from formula (3) is used to determine whether the current optimal solution is the global optimal. If it isthe global optimal, the optimization stops. If it is not, the iteration will continue until it meets the iteration stop condition, and then the optimal value will be output. The optimal solution obtained by using formula (3) is the optimal planning result that the six degree of freedom manipulator arm moving with the end trajectory [10, 11].

In conclusion, the particle swarm optimization algorithm is used to solve the model that six degree of freedom manipulator arm moves with the end trajectory. The parameters of particle swarm location and initial speed are set, and the fitness function of particle swarm is determined [12, 13]. To calculate current optimal solution and judge whether this solution is the global optimal solution can obtain optimal solution which satisfies short time consumption of mechanical arm movement, high accuracy of obstacle avoidance and strong stability of motion [14, 15]. Thus, the planning research that six degrees of freedom manipulator arm moves with end track can be completed.

3 Experimental Results and Analysis

In order to prove the feasibility of algorithm for tracking the end trajectory of six degree of freedom manipulator based on particle swarm optimization, the experimental platform was built on VC++6.0, and the object was shown in Fig. 2. The simulation time was 0–60 s. Experimental indexes were set up in three aspects: the time consumption of manipulator arm motion, the accuracy of avoiding obstacle and the stability of motion.

Fig. 2. Experimental objects

Different algorithms were applied to experimental objects, and the effectiveness of proposed algorithm was verified by three indexes above. Experimental results were as follows:

3.1   Experiment (First) Results

According to the analysis of experimental result in Fig. 3, under the same number of work pieces, the time consumption of manipulator arm in manipulator ARM motion track planning algorithm based on Cartesian space was long, and the time-consuming curve fluctuates greatly. The time consumption of manipulator arm based on particle swarm was relatively short and the time-consuming curve was relatively stable. When the number of work pieces was 30, obviously, the time consumption of manipulator ARM motion track algorithm based on particle swarm optimization was15s less than time consumption of manipulator ARM motion track planning algorithm based on Cartesian space. The experimental data show that the proposed algorithm has absolute superiority.

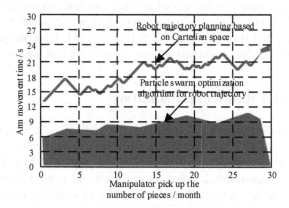

Fig. 3. Comparison of time consumption of manipulator arm motion with different algorithms

3.2   Experiment (Second) Results

From Fig. 4, we can see that the accuracy curve of obstacle avoidance of trajectory motion algorithm for manipulator arm based on Denavit-Hartenberg was first increased and then decreased, and the average accuracy of obstacle avoidance is about 50%. Thus, the reliability is low. In the preliminary stage of avoiding obstacles, the accuracy curve of robotic trajectory method based on inverse kinematics showed a steady rising trend. However, with the increase of obstacles, the accuracy curve of obstacle avoidance decreased faster, and the average accuracy of obstacle avoidance was about 45%. The accuracy curve of obstacle avoidance using algorithm of mechanical arm trajectory based on particle swarm optimization had shown a trend of stable increase, and tended to slow at about 97%, and the average obstacle avoidance accuracy was 95%. This data proves that the proposed algorithm was superior to the current algorithm in the accuracy of obstacle avoidance.

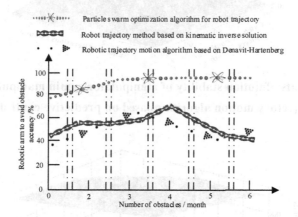

Fig. 4. Comparison of accuracy of obstacle avoidance with different algorithms

3.3 Experiment (Third) Results

According to the analysis of the experimental results in Fig. 5, the stability of the manipulator arm based on predictive control is poor. The maximum stability coefficient is 0.58. The feasibility is very low. In the trajectory motion algorithm of manipulator based on particle swarm optimization algorithm, the motion stability of manipulator arm is strong. The maximum value is 0.98. Compared with the existing algorithms, this algorithm has better motion stability. This algorithm is superior to the existing algorithms in the aspects of time consumption, obstacle avoidance accuracy and motion stability of the manipulator.

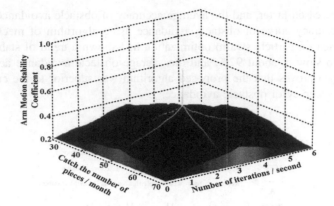

(a) Coefficients ofmotion stability of manipulator armin manipulator arm trajectory motion algorithmbased on predictive control

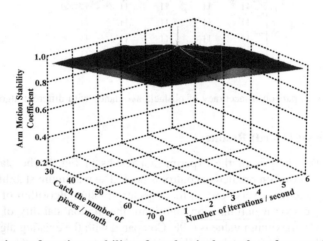

(b) Coefficient of motion stability of mechanical arm based on particle swarm optimization in manipulator track motion algorithm

Fig. 5. Comparison of motion stability of manipulator arm with different algorithms

4   Conclusions

With the constant innovation of modern technology, the current manipulator trajectory tracking algorithm is unable to satisfy the development needs of this field. An algorithm for tracking the end trajectory of six degree of freedom manipulator based on particle swarm was proposed. The inverse kinematic solution and positive kinematic solution of manipulator arm were analyzed. The motion time consumption of manipulator ARM and the accuracy and stability of obstacle avoidance were taken as constraint conditions. Then, the trajectory motion model of manipulator arm was built.

Finally, the particle swarm optimization algorithm was used to solve the model. Thus, the best planning result of manipulator arm was obtained. Experimental results show that the proposed algorithm is superior to the current algorithm in the motion time consumption of manipulator arm, the accuracy of obstacle avoidance and the stability of motion. For the development direction in the future, the following suggestions are presented:

(1) The manipulator arm is mostly applied to the industrial field. But, different trajectory planning algorithms should be used for different types of manipulator ARMs, so as to get better application results.
(2) In this research, the cost problem of trajectory planning is not considered, and it is necessary to consider this kind of problem in the future research.

References

1. Granatosky, M.C., Schmitt, D.: The mechanical origins of arm-swinging. J. Human Evol. **130**(6), 61–71 (2019)
2. Fu, Q., Wang, Z.: An obstacle avoidance trajectory planning for double manipulators. Electronic Design Eng. **25**(16), 68–72 (2017)
3. Teng, J.Y., Xu, H.B., Wang, Y., et al.: Design and simulation of motion trajectory planning for manipulator of picking robot. Comput. Simul. **34**(4), 362–367 (2017)
4. Ishikawa, A., Sakai, T., Fujii, K.: An ionic liquid gel with ultralow concentrations of tetra-arm polymers: Gelation kinetics and mechanical and ion-conducting properties. Polymer **166** (12), 38–43 (2019)
5. Levenhagen, N.P., Dadmun, M.D.: Reactive processing in extrusion-based 3d printing to improve isotropy and mechanical properties. Macromolecules **52**(17), 15–21 (2019)
6. Pekarovskiy, A., Nierhoff, T., Hirche, S., et al.: Dynamically consistent online adaptation of fast motions for robotic manipulators. IEEE Trans. Robot. **12**(99), 1–17 (2017)
7. Tiller, N.B., Campbell, I.G., Romer, L.M.: Mechanical-ventilatory responses to peak and ventilation-matched upper- versus lower-body exercise in normal subjects. Exper. Physiol. **104**(15), 25–31 (2019)
8. Garcia, A.N., Lucíola da, C., Menezes, C., Hancock, M.J, et al.: McKenzie method of mechanical diagnosis and therapy was slightly more effective than placebo for pain, but not for disability, in patients with chronic non-specific low back pain: a randomised placebo controlled trial with short and longer term follow-up. British J. Sports Med. **52**(9), 594–599 (2018)
9. Asl, H.J., Yoon, J.: Robust trajectory tracking control of cable-driven parallel robots. Nonlinear Dyn. **89**(4), 2769–2784 (2017)
10. Park, J., Karumanchi, S., Iagnemma, K.: Homotopy-based divide-and-conquer strategy for optimal trajectory planning via mixed-integer programming. IEEE Trans. Robot. **31**(5), 1101–1115 (2017)
11. Bulvas, M., Zuzana, S., Ivan, V., et al.: Prospective single-arm trial of endovascular mechanical debulking as initial therapy in patients with acute and subacute lower limb ischemia: one-year outcomes. J. Endovascular Therapy, **26**(3), 152–160 (2019)
12. Liu, S., Liu, D., Srivastava, G., et al.: Overview and methods of correlation filter algorithms in object tracking. Complex Intell. Syst. (2020) 10.1007/s40747-020-00161-4

13. Shuai, L., Gelan, Y.: Advanced hybrid information processing. Springer International Publishing, USA (2019)
14. Liu, S., Lu, M., Li, H., et al.: Prediction of gene expression patterns with generalized linear regression model. Front. Genet. **10**, 120 (2019)
15. Lu, M., Liu, S.: Nucleosome positioning based on generalized relative entropy. Soft. Comput. **23**(19), 9175–9188 (2018). https://doi.org/10.1007/s00500-018-3602-2

Automatic Track Control Method for Multi-UAV Based on Embedded System

Yu-han Jie[1] and Zong-ang Liu[2,3,4(✉)]

[1] Institute of Technology, East China Jiao Tong University,
Nanchang 330045, China
[2] No. 38 Research Institute of CETC, Hefei 230088, China
lning83@163.com
[3] Unit 91550, Dalian 116023, China
[4] National University of Defense Technology, Changsha 410073, China

Abstract. In the case of multiple UAVs, the navigation area of UAV is planned to effectively improve the accuracy of track control and ensure the navigation safety. Because there are some problems such as track deviation and delay of obstacle avoidance when using traditional methods to control the multi-UAV track, it is difficult to meet the requirements of track control accuracy and safety, an automatic control method of multi-UAV track based on embedded system is proposed. The mathematic model of UAV track control is designed based on the fuzzy algorithm, in order to obtain the route deviation parameters accurately, and the area range of UAV track is standardized according to the calculation results, and the control steps of UAV track are planned within the track range, so as to achieve the automatic control target of multi-UAV track. The experimental results show that the embedded multi-UAV track automatic control method can effectively solve the problem of large track deviation, and can avoid navigation obstacles and achieve the research goal of effective control of multi-UAV track. The experimental results show that the UAV under the control of this method can avoid obstacles accurately, solve practical problems effectively, it can effectively solve the problems of the traditional methods in track control and obstacle avoidance, it shows that the proposed method has practical application value.

Keywords: Embedded · Uav · Track control · Fuzzy algorithm

1 Introduction

UAV has the characteristics of light weight, low cost and strong adaptability. It has become a research hotspot in many countries in the world. In the current local battle dominated by information technology, UAV can achieve ideal results in reconnaissance, monitoring, guidance and strike. More and more countries also put the research and development of UAV into their military development strategy. With the rapid development of national economy, the flight area of UAV is becoming narrower and narrower, which makes it difficult to meet the requirements of track control accuracy and safety [1]. In order to improve the control accuracy of UAV's lateral track, the research results of excellent researchers at home and abroad in this field were investigated and analyzed [2].

S. Liu and L. Xia (Eds.): ADHIP 2020, LNICST 348, pp. 399–408, 2021.
https://doi.org/10.1007/978-3-030-67874-6_37

In reference [3], a three-dimensional path planning method based on the improved ispo algorithm is proposed, which applies the ispo algorithm to the path planning, and on this basis, the attraction effect between sub vectors is introduced, which effectively overcomes the defect that the algorithm is easy to fall into the local optimal solution. The simulation results show that the improved ispo algorithm has better precision and ability in track planning, but there is a certain track deviation. In reference [4], a new method of automatic acquisition of UAV's illegal navigation trajectory data is studied. The model of lidar signal and target scene's action process is established. The mapping part of each laser beam on the target scene surface is regarded as the laser footprint, and the convolution calculation of the response function of the corresponding laser footprint of the laser beam and the time distribution function of the laser radar's transmitted signal is completed, The echo signal of the corresponding target area of the laser footprint and the feedback of the laser radar signal are obtained. The distribution model of the single imaging echo peak point trajectory is established based on the feedback echo signal, and the navigation trajectory data of the UAV is obtained. For the actual navigation trajectory and the predefined navigation trajectory of the UAV, the optimized modified Hausdorff distance formula (MHD) is used to measure the matching degree of the trajectory data, According to the experience setting threshold, if the matching degree exceeds the threshold, it is considered that the actual trajectory does not match the preset trajectory of the UAV, and the corresponding actual trajectory is regarded as the illegal trajectory of the UAV, and the data of the illegal trajectory is captured. The results show that the proposed method can detect the illegal trajectory of the UAV in time, but the trajectory deviates from the actual trajectory.

In response to the above problems, the control method of UAV's lateral track was studied with embedded method. A mathematical model of track control based on fuzzy algorithm is proposed. The track control equation of UAV is solved by controlling the track of UAV, and the running state of UAV is described and controlled. In the case of multiple UAVs, the navigation area of UAV is planned, and different control schemes are proposed according to the requirements of different precision navigation control, so as to effectively improve the accuracy of track control, ensure navigation safety and fully meet the research objectives.

2 Automatic Track Control Method for Multi-UAV Based on Embedded System

2.1 Mathematical Model of UAV Track Deviation

The track of multiple UAVs is complex, so it is necessary to control the track coordinates with embedded method. In order to achieve the goal of precise track control, firstly, the track range of UAV is positioned by using the fuzzy algorithm and coordinate situation.

If z is the standard track, P and Q are the maximum and minimum DOF ranges of the standard track, and U represent the resistance during navigation, then the standard range of plane motion of UAV can be calculated by combining the fuzzy algorithm.

$$\begin{cases} x = U(\bar{a} - ct^2 - \dfrac{1}{2}m) \\ y = P(\bar{a} - ct^2 - m) \\ z = Q(\bar{a} - ct^2 - 2m) \end{cases} \tag{1}$$

In the formula, $\bar{a}$ represents the route control system, c represents the moment of inertia of the UAV's centroid, the origin is centroid m, and t represents the inertia of unmanned aerial vehicle during navigation. In the course of route control, it is necessary to input the change of navigation deviation E in time, and adjust the parameter coordinates in time by using the principles of adaptive proportional unit, integral unit and differential unit. Let φ_z represents the output of the controller, N represents variable parameters, and A represents non-linear functions. If the error variable range is $\{f_1, f_n\}$, the input and output variables are uniformly distributed, and the adaptive output characteristic formula is given.

$$\Delta N(\alpha) = E_1 \Delta \varphi_z(x) + E_2 \Delta \varphi_z(y) + E_3 \Delta \varphi_z(z)[A - N]\Delta(f_n - 1) \tag{2}$$

If T is the accuracy of proportional coefficient control system, α is the stability index of integral coefficient control system and τ is the dynamic characteristics of differential coefficient control system.

$$\begin{aligned} \Delta T(\delta) &= \Delta N(\alpha) - \Delta \varphi(\tau - 1) \\ \Delta T(\delta - 1) &= \Delta N(\alpha - 1) - \Delta \varphi(\tau - 2) \end{aligned} \tag{3}$$

So far, the design of the mathematical model of track control has been completed. According to the above attribute model, the method of effective multi-UAV track control can be optimized [5, 6].

2.2   Area Range Control of Multi-UAV Track

Combined with the above algorithm, the embedded route tracking control device is optimized to better stabilize the UAV heading and ensure the UAV navigation safety. Combining the embedded method, the route information is rearranged and classified, the UAV joint communication data is integrated, and the multi-UAV track is planned. The data rearrangement method is as follows: Assuming that there are n data UAV paths in the data integration system of multi-UAV network, K represents the transmission vector of a certain target state, and its target state model algorithm is as follows:

$$L = N_s C_n^i (SK_n + H) \tag{4}$$

In the formula, S represents the process vector with an average value of 0, C represents the state transition vector of UAV operation information transmission, H represents the impact of noise control in the process of path information transmission, and i represents the number of UAVs in operation [7–9].

C_{n-1}^i can be expressed by formula (5):

$$C_{n-1}^i = S_n H^i / L \tag{5}$$

According to the above algorithm, UAV path transmission data are rearranged to effectively control the hierarchical integration state of path management data. Further calculation shows that the regional information transmission nodes can be judged by obtaining the ranking results of information, so as to realize the hierarchical integration of data of Multi-warship joint communication module. The track distribution of regional information is shown in Fig. 1:

As shown in the Fig. 1, the UAV track automatic control system simplifies and decomposes the complex track range into three closed-loop cascade modules, ensures the rapid and accurate control and management of UAV course, defines three standard control parameters in the track range, constitutes a fuzzy adaptive controller to control the UAV track, and budget the UAV navigation deviation in order to control and manage the UAV course quickly and accurately. In order to ensure the accurate and effective control of multi-UAV track, UAV should be detected and corrected in time when deviation occurs.

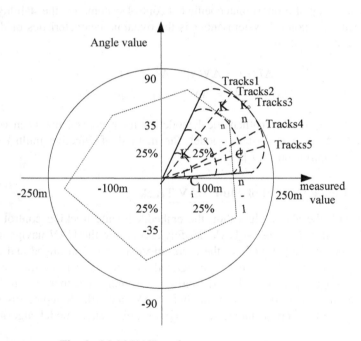

Fig. 1. Multi-UAV track area range management.

2.3   Realization of Automatic Track Control Method for Multi-UAV

In order to ensure the UAV's stable operation, the UAV's motion control system is planned [10, 11]. Accurate acquisition and accurate analysis of environmental data, combined with the multi-UAV track area environment to develop the corresponding navigation path. The route planning method of UAV is shown in Fig. 2:

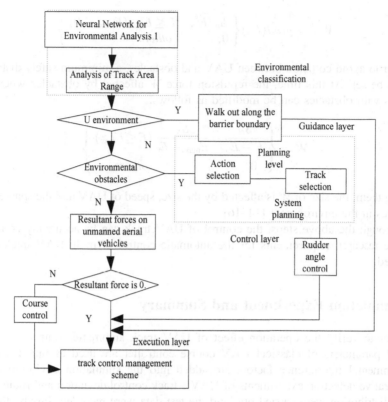

Fig. 2. UAV track control steps.

Usually, obstacles in the process of UAV moving will repel the action of UAV and form potential energy. Through investigation and analysis, it is found that the closer the UAV is to the obstacle, the bigger the potential energy is, and vice versa, the smaller the potential energy is. It can be concluded that the potential energy between the UAV and the obstacle is inversely proportional to the distance. The UAV can move to the target point quickly and accurately through the route planning step [12, 13]. It can be concluded that the potential energy between UAV and obstacle is inversely proportional to the distance. The potential energy function algorithm is as follows:

$$U_n = \left\{ \begin{array}{ll} C_{n-1}^i / S_n E_{\max}, & (S_n \neq 0, E_{\max} \neq 0) \\ 0, & others \end{array} \right\} \tag{6}$$

In the formula, n represents the distance between the point where the UAV is located and the obstacle, and S_n is the weighted coefficient. $E_{\max}$ is the maximum range of the two. When the value of E is not fixed, the repulsion force W affected by the obstacle when the UAV collides with the obstacle is:

$$W = -grad(U_n) \left\{ \begin{array}{ll} S_n/E^2, & E \leq E_{max}(E_{max} \neq 0) \\ 0, & others \end{array} \right\} \qquad (7)$$

In order to avoid collision between UAV and obstacles, a minimum safety distance of E_0 can be set. At this time, the repulsion force W affected by obstacles when UAV collides with obstacles can be modified as follows:

$$W = \left\{ \begin{array}{ll} \dfrac{S_N}{E - E_0} - \dfrac{S_N}{E_{max} - E_0}(E \leq E_{max}) \\ 0, & others \end{array} \right\} \qquad (8)$$

Among them, the size of W is affected by the size, speed of UAV and the sparseness of obstacles in the environment [14–16].

Through the above steps, the control of UAV track can be accurately completed, and the research requirements for the automatic control of multi-UAV track can be achieved.

3   Simulation Experiment and Summary

In order to verify the operation effect of UAV track automatic control system, the control parameters of classical UAV course controller are used as initial data, and environmental interference factors are added into the simulation environment. The comparative detection experiments of UAV's track control deviation and route control effect distribution were carried out, and the test data were recorded. Firstly, the comparative detection experiments are carried out under the same environment, and the track control deviation tracking data of the traditional method and the present method are recorded respectively for comparison. The specific data are shown in the Table 1 and Table 2.

Table 1. Data monitoring of trajectory control deviation tracking by traditional method.

Experiment number	Self-propelled Real-time Position			Deviation data	
	x coordinate/m	y coordinate/m	z coordinate/m	Track deviation n/m	Heading angle deviation/ψ
1	0.45	0.75	0.65	0.16	22.89°
2	0.49	1.42	0.82	0.12	18.98°
3	2.45	1.26	0.71	0.33	12.21°
4	1.71	0.94	0.64	0.21	7°
5	1.62	0.18	0.64	0.54	9.24°
6	1.96	0.87	0.51	0.41	12.95°
7	1.70	0.63	0.49	0.52	13.41°

Table 2. Embedded track control deviation tracking data monitoring.

Experiment number	Self-propelled Real-time Position			Deviation data	
	x coordinate/m	y coordinate/m	z coordinate/m	Track deviation n/m	Heading angle deviation/ψ
1	0.15	0.08	0.05	0.002	8°
2	0.29	0.07	0.02	0.07	10°
3	0.12	0.04	0.01	0.08	6°
4	0.08	0.02	0.04	0.09	0°
5	0.06	0.03	0.07	0.06	3°
6	0.07	0.06	0.09	0.05	5°
7	0.06	0.04	0.08	0.04	4°

According to the data from the Table 1 and Table 2, compared with the traditional method, the overall deviation rate loudness of the proposed control method is relatively low, and the accuracy of track control is relatively higher. This is because the method in this paper combines with fuzzy algorithm to design the mathematical model of UAV track control, using this model can accurately obtain route deviation parameters, according to the deviation parameters can effectively correct the flight path of UAV, so as to improve the accuracy of track control.

In order to further test the effect of the embedded multi-UAV track automatic control method, the experimental environment is set as the route deviation environment. At the same time, two corners are designed for the UAV, which are distributed as follows: $\varphi_z < 90°$ small corner and $\varphi_z > 90°$ big corner. In the same environment and detection time, compared with the traditional method and the method in this paper, the effect of track integration control is tested. The test results are shown in Table 3:

Table 3. Comparison of the return effect of UAV in the case of navigation deviation.

UAV Control	Proposed method	Traditional method
Data interface	RS–422	RS–422
Communication rate	9.2	9.0
Horizontal perspective	180/270	100/180
Scanning interval angle	0.9°	0.6°
Scan cycle	30 ms	40 ms
Work environment	–20 °C–70 °C	–20 °C–70 °C
Volume	145 * 170 * 140	145 * 170 * 140
Correction time	23.45 s	8.87
Overshoot	7.12 s	2.10
Speed	308 m/s	297 m/s
Angular velocity	0.2981 m/s	0.2981 m/s
Euler angles	80	60

According to the above test results, it is found that the overall control effect of this method is still higher than that of the traditional method under the same basic equipment information such as data interface and number of data. This proves that compared with the traditional method, the proposed method has better correction effect in the case of navigation deviation.

Compare the track control effects of different methods in barrier free environment, and the results are shown in Fig. 3.

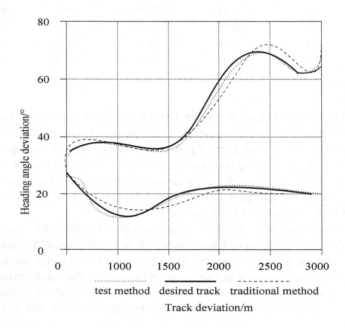

Fig. 3. Comparison of tracking accuracy in barrier free environment.

It can be seen from the analysis of Fig. 3 that in the barrier free environment, the actual navigation paths of the traditional UAV track control method and the control method in this paper are respectively compared and detected. According to the observation and detection results, there is little difference between the traditional method and the control method. It shows that in the barrier free environment, the traditional method and the method in this paper can guarantee the accuracy of the route. This is because in the process of UAV track control, the environmental factors are fully considered. On this premise, the UAV motion control system is accurately planned and the reasonable navigation path is formulated.

In the process of UAV track control, due to the complex environment, in order to ensure that the obstacles are corrected and controlled in time in the course of multi-UAV route, and to avoid deviation, the contrastive detection of the track accuracy of obstacle environment is further designed under the same link, and the test results are integrated and plotted. The specific test results are shown in the following Fig. 4.

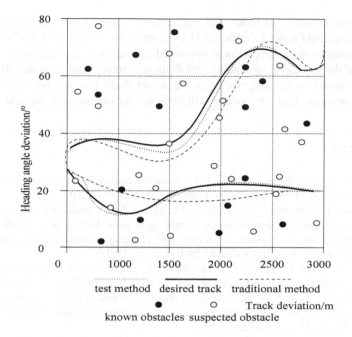

test method   desired track   traditional method
● ○   Track deviation/m
known obstacles   suspected obstacle

Fig. 4. Comparison of tracking accuracy in obstacle environment.

As shown in the Fig. 4, the traditional UAV track control method and the actual flight path of the control method in this paper are compared and detected in obstacle environment. Observed results show that, compared with the traditional method, the overall track of the control method in this paper is more consistent with the expected track. In order to avoid obstacles effectively and ensure the safety of UAV navigation, there will be small route deviation in the case of suspected track failure.

The analysis and comparison of the experimental results show that the overall control accuracy of the embedded multi-UAV trajectory automatic control method is relatively higher, which can effectively achieve the research objectives of solving the trajectory deviation and avoiding navigation obstacles. The overall control effect is much higher than the traditional method, and fully meets the research requirements.

4   Conclusion

Due to the fact that UAV is greatly influenced by the navigation environment shadow and other factors, it is easy to cause track deviation and other problems, which is not conducive to the safety and stability of UAV navigation. Therefore, an automatic control method of UAV track based on fuzzy algorithm is proposed. The track of UAV is tracked and controlled based on the indirect straight path of UAV navigation, so that the track deviation of UAV can be corrected in time. In order to verify the effectiveness of this method, a simulation experiment is designed. However, due to the limited time and the limited number of times, the experimental results show that this method has high accuracy, and the track control of UAV is much better than the traditional method.

In addition to the conclusions summarized above, there are still some areas to be improved if the method is to conduct real flight test, such as not verifying the influence of constant wind on tracking speed in the actual flight process, not considering the delay of communication, etc., which will be gradually improved in the future research.

References

1. Zhang, J., Yan, J., Zhang, P.: Multi-UAV formation control based on a novel back-stepping approach. IEEE Trans. Veh. Technol. **69**(99), 2437–2448 (2020)
2. Catapano, I., Gennarelli, G., Ludeno, G., et al.: Small Multicopter-UAV-based radar imaging: performance assessment for a single flight track. Remote Sens. **12**(5), 774 (2020)
3. Zhiyang, L., Tao, J., Yunhui, Z.: A 3D route planning method based on improved ispo algorithm. Electronics Optics Control **25**(07), 48–53 (2018)
4. Fuxian, P., Ming, L., Feng, D., et al.: Automatic acquisition method of unmanned aerial vehicle trajectory data based on Lidar. Sci. Technol. Eng. **19**(25), 225–230 (2019)
5. Hu, J., Sun, X., Liu, S., et al.: Adaptive finite-time formation tracking control for multiple nonholonomic UAV system with uncertainties and quantized input. Int. J. Adap. Control Signal Proces. **33**(1), 114–129 (2019)
6. He, Z., Gao, W., He, X., et al.: Fuzzy intelligent control method for improving flight attitude stability of plant protection quadrotor UAV. Int. J. Agric. Biol. Eng. **12**(6), 110–115 (2019)
7. Ma, J., Zhao, D., Shen, Y., et al.: Research on positioning method of low frequency oscillating source in dfig-integrated system with virtual inertia control. IEEE Trans. Sust. Energy **11**(3), 1693–1706 (2020)
8. Liu, S., Liu, D., Srivastava, G., et al.: Overview and methods of correlation filter algorithms in object tracking. Complex Intell. Syst. (2020)
9. Lu, M., Liu, S.: Nucleosome positioning based on generalized relative entropy. Soft. Comput. **23**(19), 9175–9188 (2018). https://doi.org/10.1007/s00500-018-3602-2
10. Fu, W., Liu, S., Srivastava, G.: Optimization of big data scheduling in social networks. Entropy **21**(9), 902 (2019)
11. Nguyen, H.V., Rezatofighi, H., Vo, B.N., et al.: Online UAV path planning for joint detection and tracking of multiple radio-tagged objects. IEEE Trans. Signal Process. **67**(20), 5365–5379 (2019)
12. Zhu, Z., Gong, W., Wang, L., et al.: Efficient assessment of 3D train-track-bridge interaction combining multi-time-step method and moving track technique. Engi. Struct. **183**(15), 290–302 (2019)
13. Jie, Z., Biao, W., Chaoying, T.: Flight control of quadrotor UAV based on state space model prediction algorithm. Inf. Control **47**(2), 149–155 (2018)
14. Shuai, L., Weiling, B., Nianyin, Z., et al.: A fast fractal based compression for MRI images. IEEE Access **7**, 62412–62420 (2019)
15. Liu, S., Pan, Z., Cheng, X.: A novel fast fractal image compression method based on distance clustering in high dimensional sphere surface. Fractals **25**(4), 1740004 (2017)
16. Yan, Y., Yang, J., Liu, C., et al.: On the actuator dynamics of dynamic control allocation for a small fixed-wing uav with direct lift control. IEEE Trans. Control Syst. Technol. **28**(3), 984–991 (2020)

Visual Nondestructive Rendering of 3D Animation Images Based on Large Data

Yang Zhang and Xu Zhu[✉]

Liaoning Communication University, Shenyang 110136, China
lsyyd2020@163.com, xuenne@163.com

Abstract. In the visual non-destructive rendering of three-dimensional animation images, the traditional visual non-destructive rendering method is slow, so a visual non-destructive rendering method of three-dimensional animation images based on large data is proposed. The theoretical model of pixel-by-pixel time-domain denoising process is used to denoise, and GPU is used to achieve time-domain consistent processing according to the denoising results. The non-linear Kuwahara filter is used to smooth the three-dimensional animation image, and the first-order differential operator is used to highlight the dramatically changing pixels in the image, so as to detect the edge of the image. After obtaining the distinct contour of the three-dimensional animation image, the non-destructive rendering of the three-dimensional animation image vision is realized. In order to verify the effectiveness of this method, the average rendering speed of the proposed method is 83.2%, which is significantly higher than that of the traditional method. The experimental results show that the average rendering speed of this method is the highest, the image rendering effect of this method is better, and the effectiveness of this method is verified.

Keywords: Big data · Three-dimensional animation · Image vision · Non-destructive rendering

1 Introduction

Three-dimensional animation image usually consists of point, line, surface, volume and other geometric elements and non-geometric attributes such as gray level, color, line shape, line width, etc. [1]. From the point of view of processing technology, three-dimensional animation images are mainly divided into two categories, one is composed of lines, such as engineering drawings, contour maps, curved surface wireframe drawings, and the other is similar to the shading of photographs, that is, commonly referred to as realistic images. With the emergence of raster image display, three-dimensional animation imagery has been greatly developed and widely used. With the development of three-dimensional animation imagery, three-dimensional animation imagery has been widely used. With its application, the visual non-destructive rendering of three-dimensional animation images began to develop [2].

Graphic display devices are mostly two-dimensional raster displays and dot matrix printers. From the representation of three-dimensional solid scene to the representation of two-dimensional raster and dot matrix, image rendering is called rasterization. The

S. Liu and L. Xia (Eds.): ADHIP 2020, LNICST 348, pp. 409–420, 2021.
https://doi.org/10.1007/978-3-030-67874-6_38

raster display can be regarded as a matrix of pixels. Any graph displayed on the raster display is actually a collection of pixels with one or more colors and gray levels. For a specific grating display, the number of pixels is limited, the color and gray level of the pixels are limited, and the pixels are of size, so the grating graphics are only approximate actual graphics. How to make the grating graphics approximate the actual graphics perfectly is the content of the grating graphics to be studied [3, 4].

At present, the commonly used highlight rendering methods are depth of field rendering algorithm based on hierarchical anisotropic filtering [5] and 3D animation image texture real-time rendering system [6]. For the input depth map, the pyramid of depth map is constructed, and the discontinuous region is repaired. Combined with the diffusion pattern distribution model, the fuzzy radius parameters of each point in the scene are determined. According to the depth information, the depth map is rendered layer by layer. The radius of the filter core of each layer is consistent with the radius of the diffusion circle. Finally, the separated anisotropic Gaussian filter is used to quickly get the rendering results. The latter first designs the hardware system of the rendering system, which is composed of image client, image management node, storage node and computing node. By drawing pixels of the corresponding coordinate points of the texture of the rendered image model, the purpose of rendering the texture of the animation image is achieved. Then, the bilinear algorithm is used to calculate the texture rendering of 3D animation image.

Due to the slow speed of traditional visual non-destructive rendering method for three-dimensional animation images, a new visual non-destructive rendering method for three-dimensional animation images based on large data is proposed.

2 Visual Nondestructive Rendering of 3D Animation Images Based on Large Data

2.1 Time Domain Consistency Processing

Firstly, time-domain consistency processing is used to reduce flicker in three-dimensional animation images, and the consistency of three-dimensional animation images in time-domain, i.e. between frames, is obtained. Because noise in time domain exists between different frames, it can not be eliminated by processing a single image. Considering the real-time requirement and the characteristics of graphics hardware operation, a new solution based on large data is proposed, which achieves ideal results with the help of graphics hardware, and has certain adaptive characteristics. Firstly, the theoretical model of the pixel-by-pixel time-domain denoising process is used to denoise the image. It is assumed that the color values of each unchanged position in the three-dimensional animated image obey normal distribution, $I \sim N(\mu, \sigma^2)$. Using I_n to represent the color of (x, y) position in the n frame, suppose that a position in the three-dimensional animated image does not change from the m frame to the n frame, that is, $I_m, I_{m+1}, \ldots$. If I_n is a sample of I, the theoretical model is as follows:

$$\begin{cases} \bar{\mu} = \dfrac{1}{n-m+1}\displaystyle\sum_{i=m}^{n} I_i \\[4mm] S^2 = \begin{cases} S_0^2 & m=n \\[2mm] \frac{1}{n-m}\displaystyle\sum_{i=m}^{n} I_i - \bar{\mu} & m<n \end{cases} \\[4mm] \tau = \gamma \times S^2 \end{cases} \tag{1}$$

As a basis for judging whether the pixel changes in frame $n+1$: If the deviation of frame $n+1$ at this position is less than τ, it is considered that there is no change, the output color is $\bar{\mu}$. Otherwise, it is considered that the position changes in frame $n+1$ and output I_{n+1} Among them, S_0^2 is a constant, which is the demarcation point of color change caused by noise and change, and can be obtained by learning; S^2 represents the value of noise change; γ is greater than 0, which is used to control the coefficients. It needs to be given according to the performance of the equipment and the characteristics of three-dimensional animation images. The theoretical model is used to define the position color.

$$\mu_{n+1} = I_{n+1} \tag{2}$$

In the formula, μ_{n+1} represents the final position color and denoises again based on the final position color:

$$S_{n+1}^2 = \frac{S_n^2 \times (N-1) + \mu_{n+1}^2}{N} \tag{3}$$

If there is noise in the position, as the processing proceeds, μ_n and S_n^2 will gradually calculate the variance of the color and noise of the position, which can be used as a basis for judging whether the pixel in a new frame has changes other than noise. It is easy to see that these two parameters have certain adaptability to the relatively slow motion and change of objects in three-dimensional animation images. Although the motion or change of objects in three-dimensional animated images will also affect μ_n and S_n^2, the two parameters will be automatically corrected as processing proceeds. The size of N reflects the sensitivity to change [7]. In addition, the selection of N also needs to consider the frame frequency of the three-dimensional animation image.

After noise reduction, GPU is used to achieve time-domain consistent processing. Firstly, the frame image of the three-dimensional animation image is loaded into the texture in the main program, and then processed by GPU line operation. At the same time, Render To Surface is used to save the operation results to the texture cache for the next operation.

2.2   Image Smoothing Processing

After the time-domain consistent processing, it is necessary to smooth the three-dimensional animation image. By smoothing the image, a large range of color consistency can be obtained, and a relatively uniform color in space can be obtained, so as to achieve a better smoothing coloring effect [8]. At the same time, smoothing processing has important local operation characteristics, which can be better processed by parallel processing. In order to reduce the loss of boundary information, a non-linear Kuwahara filter is used.

The principle of Kuwahara filtering is to select the average color of the most gentle position near each pixel as the color of the current pixel.

Firstly, take a template of size $J \times K$, where J and K belong to set $\{i \in Z | i \bmod 4 = 1, \ i > 4\}$, and then divide the template into four $[(J+1)/2] \times [(K+1)/2]$ size regions of size 1, 2, 3 and 4. There are overlaps of $[(J+1)/2] \times 1$ or $1 \times [(K+1)/2]$ between two adjacent regions, and one common pixel in the four regions is located at the center of the template. Moving the template over the whole image range, the average and variance of each small area are calculated for each location, and then the least variance of the four regions is determined by comparison, and the $\bar{I}$ of the region is used as the color of the central position pixels of the template.

After filtering, the noise in the smoothed area will be greatly reduced, the color will be more uniform, and the region information in the image will be well preserved. In order to get as close as possible to the effect of smooth coloring, four times of smoothing were carried out to obtain a larger range of color consistency [9]. In the implementation, a single 5×5 Kuwahara filter needs to be completed in two steps: Firstly, the mean and variance of 3×3 region around each pixel are calculated, and the results are saved to the central pixel position of a small region. After calculating the mean and variance, the mean and variance of the four vertices in 3×3 rectangular region near each pixel are checked, and the mean of the pixel with the smallest variance is obtained as the result of this filtering. There are two reasons for this: (1) Operational efficiency: if two steps are completed in the same process, each 3×3 region will be calculated four times, and only one time is needed to divide into two processes; (2) The requirement of Pixel Shader program for program size: the number of instructions in a ps_2_0 program can not exceed 256, which makes it difficult to complete the whole algorithm.

After calculating the mean and variance, the mean and variance of the four vertices in 3×3 rectangular region near each pixel are checked, and the mean of the pixel with the smallest variance is obtained as the result of this filtering. The specific image smoothing process is shown in Fig. 1.

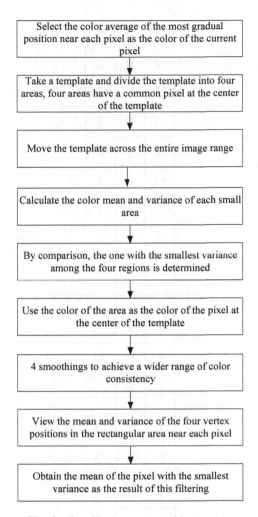

Fig. 1. Specific image smoothing process

2.3 Image Edge Detection

After the image smoothing process is completed, the image needs to be edge-detected to obtain a sharp outline of the three-dimensional animated image. The first-order differential operator is used to highlight the sharply changing pixels in the image [10]. First, calculate the I_x and I_y separately using two Sobel operators, and then take $mag(\nabla I)$ as the final result. Two Sobel operators are the horizontal Sobel operator and the vertical Sobel operator, respectively. The schematic diagram is shown in Fig. 2.

-1	0	1
-2	0	2
-1	0	1
-1	0	1
-2	0	2
-1	0	1
-1	0	1
-2	0	2
-1	0	1

(a) Horizontal Sobel operator

1	0	-1
2	0	-2
1	0	-1
1	0	-1
2	0	-2
1	0	-1
1	0	-1
1	0	-2
1	0	-1

(b) Vertical Sobel operator

Fig. 2. Schematic diagram

2.4   Image Visual Lossless Rendering Implementation

After the image edge detection is completed and the sharp outline of the three-dimensional animated image is obtained, the three-dimensional animated image is visually losslessly rendered based on the big data [11, 12]. First, real-time blanking is performed, and the three-dimensional information is transformed by a projection transformation on a two-dimensional display surface. Since the projection transformation loses depth information, it often leads to the ambiguity of the image. To eliminate ambiguity, you must eliminate occluded invisible lines or faces when drawing, customarily called eliminating hidden lines and hidden faces, or simply blanking. The projected image obtained by blanking is called the real image of the object [13]. The object of blanking is a three-dimensional object. The simplest representation of a three-dimensional object is represented by a planar polygon on the surface. The blanking result is related to the observation object and also to the viewpoint.

Then the level details are simplified. The drawing complexity of the 3D scene is very high. A complex scene may contain dozens or even millions of polygons. It is very difficult to image the complex scene. Hierarchical detail simplification is to improve the speed of image drawing by reducing the complexity of the scene.

Simplifying the level of detail requires the use of hierarchical detail display simplification technology, which reduces the geometric complexity of the scene by simplifying the surface details of the scene one by one without affecting the visual effect of the picture, thereby improving the efficiency of the rendering algorithm. This technique typically establishes geometric models of several different approximation degrees for a primitive polyhedral model. Compared with the original model, each model retains a certain level of detail. When observing objects from close proximity, a fine model can be used. When observing objects from a distance, a rougher model is used. In complex scenes, the complexity of the scene can be reduced, and the speed of image generation can be greatly improved. This is the basic principle of hierarchical detail display and simplification technology [14, 15].

However, it should be noted that when the viewpoint changes continuously, there is a significant jump between the two different levels of the model, and it is necessary to form a smooth visual transition between the adjacent levels of the model, that is, the geometric transition. The generated sequence of photorealistic images is visually smooth. The study of hierarchical detail techniques focuses on how to model the different levels of detail of the original mesh model and how to create geometric transitions between adjacent hierarchical polygon mesh models.

Hierarchical detail display and simplification is based on the geometric model of the object scene. By reducing the geometric complexity of the scene, that is, reducing the number of scene patches that the realistic graphics algorithm needs to render, the efficiency of rendering the realistic image is improved. Requirements. However, for 3D animated images, a technique capable of real-time realistic graphics rendering is required, and this technique is required to be applied to a general computer [16–18].

In recent years, technologies that meet this requirement have begun to emerge, that is, image-based rendering techniques. It starts from some pre-generated realistic images and generates realistic images at different viewpoints through certain operations such as interpolation, blending, and deformation. After generating realistic images at different viewpoints, it is necessary to render non-photorealistic images of these realistic images, including scientific data visualization processing and artistic style rendering processing [19].

Visualization of scientific data pays more attention to highlighting important information and neglecting secondary information so that the most important data can be more clearly expressed; artistic style rendering processing pays more attention to the personalization and artistic expression of 3D animated images, combined with scientific data. Visual processing and artistic style rendering processing realize visual lossless rendering of 3D animated images.

3   Experimental Research

In order to detect the visual lossless rendering method of three-dimensional animated images based on big data proposed in this paper, a comparative experiment was designed.

3.1   Experimental Parameters

The parameters of this experiment are shown in Table 1:

Table 1. Experimental parameters

Project	Data	Environment
Data sources	3D animated image library	Software environment: 3D animated image support database
Hardware accelerated frame rate	18.54 fps	
Algorithm frame rate	0.06 fps	
Experiment platform	MATLAB R2014a	
Memory	256 MB RAM	Hardware environment: NVIDIA GeForce FX 5600 graphics card, Intel(R) Celeron(R) CPU 2.40 GHz processor
Rendering scene	Sponza scenes	
Data source approach	Get the actual parameters	
Experiment process	Visually lossless rendering of 3D animated images	
Evaluation basis	Rendering speed	
Operating system	Microsoft Windows XP Professional (5.1,Build 2600)	

3.2   Experimental Process

The 3D animated image is supported by the database to collect the experimental 3D animated images, and the 3D animated images are subjected to time domain uniform processing, image smoothing processing, and image edge detection, thereby realizing image visual lossless rendering and comparing the rendering speed. In order to ensure the validity of the experiment, image depth of field rendering algorithm based on hierarchical anisotropic filtering (Method 1), real time rendering system of 3D animation image texture (Method 2) is compared with the big data-based three-dimensional animated image visual lossless rendering method proposed in this paper.

3.3   Experimental Results

The comparison results of rendering speed of different methods are shown in Fig. 3.

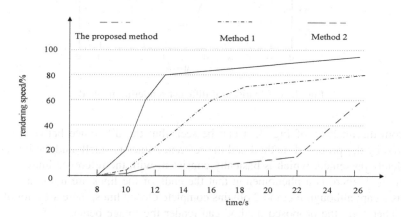

Fig. 3. Rendering speed comparison

It can be seen from Fig. 3 that the maximum speed of visually non-destructive rendering of 3D animation images using Method 1 is about 80%; the maximum speed of visually non-destructive rendering of 3D animation images using Method 2 is about 60%; and the use of 3D based on big data Animated image visual lossless rendering method The highest rendering speed for visually lossless rendering of 3D animated images has reached more than 90%. By comparing the average rendering speed, it can be seen that the average rendering speed of the 3D animation image visual lossless rendering method based on big data is higher than the existing method, which proves the superiority of the method. This is because the proposed method uses pixel by pixel time domain noise reduction to deal with the noise in the image, and uses GPU to achieve time-domain consistent processing according to the denoising results, which reduces the noise interference on image rendering and improves the rendering speed.

In order to further verify the effectiveness of the proposed method, taking the line elements in the 3D animation image as the object, different methods are used to render the lines in the 3D animation image, and the results are shown in Fig. 4.

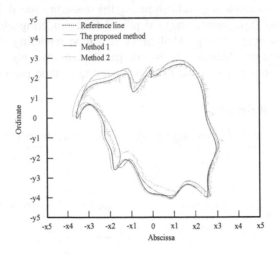

Fig. 4. Comparison of different rendering methods

From the analysis of Fig. 4, it can be seen that the difference between the lines rendered by the proposed method and the reference lines is small, and the lines can be completely presented, which is basically consistent with the reference lines. However, in method 1, there is a phenomenon that the ends of the lines are not connected and there is a gap; although method 2 obtains complete closed lines, there are a lot of rough lines. Therefore, the proposed method can render the image better.

4 Conclusion

Image rendering is the process of transforming three-dimensional light transfer processing into a two-dimensional image. Scenes and entities are expressed in three-dimensional form, which is closer to the real world and easy to manipulate and change. Before rendering the 3D animated image, you need to prepare 3D geometric model information, 3D animation to define information and material information. The 3D geometric model is obtained by 3D scanning, 3D interactive geometric modeling and 3D model library to obtain 3D animation definitions. Motion design, motion capture, motion calculation and dynamic deformation are used to obtain materials from scanned photos, computer calculated images and human paintings. Get it out of the picture. The work to be done in image rendering is to generate images by geometric transformation, projection transformation, perspective transformation and window clipping, and then through the acquired material and light and shadow information. After the image is rendered, the image information is output to an image file or a video file, or to a frame buffer of the display device to complete the graphics generation. The experimental results show that the visual lossless rendering method based on big data and 3D animated image can reduce the amount of calculation and improve the rendering speed, and the image rendering effect is good.

References

1. Song, L.: The early stage of the visual effects of movies - the concept and skills of the early stage shooting of the new visual effects of movies. Modern TV Technol. **10**(11), 150–155 (2017)
2. Morikawa, T., et al.: Image processing analysis of oral cancer, oral potentially malignant disorders, and other oral diseases using optical instruments. Int. J. Oral Maxillofac. Surg. **49** (4), 515–521 (2020)
3. Jinming, L.: Optimized simulation of feature point matching for 3D reconstruction of multi-vision animation images. Comput. Simul. **34**(39), 341–344 (2017)
4. Xixi, W.: Three-dimensional visual representation method in virtual display design of metallurgical automation. Autom. Instrum. **21**(22), 147–149 (2018)
5. Zhiheng, O.Y., et al.: A layered image depth of field rendering algorithm using anisotropic filtering. Optical Technique **044**(004), 469–475 (2018)
6. Suran, K., Junping, Y.: Design of image texture real-time rendering system for three-dimensional animation. Modern Electron. Technique **041**(005), 102–105 (2018)
7. Liang, S., Dawei, L., Liang, L., et al.: Design and implementation of disparity visualization adjustment method for three-dimensional animation production. J. Comput. Aided Des. Graph. **29**(37), 1245–1255 (2017)
8. Liu, S., Sun, G., Fu, W. (eds.): eLEOT 2020. LNICST, vol. 339. Springer, Cham (2020). https://doi.org/10.1007/978-3-030-63952-5
9. Qian, D., Jingshuang, W.: Talking about the script creation and technical application of three-dimensional animation film and television works. Tomorrow Fashion **10**(12), 361–361 (2018)
10. Lu, M., Liu, S.: Nucleosome positioning based on generalized relative entropy. Soft. Comput. **23**(19), 9175–9188 (2018). https://doi.org/10.1007/s00500-018-3602-2
11. Deming, Z.: A brief analysis of the real reproduction of three-dimensional animation in the historical documentary "palace museum". J. Res. Guide **18**(25), 282–282 (2017)
12. Dongli, X.: A brief talk on the visual language performance of graphic design elements in animation. Time Educ. **14**(15), 198–198 (2017)
13. Bing, L., Shugang, L.: Efficient polarization direction measurement by utilizing the polarization axis finder and digital image processing. Opt. Lett. **43**(12), 2969–2972 (2018)
14. Megibow, A.J., Kambadakone, A., Ananthakrishnan, L.: Dual-energy computed tomography: image acquisition, processing, and workflow. Radiologic Clin. North Am. **56**(4), 507–520 (2018)
15. Engelkes, K., Friedrich, F., Hammel, J.U., Haas, A.: A simple setup for episcopic microtomy and a digital image processing workflow to acquire high-quality volume data and 3D surface models of small vertebrates. Zoomorphology **137**(1), 213–228 (2017). https://doi.org/10.1007/s00435-017-0386-3
16. Xiaoyuan, Y., Jingkai, W., Ridong, Z.: Random walks for synthetic aperture radar image fusion in framelet domain. IEEE Trans. Image Process. **27**(2), 851 (2018)

17. Mujika, K.M., Méndez, J.A.J., de Miguel, A.F.: Advantages and disadvantages in image processing with free software in radiology. J. Med. Syst. **42**(3), 1–7 (2018). https://doi.org/10.1007/s10916-017-0888-z
18. Haiyang, L., Xiaofeng, H., Xu, L.: Remote real-time rendering system based on graphics cluster. J. Syst. Simul. **31**(005), 886–892 (2019)
19. Liu, S., Pan, Z., Cheng, X.: A novel fast fractal image compression method based on distance clustering in high dimensional sphere surface. Fractals **25**(4), 1740004 (2017)

Visual Reconstruction of Interactive Animation Interface Based on Web Technology

Xu Zhu$^{(\boxtimes)}$ and Yang Zhang

Liaoning Communication University, Shenyang 110136, China
xuenne@163.com

Abstract. The human perception that more than 80% of the external information is visually acquired, therefore, in the interactive animation interface
design, the visual effect is very important. In this background, an interactive
animation interface visual reconstruction method based on Web technology is
proposed. The method is mainly described by two aspects, firstly, the related
description is carried out on the Web technology, and then the visual reconstruction of the interactive animation interface is realized by using the technology, and the method comprises the visual feature extraction, the visual
feature matching and the visual feature 3D reconstruction. The results show that,
after the visual reconstruction, the visual effect of the interactive animation
interface is improved, and the visual existence in the design of the interactive
animation interface is solved.

Keywords: Web technology · Interactive animation interface · Visual
reconstruction

1 Introduction

In the 1970s, the American futurist Alvin Towler put forward the view that "the service
industry will eventually outpace the manufacturing industry and the experience production will surpass the service industry" in "the impact of the Future". The development of the "Experience Economy", it is the improvement of the degree of human
civilization, how to meet the growing material and cultural needs of people, and to
improve the interactive experience is an important research subject of the present
designer [1]. Interactive animation participation interface design provides users with a
self-help experience, in a more free, natural, friendly, multi-channel information
transmission, through the visitors' multi-sensory information transmission. In the present day of information explosion, the information transfer mode is rapidly inserted into
the design field with the characteristics of interactivity, nonlinearity, entertainment,
accuracy and the like, so as to construct between the person and the person, between
the person and the product, between the product and the product, The interaction
network between people and the environment increases the effectiveness of product
communication. Reflects the design of "people"-oriented concept, around the needs of
people, to meet the needs of people. Animation is realized in a variety of ways, such as
traditional hand-painting, three-dimensional production, physical animation, and so on.
In addition, some network editing software flash, dreamweaver, toombom can achieve

S. Liu and L. Xia (Eds.): ADHIP 2020, LNICST 348, pp. 421–432, 2021.
https://doi.org/10.1007/978-3-030-67874-6_39

animation effect according to the need. However, the interactive animation interface designed by these techniques is not very good in visual effect, and user experience is poor. Aiming at the above problems, a visual reconstruction method of interactive animation interface based on Web technology is designed in order to enhance the visual characteristics of interactive animation interface design [2]. Methods the application of Web technology was analyzed firstly, and then the visual reconstruction of interactive animation interface was described. Finally, the visual effect of interactive animation interface is improved after visual reconstruction using this method.

2   Visual Reconstruction of Interactive Animation Interface based on Web

Interactive animation refers to an animation that supports event response and interaction in the play of an animated work, that is, a certain control can be accepted when the animation is played. This control may be a certain operation of the animation player or may be an operation prepared in advance at the time of the animation production. This interactivity provides the audience with the means to participate in and control the content of the animation, so that the audience can change from passive acceptance to active selection. Interactive design is the design of the user and the product interaction process, the way, and so on [3]. Interactive media systems are more acceptable to users, more satisfying, easier to learn and master, and more cooperative. Not only can obviously improve the product friendliness. So that the original monotonous, rigid information transfer system interface becomes more gorgeous, has affinity. More importantly, it can increase the efficiency of information transmission, and achieve a better effect of information promotion.

WEB technology refers to the development of Internet application technology, generally including WEB server technology and WEB client technology. The animation produced by web technology has the characteristics of short and small, so it is widely used in the design of web page animation. In order to become the current web page animation design one of the most popular technology.

The visual reconstruction of interactive animation interface is aimed at transmitting visual and natural information through the picture. Therefore, some necessary design principles should be obeyed to ensure the visual display of the interface. They include clarity, simplicity, novelty, beauty, kindness, efficiency, and consistency [4].

2.1   Web Technology

WWW is the abbreviation for Word Wide Web, also known as Web. And is a network information service based on a TCP/ IP protocol. The Internet is a network connecting the global computer network, which is used to share the resources of the global computer. The Internet and the Web are two different concepts, and the Internet is the basic platform of the Web, and the Web is an application layer service on the Internet platform. It allows computer users to locate and read information resources such as text, graphics, animation, audio and video from all corners of the world. These resources can be linked through hyperlinks, logically forming a huge "information network" around the world.

So far, there are four techniques available for visual reconstruction of interactive animation interfaces: CGI (Common Gateway Interface), ASP (Active XServerPage), PHP(Personal Home Page) and JSP (Java ServerPage). But if we want to realize the dynamic web page under the existing technology, we can only adopt CGI, because at present, the web technology does not support the visual reconstruction such as ASP, PHP and so on [5].

CGI provides a channel for Web servers to execute external programs, a server-side technology that makes browsers and servers interactive. CGI programs belong to an external program and need to be compiled into executable files to run on the server side. Its application structure is shown in Fig. 1.

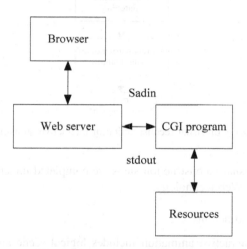

Fig. 1. CGI program running flow

The browser sends the interactive animation interface-related data to the Web server, the Web server sends the data to the CGI program using STDIN, after executing the CGI program, you may access some documents that store the data, and finally use STDOUT to output the visual refactoring result file. Web server sent back to the browser to display to the user.

CGI programs can be written in any programming language, such as shell scripting language, Perl, Fortran, Pascal, C language, etc. But UC Linux does not support the language of Perl, Fortran, and so on, and the C language is not well-related to the platform, so we choose to use the C language to write CGI programs. In addition, the web does not support the database, so the data that needs to be saved can only be saved to the file, and the CGI query data is also a query of the files, rather than accessing the database.

2.2   Implementation of visual reconstruction of interactive animation interface

In that follow, the visual reconstruction flow of the interactive animation interface is realized by the above-mentioned Web technology, as shown in Fig. 2 below.

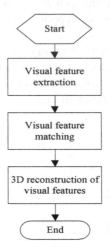

Fig. 2. Process of visual reconstruction of interactive animation interface

All the above visual reconstruction steps are completed under the guidance of the program written by Web technology.

(1)   Visual feature extraction

The scene of interactive animation includes logical scene and visual scene. The logical scene reflects the content information of animation from the logical feature, and the visual scene reflects the content information of animation from the visual feature. Therefore, visual feature extraction is the basis of visual reconstruction. At present, there are two main methods to extract feature points.

Methods based on contour lines: this method first extracts the edge of the image from the interactive animation interface, then searches for the maximum curvature point on the chain composed of edges, or approximates the edge with polygons, and then calculates the vertices of polygons as feature points [6]. Asada and Brady extract the feature points from the edge profile curve and, in order to get robust results, they also introduce a multi-scale framework. This approach is very similar to the method proposed by Mok 'harian and Machworth using the deformation point on the curve. Mediumi and Yasumot. The B-spline function is used to approximate the contour curve, and the characteristic points are the places where the curvature changes most on the B-spline number function. Horaud first extracts the part of the line from the contour line, and these lines are grouped according to certain rules. The intersection point of the line in each group is the characteristic value.

Method based on gray value. This kind of method firstly defines an operator and extracts the feature points by searching the extreme value of the operator on the grayscale changing image. In 1978, Beaude proposed a method of detection based on the rotation invariant. Dreschler and Nagel used the Gaussian curvature principle in the detection process. Kitchen and Rosenfeld proposed a method of finding a feature point along the gradient change direction of the edge of the interactive animation interface. Noek attempts to give a theoretical formula for corner detection with differential geometry, and he plans the principle of detection under the Plessey's algorithm [7]. In 1988, Harris and Stephen proposed the Harris operator.

Here we collect the Harris operator to complete the point detection in the feature. First, the average gradient square matrix of each pixel in the image is calculated.

$$S(x,y) = \begin{bmatrix} \frac{I(x,y)^2}{x} & \frac{I(x,y)}{x}\frac{I(x,y)}{y} \\ \frac{I(x,y)}{x}\frac{I(x,y)}{y} & \frac{I(x,y)^2}{y} \end{bmatrix} \tag{1}$$

Of which, $I(x,y)$ is coordinate (x,y) in the ima The grayscale of the point;

In order to quickly find the gray pixels of the image, this paper defines an index GI to judge the pixel gray value in the logarithmic range, and the expression is:

$$GI(p) = \sqrt{\frac{1}{3}\sum_{c\in\{r,g,b\}} \frac{\left(\Theta I_{\log}^c(p) - \Theta \bar{I}_{\log}(p)\right)^2}{S(x,y)}} \tag{2}$$

In the formula: $\Theta I_{\log}^c(p)$ represents the local area operator for identifying gray pixels in the logarithmic range, namely:

$$\Theta I_{\log}^c(p) = \left(\sqrt{\sum_{q\in\omega(p)} I_{\log}^c(q) - I_{\log}^{-c}(p)}\right) \tag{3}$$

Among them, ωp represents the local neighborhood with p as the center and size 3×3, $I_{\log}^c(q)$ represents the result of taking the logarithm of the $c(c\in\{r,g,b\})$ component $I^c(p)$ of p, $I_{\log}^c(q)$ is the partial mean of $I_{\log}^c(p)$ in neighborhood $\omega(p)$. Assuming $\Theta I_{\log}^r(p) = \Theta I_{\log}^g(p) = \Theta I_{\log}^b(p)$ and $GI(p) = 0$, p is a standard gray pixel.

In order to remove the pixels with extremely low light, modify GI to obtain the following formula:

$$GI^*(p) = \frac{GI(p)}{E(p) + \varepsilon} \tag{4}$$

In the formula, $E(p) = \left(I^r(p) + I^g(p) + I^b(p)\right)/3$ represents the brightness value of pixel p, and ε represents a small positive number, which can avoid the situation where the denominator is zero. Therefore, it can be concluded that the pixel $E(p)$ in the area with extremely low light is also low, and the GI^* value will rise, and it will not be considered as a gray pixel.

If the two eigenvalues of the mean gradient square matrix corresponding to a point are larger, then a small movement near the point will result in a larger gray level change, which means that the point is a corner point. Corner response functions are:

$$R = \det S(x, y) GI^*(p) - k[trace[S(x, y)]]^2 \tag{5}$$

Where, k Set to 0. 04 (recommended by Harris).

Any pixel satisfying R greater than a certain closed value T is considered to be a feature point when the feature points are extracted by the formula above. The closed-value T depends on the attributes of the actual image, such as size, texture, etc., but because T has no intuitive physical meaning, its specific value is difficult to determine, Therefore, the method of indirectly determining T is adopted in the experiment: we only need to give the maximum number of possible feature points to be extracted in the image, that is, Smax, matching program to sort the possible feature points according to R value. Then some pixels with maximum R value are selected as feature points according to Smax. When you actually match, you should constantly adjust the Smax according to the matching result [8].

(2) Visual Feature Matching

The goal of the initial match is to determine a set of candidate matching pairs. Candidate matching pairs can contain a large number of mismatches, all of which will be eliminated in subsequent robust matching processes as well as in the reconstruction process. Binocular stereo matching is based on a point matching criterion, according to some characteristics of the visual image. At present, visual feature matching mainly includes gray-scale correlation-based matching, feature-based matching and model-based matching. Among them, gray-scale correlation-based matching is a classical method of image matching. Based on the gray-scale information, the visual interface image is statistically analyzed, and gray-scale correlation and similarity are used as correlation matching decision. Search and match according to one or more similarity measures. In this chapter, the feature point matching method based on gray-scale correlation is adopted [9].

The basic idea of matching based on gray scale correlation is as follows: firstly, geometric transformation is done to the visual image of the registration animation interface, and then an objective function is defined according to the statistical characteristics of the gray level information, which is used as a similarity measure between the reference image and the transformed image. So that the registration parameters are obtained at the extreme value of the objective function, which is used as the criterion of registration and the objective function of the optimization of registration parameters, so that the registration problem can be transformed into the extreme value problem of multivariate functions. Finally, the correct geometric transformation parameters are obtained by a certain optimization method, as shown in Fig. 3 below.

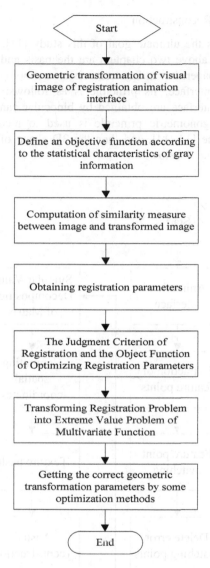

Fig. 3. Visual feature matching based on Gray-scale correlation

The performance of this kind of algorithm mainly depends on the similarity measure and the choice of search strategy. The selection of matching window size is also a problem that must be considered by this kind of method. Large windows may be mismatched, and small windows can not cover enough intensity changes. Therefore, the size of the matching region can be adjusted adaptively to achieve better matching results [10].

(3) Visual Feature 3D Reconstruction

Visual reconstruction is the ultimate goal of this study [11]. Feature extraction and stereo matching in the above two chapters are the basis and precondition of visual reconstruction. In this paper, binocular stereo vision technology is used to reconstruct interactive animation interface. Its principle is as follows: firstly, two images of interactive animation interface are obtained by binocular camera, based on parallax principle [12]. The trigonometric principle is used to recover the 3D geometric information of the scene [13, 14], and then the 3D shape of the interface is reconstructed, as shown in Fig. 4 below.

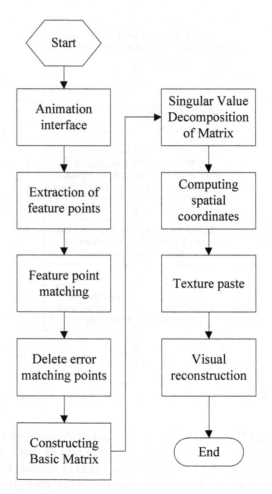

Fig. 4. Process of 3D reconstruction of visual features

3   Experimental Results and Analysis

The purpose of this study is to improve the visual quality of the interactive animation interface, which is more humanized and better reflected in the interactive performance. So in this simulation experiment, the visual effect of the interface before and after visual reconstruction is taken as the object, and the visual effect of the interface before and after visual reconstruction is taken as the object. Carry out experimental analysis.

Fig.5 below shows the interactive animation interface for visual reconstruction.

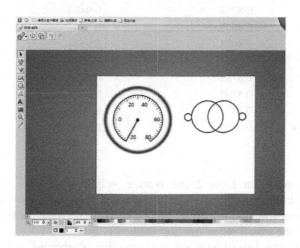

Fig. 5. The interactive animation interface to be visually reconstructed

Now, using the method mentioned in the above-mentioned article to perform the visual reconstruction, the visual effect evaluation is carried out on the two interfaces before and after the visual reconstruction by the expert evaluation method, and the result of the evaluation is shown in Table 1 below.

Table 1. Expert Assessment score (score).

Expert	Before Visual Reconstruction	After Visual Reconstruction
1	92	98
2	89	97
3	88	97
4	90	96
5	88	98
6	88	97
7	90	98
8	91	96
9	92	97
10	89	97

In order to reduce the subjectivity of the expert screen evaluation, the weight conversion calculation is carried out, and the results are shown in Table 2 below.

Table 2. Evaluation results weights

Expert	Before Visual Reconstruction	After Visual Reconstruction
1	0.8798	0.9547
2	0.8567	0.9514
3	0.8642	0.9514
4	0.8754	0.9474
5	0.8642	0.9547
6	0.8642	0.9514
7	0.8754	0.9547
8	0.8788	0.9474
9	0.8798	0.9514
10	0.8567	0.9514
Average value	0.8695	0.9516

As can be seen from Table 2, the average weight of interface visual effect after visual reconstruction is 0.9516, while the average weight of interface visual effect before visual reconstruction is 0.8695. The comparison between the two shows that the performance of the visual reconstruction method is better.

On this basis, the time of interface reconstruction is recorded and counted, and the statistical results are shown in Fig. 6.

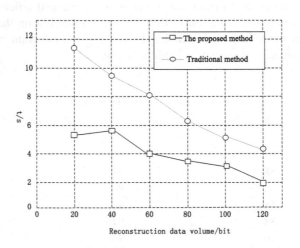

Fig. 6. Comparison of Interface Reconstruction

As shown in Fig. 6, the proposed method has short time, obvious advantages and high practical application.

4   Conclusion

To sum up, visual reconstruction has always been one of the focal points and hot spots in interactive animation interface research. In order to generate a synthetic environment for known objects and virtual objects, the research on reconstruction methods is of great academic significance and application value both in computer vision and in engineering applications. In order to solve the problem of poor visual effect of inter-active animation interface, a visual reconstruction method of interactive animation interface based on Web technology is proposed in this paper. Finally, after the visual reconstruction of interactive animation interface with web technology, the visual effect is better, which achieves the purpose of this study.

References

1. Anyifei. Application of Dual-mode Interface Based on Three Library in Interactive Web3D. Electr. Technol. Softw. Eng. **132**(10), 62–65 (2018)
2. Rui, W., Liang, H., Peng, B., et al.: Research on interactive 3D geographic scene construction method based on WebGL. Map. Spatial Geogr. Inf. **226**(2), 62–64 (2018)
3. Mengye, L., Shuai, L.: Nucleosome positioning based on generalized relative entropy. Soft. Comput. **23**, 9175–9188 (2019)
4. Li S., Nan, Y.D., et al.: Research on web dynamic page hierarchical reconstruction in complex context. Comput. Digital Eng. 45(11), 2218–2222 (2017)
5. Liu, S., Liu, D., Srivastava, G., et al.: Overview and methods of correlation filter algorithms in object tracking. Complex Intell. Syst. (2020). https://doi.org/10.1007/s40747-020-00161-4
6. Jinglei, H., Yongqiang, Z., Haimeng, Z., et al.: Three-dimensional reconstruction of highly reflective and textureless targets in polarized multispectral machine vision. J. Surv. Map. **47** (6), 130–138 (2018)
7. Pingxi, L.Y., Xinming, D., et al.: Three-dimensional reconstruction of textureless high reflective targets based on polarized binocular vision. J. Infrared Millimeter Wave **36** (4), 432–438 (2017)
8. Marlowe, Y.B., Ting, D.: Research on three-dimensional reconstruction simulation of human identity authentication visual image. Comput. Simul. **34**(9), 288–291 (2017)
9. Baiyue, W.M., Hao, Z., et al.: Cotton three-dimensional reconstruction technology based on binocular vision. Mech. Design Manuf. Eng. **46** (11), 150–154 (2017)
10. Wenwen, L., Shanxi, D.: Optimizing the foreground vision of 3-D free-form stereo display based on double-view reconstruction. Chinese J. Image Graph. **12**(6), 1119–1123 (2018)
11. Zheng, P., Shuai, L., Arun, S., Khan, M.: Visual attention feature (VAF): a novel strategy for visual tracking based on cloud platform in intelligent surveillance systems. J. Parallel Distrib. Comput. **120**, 182–194 (2018)
12. Di, W., Hua, L., Xiang, C.: A miniature binocular endoscope with local feature matching and stereo matching for 3D measurement and 3D reconstruction. Sensors **18**(7), 2243 (2018)

13. Li, C., Lu, B., Zhang, Y., et al.: 3D Reconstruction of indoor scenes via image registration. Neural Process. Lett. **48**(3), 1281–1304 (2018)
14. Fu, W., Liu, S., Srivastava, G.: Optimization of big data scheduling in social networks. Entropy **21**(9), 902 (2019)

Micro Image Surface Defect Detection Technology Based on Machine Vision Big Data Analysis

Chao Su[1(✉)], Jin-lei Hu[1], Dong Hua[2], Pei-yi Cui[2],
and Guang-yong Ji[3]

[1] Qingyuan Power Supply Bureau of Guangdong, Power Grid Co., Ltd.,
Qingyuan 511500, China
zour20@2980.com
[2] Guangdong Jun and Hua Energy Technology Co., Ltd.,
Guangzhou 510663, China
[3] Yantai Vocational College of Culture and Tourism, Yantai 264003, China

Abstract. The traditional micro image surface defect detection system had slower running speed and less detection precision, which made the detection system operate inefficient and could not meet the requirements of small image surface defect detection. To this end, the optimization design of the micro image surface defect detection system based on machine vision-based big data analysis was carried out. The system design was optimized with MATLAB 7.0 programming environment; MATLAB technology was used to process small images to visualize calculation results and programming; The filtering of the micro image was detected by the method of spatial domain filtering to complete the detection task of the surface defect of the micro image. The design method was validated and the test data showed that the micro image surface defect detection system ran faster and the detection was more precise. The detection accuracy was 92% and the detection quality was high.

Keywords: Machine vision · Micro image · Surface defect · Big data · Detection technology

1 Introduction

With the continuous expansion of machine vision products, the application in the field of machine vision has become more and more extensive, and the visual performance has been improved as never before. Machine vision is a multi-disciplinary interdisciplinary subject involving image processing, artificial intelligence, pattern recognition, etc. The industries applied include industry, agriculture, military and defense, traffic management, remote sensing image analysis [1, 2]. For example: defect detection and measurement of parts in industry, detection of agricultural product quality, military sonar imaging, identification of vehicles or license plates in traffic management. Machine vision applications are the most common in the industrial industry. Machine vision systems can replace manual inspection, identification and classification of smaller parts, control their production processes, and improve product quality and

S. Liu and L. Xia (Eds.): ADHIP 2020, LNICST 348, pp. 433–441, 2021.
https://doi.org/10.1007/978-3-030-67874-6_40

production efficiency. Therefore, it is of great significance to study the micro image surface defect detection technology under machine vision big data analysis.

2 Inspection System Overall Design

2.1 Overall Design of Micro Image Surface Defect Detection System

A complete machine vision based defect detection system captures images through an image acquisition system. The image processing module receives the image and processes the image, converts it into a digital signal, and then performs preprocessing, image enhancement, defect segmentation, feature extraction on the digital image, and finally realizes classification of product defects. The research process is shown in Fig. 1.

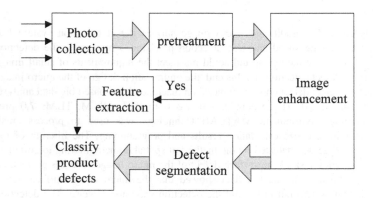

Fig. 1. Overall flow chart of micro image surface defect detection system

The overall design of the micro image surface defect detection system mainly includes two parts: hardware structure design and software algorithm design. According to the function division, the whole system can be divided into four modules: industrial camera, optical lighting system, image processing module and automatic detection result module [3, 4]. Among them, the industrial camera and the optical lighting system belong to the hardware component, and the image processing module and the automatic detection result module belong to the software component.

The object to be measured is placed on the stage, and the image of the part is acquired by the CCD camera through the uniform illumination of the LED light source, the collected image is transmitted to a computer for image processing, and the characteristic parameters are extracted to realize the identification and classificationof the defective image, and finally complete the detection task of the system [5].

2.2 System Hardware Design

Image sensors, also called photosensitive elements, are widely used in digital cameras and other electro-optical devices. Image sensors which are currently used most are CCD and CMOS.

CMOS sensors are semiconductors made from two chemical elements, silicon and germanium, which perform their functions due to positive and negative charge transistors on CMOS. In industrial cameras, CCDs are the most widely used, and CMOS is gradually being applied to HD surveillance cameras. Because CMOS has the characteristics of high integration, low power consumption, fast speed, etc. With the rapid development of CMOS technology, the effect of CMOS is getting closer to the effect of CCD, and there is a tendency to gradually replace CCD [6].

The advantage of CCD device is that it has the advantages of high integration, low power consumption, small pixel and low noise, which makes CCD widely used in three fields of camera, signal processing and storage. Compared with CMOS, first of all, CCD can convert light into electric charge and transfer charge storage, and can also take out the stored charge to make the voltage fluctuate. It is the ideal camera original. Secondly, in the case of the same pixel, the size of the photosensitive device is also the same, and the sensitivity of the CCD device is higher than that of CMOS. Thirdly, under the same conditions, the CCD sensor can use a large area, can receive and output a strong photoelectric signal, the acquired image has high definition, less noise, and the seismic effect is also very good [7].

The micro image surface defect detection system designed in this paper uses the BB-500GE industrial digital camera developed by JAJ Company of Denmark to obtain images. It features high resolution, high precision, low noise, etc. It uses Gigabit Ethernet interface processing functions, and is easy to install and use. It is the first choice for indoor and outdoor industrial inspection applications. Compared with foreign products of the same grade, this camera has obvious price advantage and is widely used.

In machine vision systems, the main goal of the illumination system is to select the light source that fits the system and project the light onto the object to be measured, mainly to highlight the contrast between the target and the background. A good lighting system can avoid problems such as flowering, glare, overexposure, etc., and can improve the resolution of the entire system. As it can improve the accuracy and stability of the system and simplify the operation of the software, so the lighting system is the key to obtaining high quality images [8].

After the light source is determined, the correct illumination mode should be selected according to the characteristics of the object to be measured. When used in a specific environment, the choice of light source needs to be considered: the characteristics of the object, the state of motion, the application environment, and the type of camera used. There are four lighting methods, the first one is forward illumination [9, 10]. Forward illumination is to place the light source on the front of the object and the camera. Using the reflection principle of light on the surface of the object to be measured, the scratches, defects and important details of the surface of the object can be well displayed, which is convenient for subsequent experiments and research, forward lighting is the preferred method of illumination for most applications.

The second is backward illumination. Backward illumination is to place the object on the middle of the camera and the light source, which can clearly outline the edge of the object to be measured, and is suitable for observing the shape of the image.

The third is structured light illumination. The structured light illumination is a distortion generated by a line source or a grating projected onto the object to be measured, and the three-dimensional information of the measured object is calculated.

The fourth is stroboscopic illumination. The illumination is to illuminate a high frequency light pulse onto the object, requiring the camera to be synchronized with the light source.

2.3   Software Design

The defect detection process based on machine vision includes preprocessing, segmentation, defect feature extraction and classification of images. As shown in Fig. 2, The collected micro images are processed by the defect detection algorithm, and finally the surface defects of the micro images are detected, thereby achieving the purpose of detecting and controlling the micro image quality. However, the implementation of the defect detection algorithm requires a powerful development environment. The micro image surface defect detection designed in this paper is based on the MATLAB7.0 programming environment.

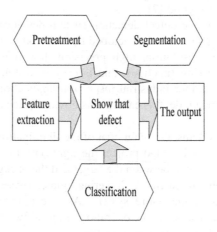

Fig. 2. Machine vision based defect inspection system diagram

MATLAB is a high-tech computing environment released by Mathworks Corporation of the United States. It is a combination of the word matrix & laboratory, meaning a matrix factory. MATLAB has perfect image processing capabilities to enable visualization of calculation results and programming; MATLAB also has a feature-rich toolbox, and can also directly call C, C++, and JAVA to meet the different needs of users. MATLAB's ability to handle matrix operations determines its unique advantages in digital image processing, and MATLAB's analysis toolbox also supports indexed images, grayscale images, image filtering and other functions, which is of great help to the development of this topic. The software part of the surface defect detection system for micro images mainly consists of image preprocessing module, defect segmentation module, defect feature extraction and selection module.

The captured image is often plagued by noise due to interference from the light source, camera and environment [11, 12]. During the detection process, noise and defects are very confusing, reducing system performance, affecting the results of the experiment, and also having a serious impact on subsequent feature extraction processes [13, 14]. Therefore, the acquired image is filtered before image processing. Commonly used filtering methods include median filtering and averaging filtering.

Defect detection technology mainly studies the detection and classification of defects on the gear surface. Image segmentation technology divides defects from images and lays the foundation for subsequent classification of defects. Common methods for image segmentation include region-based detection methods and edge-based detection methods.

Describing the characteristics of the target area in the image can be started from the following aspects: gray statistical features, geometric features, algebraic features, intuitive features, topological features and transform domain features, As shown in Fig. 3. Feature extraction is to extract the attributes and parameters that can describe the characteristics of the target from the target, identify different defects, and extract the different parameter characteristics, which feature to use depends on the problem.

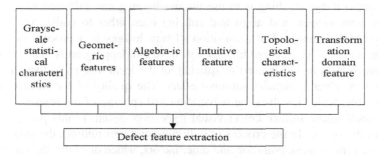

Fig. 3. Defect feature extraction map

The defect classification is mainly to realize the automatic extraction and classification of the defects extracted from the image according to its own characteristics, which is mainly realized based on the processing of image processing and feature extraction. Commonly used defect feature recognition methods include artificial neural network pattern recognition method and fuzzy pattern recognition method. Which defect identification method to choose depends on the specific characteristics of the application and the defect.

3 Key Technology for Detection

The result of scanning an image according to a rectangular scan grid is that a one-dimensional integer matrix corresponding to the image is generated. The position of each element (pixel) in the matrix is determined by the order of scanning, and the gray value of each pixel is determined by the sample, an integer representation of the gray value of each pixel is obtained after quantized. Therefore, the result of image acquisition is to digitize a continuous image of nature and finally obtain a digital image. It generally has two common representations: the first is an array representation of grayscale images, as follows:

Let continuous images $f(x, y)$ be sampled at equal intervals and arranged in $M \times N$ array (generally square matrix $N \times N$) as shown below,

$$f(x,y) = \begin{pmatrix} f(0,0) \cdots f(0,N-1) \\ f(1,0) \cdots f(1,N-1) \\ f(N-1,0) \cdots f(N-1,N-1) \end{pmatrix} \tag{1}$$

The second is binary image representation. In digital image processing, in order to reduce the amount of calculation, the grayscale image is often converted into binary image processing. The so-called binary image is only two gray levels of black and white, that is, the pixel gray level is 1 or 0.

The micro image stitching is to use the overlapping part information between the collected local images, and align and splicing each other to realize the full scene generation global image. The acquisition of tiny images is a process of collecting sequence images. The adjacent images will have overlapping parts. The overlapping partial image information is used to splicing all the local images to form a complete image, which gives a visually intuitive effect. The quality of the image surface is clearly visible, which involves the registration and splicing of the image.

The small image surface defect visual inspection system in this paper uses CCD camera as the sensor. In the process of collecting and transmitting, the small image is affected by various noise pollution and other factors, which degrades the image quality and affects the reliability of the detection result. Therefore, it is necessary to perform accurate filtering processing. After these processes, the quality of the output image is improved to a considerable extent, and the edges of the image are more intuitive, which not only improves the visual effect of the image, but also facilitates the computer to analyze, process and recognize the image.

The spatial domain filtering method is implemented in the image space by means of a template for neighborhood operation. A template is a one-dimensional array, and the value of each element in the template determines the functionality of the template. The template operation implements a neighborhood operation, that is, the result of a certain pixel point is not only related to the gray level of the pixel, but also related to the gray level of the neighboring pixel point. The description of the template operation in mathematical operations is a convolution (or cross-correlation) operation. Template operation is an operation method often used in digital image processing. Image filtering, sharpening and refinement, and edge detection are all used in template operations. The filtering is divided into two types: linear and nonlinear. The simplest linear filter is a local mean filter. The algorithm is that the gray value of each pixel $f(k,l)$ is replaced by the mean, $h(i,j)$ of the gray values of the points in the local neighborhood indicated by its template. The mean calculation formula is as follows:

$$h(i,j) = \frac{1}{M} \sum_{k,j \in N} f(k,l) \tag{2}$$

$$h = \frac{|i,j|^2}{2} \tag{3}$$

Where M is the total number of pixels contained in the neighborhood N; $f(k,l)$ is the gray value at the position (k,l) in the neighborhood N.

4 Experimental Results and Analysis

In order to ensure the effectiveness of the micro image surface defect detection system proposed in this paper, experimental demonstration is carried out. In the experiment, the image of 1–5 in the micro picture feature library was selected as the verification sample of the model. The experimental data is shown in Table 1 and Table 2.

Table 1. Error verification table and error results

	M1	M2	M3	M4	Desired output	error
1	1.03	1.10	−0.97	0.19	1	−0.62
2	1.04	1.18	0.03	0.01	1	0.06
3	1.02	1.17	−0.17	0.12	1	−0.13
4	1.06	1.42	0.27	0.03	1	−0.01
5	1.03	1.21	−0.41	0.21	1	−0.07

Table 2. Verification sample situation statistics

The total number of samples			
Number of qualified 4		Unqualified number 1	
Qualified	Unqualified	Qualified	Unqualified
3	1	0	1

It can be seen that if the relative error range of the qualified product is set to 12%, and all the others are unqualified (including the unrecognized all the records are unqualified), then, among the five tiny bearing pictures in the sample, one of the total of four qualified was judged as unqualified; All with one originally failed are judged as unqualified. The statistical table is shown in Table 2. The correct recognition rate of the model is 92%, which fully demonstrates that the micro image surface defect detection system under machine vision big data analysis has high accuracy, reliability and safety.

On this basis, the detection error is taken as the test index, and the comparison results are shown in Fig. 4.

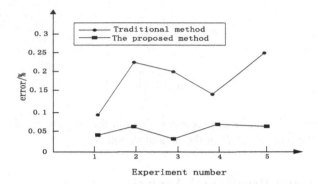

Fig. 4. Comparison of error of different methods

As shown in the figure, the detection error of the traditional method fluctuates greatly, about 0.2, while the detection error of the proposed method is about 0.05, which has obvious advantages and high practical application.

Acknowledgements. Research and development of fault warning system for transmission equipment based on multi-source image feature matching of UAV(031800KK52160007/GDK JQQ20161094).

References

1. Zhen-Xing, L.I., Zhang, Y.P., Ding, Z.L.: Work-piece surface tiny defects detection algorithm based on pixel level image segmentation and marker mapping. Modular Mach. Tool Autom. Manufact. Tech. **19**(1), 102–112 (2018)
2. Nakajima, R., Asano, Y., Hida, T., et al.: The relationship between dirt levels of inspection surface and defect detection in visual inspection utilizing peripheral vision. Ind. Eng. Manage. Syst. **17**(1), 102–112 (2018)
3. Liu, Y., Liheng, B.I.: Surface defect detection of new energy automobile solar panels based on improved particle swarm optimization algorithm. Res. Explor. Lab. **17**(1), 102–112 (2018)
4. Ge, J., Li, W., Chen, G., et al.: Investigation of optimal time-domain feature for non-surface defect detection through a pulsed alternating current field measurement technique. Meas. Sci. Technol. **29**(1), 134–142 (2018)
5. Fu, X.B.: Simulation and experimental verification of ultrasonic phased array imaging technology for electrofusion joint defect detection in polyethylene pipes. J. Mech. Electr. Eng. **10**(21), 104–108 (2018)
6. Lou, W., Shen, C., Zhu, Z., et al.: Internal defect detection in ferromagnetic material equipment based on low frequency electromagnetic technique in 20# steel plate. IEEE Sens. J. **99**, 1 (2018)

7. Zhang, K., Wang, M., Zhang, J.: Surface defect detection system for optical fiber devices for image transmission based on machine vision. J. Test Meas. Technol. **21**, 132–137 (2017)
8. Zhen-Xing, L.I., Zhang, Y.P., Ding, Z.L.: Work-piece surface tiny defects detection algorithm based on pixel level image segmentation and marker mapping. Modular Mach. Tool Autom. Manufact. Tech. **16**, 243–248 (2018)
9. Liu, K., Luo, N., Li, A., et al.: A new self-reference image decomposition algorithm for strip steel surface defect detection. IEEE Trans. Instrum. Meas. **69**(7), 4732–4741 (2020)
10. Fu, W., Liu, S., Srivastava, G.: Optimization of big data scheduling in social networks. Entropy **21**(9), 902 (2019)
11. Laraqui, A., Saaidi, A., Satori, K.: MSIP: Multi-scale image pre-processing method applied in image mosaic. Multi. Tools Appl. **77**(6), 7517–7537 (2018)
12. Liu, S., Bai, W., Zeng, N., Wang, S.: A Fast fractal based compression for MRI images. IEEE Access **7**, 62412–62420 (2019)
13. Harkaitz, E., Oskar, C., Iciar, M.: The shannon entropy trend of a fish system estimated by a machine vision approach seems to reflect the molar Se: Hg ratio of its feed. Entropy **20**(2), 90 (2018)
14. Liu, S., Li, Z., Zhang, Y., et al.: Introduction of key problems in long-distance learning and training. Mob. Netw. Appl. **24**(1), 1–4 (2019)

Strength Detection Method for Subway Vehicle Bogie Frame in Big Data Environment

Wang Shi[1(✉)], Hu Hai-tao[1], Zhou Ye-ming[1], Wang Yu-guang[1], Zhao Wei[1], and Jin Feng[2]

[1] Crrc Qingdao Sifang Co., Ltd., Qingdao 266000, China
xuenxuen20@163.com
[2] Information and Communction College National University of Defense Technology, Xi'an 710106, China

Abstract. Aiming at the problem that the accuracy of the frame strength detection method is not high, the strength detection method of the subway vehicle bogie frame is studied in the big data environment. Firstly, the new structure of the subway vehicle steering frame is taken as the research object. The CATIA software is used to carry out the solid modeling of the subway steering frame to construct the frame 3D model to obtain the frame strength detection parameters. Then, the frame test rig is built, and the strength test of the frame test rig is carried out by the finite element model of the frame strength detection to realize the strength detection of the subway vehicle bogie frame. The simulation experiment is carried out to verify the detection accuracy of the strength detection method of the metro vehicle bogie frame. The experimental comparison shows that the strength detection method of the metro vehicle bogie frame is higher than the traditional frame strength detection method.

Keywords: Big data environment · Subway vehicles · Bogies · Frame strength detection · CATIA software

1 Introduction

Since the structural reliability of rail vehicles is an important factor to ensure the safe operation of rail vehicles, and the running of rail vehicles has a tendency to develop at a high speed, its safety has received more and more attention [1]. However, in the current design, traditional quantitative design methods are still used, which often cannot meet the requirements of actual strength testing of vehicle parts. As an important part of the vehicle system, the reliability of the subway steering frame is extremely important for ensuring the safe operation of the subway vehicle, which is directly related to passenger safety and comfort [2]. During the operation of the subway vehicle, the bogie is subjected to various loads while ensuring that the vehicle has good dynamic performance. The bogie frame is an important component of the bogie. It integrates the wheel axle box device, the spring damper device, the traction drive device, the basic brake device, the vehicle body suspension device, etc., to bear the load and transmit the load. Whether the structure of the structure is reasonable or not directly affects the running quality, power performance and driving safety of the vehicle. The strength and rigidity

© ICST Institute for Computer Sciences, Social Informatics and Telecommunications Engineering 2021
Published by Springer Nature Switzerland AG 2021. All Rights Reserved
S. Liu and L. Xia (Eds.): ADHIP 2020, LNICST 348, pp. 442–451, 2021.
https://doi.org/10.1007/978-3-030-67874-6_41

of the frame will affect the service life of the metro vehicle bogie, the smooth running of the vehicle and the comfort of the passengers. It is also related to the economics of the subway operation.

At present, researchers have done a lot of research on the structural strength detection methods of metro vehicles. Reference [3] combined with the characteristics of nondestructive testing in strength test, the nondestructive testing scheme and damage assessment method in the strength test of CFRP structure are proposed by using the method of typical signal analysis. Finally, the effectiveness of the detection scheme and evaluation method is verified by experiments, but the accuracy of the test results is not high. Reference [4] analyzes the evaluation method for welding fatigue strength of bogie frame of locomotive and rolling stock. This method calculates and analyzes the welding fatigue strength of a B0 bogie frame under simulated operation conditions. By extracting the node stress at 2 mm from the outside of weld toe, the main weld of bogie frame and suspension seat of driving device, primary vertical shock absorber seat and brake seat are respectively analyzed, the fatigue strength of welding seam of traction seat is checked. The experimental results show that this method can obtain the strength coefficient of each component structure, but the detection results are accurate.

Therefore, it is particularly important to detect the strength of the subway frame bogie frame. The development of modern finite element technology and computer technology in the big data environment has laid a good foundation for the promotion of structural strength detection in the field of vehicle engineering [5].

2 Strength Detection Method for Subway Vehicle Bogie Frame

2.1 Structural Strength Detection Parameter Selection

There are two main structures of metro vehicle steering frame in China. Among them, Qingdao No. 11 and Chengdu No. 10 are old structures. The new structure is mainly changed in the upper structure of the steering frame mounting arm and the circular arc structure of the lower cover [6]. The new structure tangentially cuts the side arc of the positioning arm seat with the arc of the lower cover. This change reduces the stress on the most dangerous part of the frame and increases the safety of the frame [7, 8]. This paper is to study the new structure of the subway vehicle steering frame.

The CATIA software is used to carry out the solid modeling of the subway steering frame, starting with the 2D drawing design. Then use the methods of stretching, rotating, digging, slotting, scanning, etc. to create physical signs, and use the editing commands such as mirroring, copying, and array to edit the physical features. A 3D solid model is a combination of a number of feature commands. The 3D solid model is shown in Fig. 1:

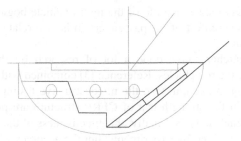

Fig. 1. Architecture 3D solid model.

It can be seen from the three-dimensional solid model that the main components of the frame are side beams, beams, longitudinal beams, pallets, spring bearings, etc. The main connection form is welded structure [9, 10]. The side sill is a variable-section hollow box-type welded structure. The middle and lower concave fish-belt design is adopted. The upper and lower covers are all steel plates with a thickness of 12 mm, and the lower surface of the side sill is also welded with a pallet. The beam is a seamless steel pipe structure, the surface is pickled and phosphated, the inner cavity is used as an additional air chamber of the air spring, and a through connection is adopted between the beam and the side beam. A sleeve is provided at the joint for reinforcement, and in order to increase the torsional strength of the entire frame, two longitudinal members are welded to the end of the beam adjacent to the side members. The thickness of other welded steel plates is also basically between 10 mm and 20 mm.

2.2   Build a Frame Test Bench

The frame test bench mainly includes a test bench base, a gantry, a vertical actuator, a longitudinal actuator, a lateral actuator, and a support restraint unit. First, the base of the entire test rig is built. The pedestal mainly provides a platform for the loading and restraining structures in the virtual test. The size depends on the size of the frame, and the size is roughly determined to be 3550 × 2930 × 50 mm. The size of the base thus established can meet the requirements of the virtual test of the subway bogie [11, 12]. The gantry is a 200 mm "U" shaped structural member. The "U"-shaped structure has a width of 2700 mm, which satisfies the width requirement of the bogie wheelbase and is 1725 mm high. The function of the gantry is mainly to withstand the reaction of the load loaded on the bogie and to establish a vertical actuator thereon to power the vertical load on the bogie frame. The vertical actuator is fixed on the gantry and is simplified into two cylinders with a diameter of 180 mm and a height of 200 mm; The support restraint unit is a relatively important structure of the virtual test rig. There are eight constrained support units, which can be simplified into a cylinder having a diameter of 230 mm, the height being determined according to the structure of the bogie; The position of the transverse actuator is on the lateral baffle of the bogie, because the constrained position of the bogie is still a certain height from the base. Therefore, an actuator receiving member is to be established when the lateral actuator is established to satisfy the load acting position and the actuator position; The establishment of the longitudinal actuator is basically the same as that of the transverse actuator, except that there are four sets of brake hanger vertical plates. The framework simulation test bench based on the above data and requirements is shown in Fig. 2.

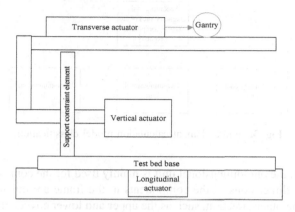

Fig. 2. Frame simulation test bench.

2.3   Finite Element Frame Strength Detection

A finite element model is a discrete model of a structure or component that contains the physical properties of the structure, the displacement and the boundary conditions of the force. Therefore, the establishment process of the framework finite element model includes selecting the unit type that expresses the structural characteristics, dividing the reasonable grid distribution, defining the material and attribute information, and defining constraints and loads on the nodes or geometry to be generated [13]. The basic principles that must be followed in the establishment of a finite element model in a big data environment are: On the one hand, the accuracy of the calculation results is guaranteed; on the other hand, the scale of the model is moderately controlled, thereby improving the calculation efficiency.

2.3.1   Mesh Generation

The finite element pre-processing software HyperMesh is used to process the three-dimensional solid model of the framework to establish a finite element model of the bogie frame. The finite element pre-processing software HyperMesh's meshing function is recognized by the industry and has interfaces with mainstream CAD software such as CATIA, PRO/E, UG, IGES, STEP, etc. The rationality of the finite element calculation model is largely determined by the form of the grid. According to the characteristics of the actual component geometry, it can be divided into categories, as shown in Fig. 3.

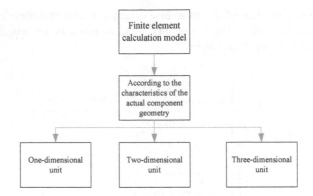

Fig. 3. Finite element calculation model classification.

In this paper, the one-dimensional unit is mainly used for the connection of components; when the dimensions of the components in the frame are much larger than the dimensions in the other direction, such as the upper and lower covers of the side beams, the shell-shell unit is used for simulation; When the dimensions of the structural members in the three directions are not much different, basically another order of magnitude, such as a spring bearing, this article uses a three-dimensional unit to simulate.

Before dividing the shell-and-shell unit, the inner surface of each member is extracted. The thickness of the member is represented by a numerical value instead of geometric representation. After the middle surface is extracted, the mesh of the shell-and-shell unit is divided on the middle surface [14]. When dividing the grid, pay attention to the size of the control grid and the density of the grid. The size of the grid and the density of the grid directly determine the number of elements of the finite element model, which in turn affects the calculation time and calculation accuracy. For components that require a large gradient of strength analysis data, a dense mesh simulation is used, and other components can be moderately sparsely meshed. With the development of computer technology, the computational storage capacity is also becoming less and less restrictive for the number of grids and calculation time in finite element analysis [15]. Based on the structural characteristics of the integrated structure and the configuration of the computer, the mesh size of the component that can withstand large loads on the frame is 10 mm, and the other components are divided by 20 mm mesh size. In the process of dividing the two-dimensional unit, the plate-shell unit is mainly composed of a quadrilateral plate unit, and the triangular plate unit is supplemented by a combination of automatic mesh division and manual mesh division. For the supporting members of the frame, such as the spring seat and the spring seat plate, the dimensions in the three directions are basically the same order of magnitude, and the difference is not large, so the hexahedral element is used for simulation.

2.3.2 Grid Quality Check

In the process of meshing, mesh deformation is easy to occur. If the deformation exceeds a certain limit, the calculation accuracy will decrease significantly with the degree of deformation, and the cell with poor mesh quality will not pass the mesh quality check. The mesh quality of the unit is directly related to the accuracy and convergence of the finite element model calculation analysis results. The general mesh quality check is accompanied by meshing simultaneously. Grid quality inspection includes: unit continuity check, unit normal direction check, repeat unit check, and unit quality check. Software HyperMesh provides tools for checking grid quality based on user-defined values—Check Elems and Quality Index panels. The Check Elems panel can be used to check the free end of a one-dimensional unit, the minimum and maximum internal angles of a two-dimensional unit, and the Jacobian value, the warpage and distortion of a three-dimensional unit. The Quality Index panel provides a comprehensive quality assessment of grid quality, and quality indicators can be defined by the user.

3   Implementation of Frame Strength Detection

3.1   Strength Check

Before the calculation, the structural finite element of the frame test rig shall be checked for stiffness and strength by the finite element model of the frame strength test. Only when the stress is less than the allowable stress of the material and tends to be 0 deformation, the whole frame test bench can be avoided to have a great influence on the strength test result of the bogie frame. Therefore, the structural finite element analysis of the frame test rig is first performed before the strength measurement of the bogie frame is performed.

In this paper, the turning brake condition is selected. Under this condition, the bogie frame receives vertical load, lateral load, additional vertical load, longitudinal inertial force, and additional vertical load. Each load is distributed to the components of the gantry, wherein the vertical brakes are subjected to loads of 100250 N and 129659 N, respectively; the transverse actuators are subjected to loads of 31856 N; and the longitudinal actuators are subjected to loads of 20916 N; The eight support units are required to bear the quality of the bogie and the dynamic load of the car during operation. The load sizes are 25062.5 N and 32414.75 N respectively. The main components of the test bench stiffness test bench are made of Q235 series steel, which has lower cost and reliability. The ribs are welded between the column and the base to enhance the structural strength. The allowable stress of Q235 series steel is 245 MPa, and the bench design meets the requirements as long as the calculation result is less than its allowable stress. Through the finite element model of frame strength detection, the maximum stress is 122 MPa, which is less than the allowable stress of the material. The structural strength meets the design requirements, while the maximum deformation of the gantry is 0.855 mm, and the deformation is small, meeting the design requirements.

3.2   Intensity Detection

The stiffness determines the deformation of the component, and the strength has a great relationship with the stress of the component. Therefore, it is especially important to detect the strength and stiffness of the frame in the strength detection of the bogie frame. C-B modal analysis and stiffness and strength valency check of the frame test bench are carried out by the finite element model of the frame strength test, and the C-B modal analysis can greatly reduce the calculation amount. On the other hand, C-B modal analysis also realizes the rigid-flexible coupling and flexible body constraints of the bogie frame and the virtual test rig. Finally, the stiffness and strength of the frame are detected by the modal analysis results.

From the basic knowledge of multi-body dynamics, the expression of kinetic energy T can be written in the following form:

$$T = \frac{1}{2} \xi^T M(\xi) \xi \tag{1}$$

Where, ξ is the rigid-flexible coupling degree and M is the mass matrix, which consists of a 3×3 matrix, the matrix form is as follows:

$$M(\xi) = \begin{bmatrix} M_{tt} & M_{tr} & M_{tm} \\ M_{tr} & M_{rr} & M_{rm} \\ M_{tm} & M_{rm} & M_{rr} \end{bmatrix} \tag{2}$$

Where, t is the displacement, r is the rotation, and m is the modality of freedom. The rigid-flexible inertial coupling is determined by the above mass matrix, and can be divided into four rigid-flexible coupling modes according to the expression of the mass matrix: rigid coupling, static coupling, partial dynamic coupling, full dynamic coupling. The commonly used rigid-flexible coupling simulation analysis is basically a method of applying partial dynamic coupling. Deriving the equation of motion of a flexible body from the Lagrangian equation yields:

$$\xi = [X \Omega q] \tag{3}$$

Where, ξ is the rigid-flexible coupling, X is the displacement coordinates; Ω is the euler angle coordinates; q is the modal coordinates. From Eq. 3, the actual constraints of the flexible body can be divided into two categories: primary constraints and secondary constraints. Through the detection of the finite element model of the frame strength detection, it is known that the frame test rig realizes the constraint of rigid-flexible coupling and flexible body.

Before the C-B modal analysis, the boundary conditions of the bogie frame are constrained. The correctness of the boundary conditions is significant for the finite element analysis. To establish the boundary conditions, the actual working conditions are first quantified, and the quantified working conditions are defined as the boundary conditions in the model. Therefore, the constraints should be simulated realistically according to the actual stress conditions of the framework.

This type of subway bogie frame is supported by 8 spring supports at the bottom of the side members, so these supports are constrained. The side beam is simulated by the shell element, and any unit node has 6 degrees of freedom, which are translational freedom in three directions along the x, y, and z axes and freedom of rotation in three directions around the x, y, and z axes. The spring support and spring support plate are simulated by hexahedral elements, which have 3 degrees of freedom, which are the translational degrees of freedom along the x, y, and z axes. Therefore, only the translational freedom of the three supports along the x, y, and z axes can be constrained. The intensity calculation results are extracted and edited by C-B modal analysis, and the corresponding loads are applied according to different working conditions and calculated. The calculation results show that the maximum stress of the frame is 77.1 MPa, 99.9 MPa, 110 MPa, 109 MPa and 109 MPa under full load static, full load, cornering, braking and turning braking conditions. The frame material is 16Mn R and its allowable stress is 345 MPa. It can be concluded that the maximum stress value does not exceed the elastic limit of the material, indicating that the strength of the frame meets the design requirements; Under the five working conditions, the maximum deformation of the frame is only 0.137 mm, which indicates that the structural strength of the steering frame of this type of subway vehicle meets the design requirements.

4 Simulation Test

In order to ensure the effectiveness of the strength detection method of the steering frame of the subway vehicle, a simulation experiment was designed. During the experiment, a metro vehicle steering frame was taken as the experimental object, and the frame test bench was built and the strength detection of the subway vehicle bogie frame was carried out by the frame strength detection finite element model. In order to ensure the validity of the experiment, the traditional frame strength detection method is used to compare the accuracy of the frame strength detection with the strength detection method of the subway vehicle steering frame designed in this paper, and the test results are observed. The results of the strength detection of the subway vehicle bogie frame using the conventional frame strength detecting method are shown in Table 1. The results of the strength detection of the subway vehicle bogie frame using the strength detection method of the subway vehicle bogie frame designed in this paper are shown in Table 2.

Table 1. Strength measurement of traditional frame strength test method.

Location	Unit node ID	Stress value (MPa)
Frame right side beam assembly upper cover and guide post joint	12095	109
Frame left joint assembly upper cover and guide post joint	18163	104
Frame left joint assembly upper cover and guide post joint	18164	101
Frame right side beam assembly upper cover and guide post joint	12097	99.2
Frame left joint assembly upper cover and guide post joint	18162	97.1
The left side beam upper cover is connected to the outer edge of the air spring seat plate	12056	93.1
The right side beam upper cover is connected to the outer edge of the air spring seat plate	49415	92.3
The left side beam upper cover is connected to the outer edge of the air spring seat plate	12053	91.8
Accuracy		83.6%

Table 2. Strength detection method for strength measurement of subway vehicle bogie frame.

Location	Unit node ID	Stress value (MPa)
Frame right side beam assembly upper cover and guide post joint	18160	109
Frame left joint assembly upper cover and guide post joint	18162	108
Frame left joint assembly upper cover and guide post joint	12092	105
Frame right side beam assembly upper cover and guide post joint	12095	93.6
Frame left joint assembly upper cover and guide post joint	49410	93.5
The left side beam upper cover is connected to the outer edge of the air spring seat plate	12053	87.9
The right side beam upper cover is connected to the outer edge of the air spring seat plate	49415	87.5
The left side beam upper cover is connected to the outer edge of the air spring seat plate	39089	87.0
Accuracy		98.2%

It can be seen from Table 1 and Table 2 that the strength detection method of the bogie frame of metro vehicles is more accurate than that of the traditional frame strength detection method, which shows that the method in this paper is more reliable. This is because this method is based on CATIA software to establish a three-dimensional model of the frame, through which the strength detection parameters of the frame can be obtained. According to the obtained parameters, combined with the finite element model, the strength test of the frame test-bed is carried out to realize the strength detection of the metro vehicle bogie frame, so as to improve the accuracy of the detection results and the detection performance of the detection method.

5   Conclusion

A large number of studies have shown that the damage form of the vehicle bogie is mostly the fatigue failure caused by random load. The structural strength testing of bogies has become one of the necessary processes in the early development of new products for bogies. The effective frame strength detection method to detect the strength of the bogie frame is an important means of structural design of the bogie frame. The strength of the bogie frame is the core indicator of the frame design. Therefore, in order to ensure the safety of subway vehicles, it is very important to detect the steering frame of subway vehicles. In this paper, the strength detection method of metro vehicle bogie frame is proposed in the big data environment. Taking the new structure of metro vehicle bogie as the research object, the three-dimensional frame model is established based on CATIA software, and the frame strength detection parameters are obtained. The finite element model is used to test the strength of the frame test-bed to realize the strength detection of the metro vehicle bogie frame. The experimental results show that the detection accuracy of this method is high, which shows that the method has strong practical application.

References

1. Tao, L., Ding, P., Shi, C., et al.: Shaking table test on seismic response characteristics of prefabricated subway station structure. Tunn. Undergr. Space Technol. **91**(9), 102994 (2019)
2. Keyu, Z., Zikai, H.: Failure analysis and strength testing of bone screws. J. Med. Biomech. **33**(03), 280–284 (2018)
3. Dan, W., Ning, N., Pengfei, Y., et al.: Non-destructive testing technology for carbon fiber reinforced polymer composite in the aircraft structure strength test. Sci. Technol. Eng. **18** (18), 313–322 (2018)
4. Cao, J.W., Xu, C.B.: Application of engineering method for welding fatigue strength of locomotive bogie frame. Chin. J. Constr. Mach. **016**(001), 81–86,94 (2018)
5. Yao, S., Xiao, X., Xu, P., et al.: The impact performance of honeycomb-filled structures under eccentric loading for subway vehicles. Thin-Walled Struct. **123**(2), 360–370 (2018)
6. Guoxin, Z., Yuhuan, L., Su, H., et al.: Development of dynamic vehicle exhaust emission inventory based on traffic big data: a case study of Guangzhou inner ring. Environ. Pollut. Control **40**(06), 723–727 (2018)
7. Shengwa, L., Junming, S., Xiang, G., et al.: Analysis and establishment of drilling speed prediction model for drilling machinery based on artificial neural networks. Comput. Sci. **46** (1), 605–608 (2019)
8. Dan, S., Shengyan, L., Yanting, A., et al.: Dynamic characteristics of journal bearing and stability of rotor system based on work of oil film force. J. Drainage Irrig. Mach. Eng. **37**(8), 699–704 (2019)
9. Shuai, L., Weiling, B., Nianyin, Z., et al.: A fast fractal based compression for MRI images. IEEE Access **7**, 62412–62420 (2019)
10. Junping, H., Yukai, Q., Kejun, L., et al.: Stability analysis of the mast of continuous flight auger driller with double braces under direct-drill-lifting condition. Chin. J. Eng. Des. **25** (02), 151–158 (2018)
11. Connick, M.J., Beckman, E., Vanlandewijck, Y., et al.: Cluster analysis of novel isometric strength measures produces a valid and evidence-based classification structure for wheelchair track racing. Br. J. Sports Med. **52**(17), 1123 (2018)
12. Liu, S., Liu, G., Zhou, H.: A robust parallel object tracking method for illumination variations. Mob. Netw. Appl. **24**(1), 5–17 (2018). https://doi.org/10.1007/s11036-018-1134-8
13. Lu, M., Liu, S.: Nucleosome positioning based on generalized relative entropy. Soft. Comput. **23**(19), 9175–9188 (2018). https://doi.org/10.1007/s00500-018-3602-2
14. Xuyun, F.U., Luo, H., Zhong, S., et al.: Aircraft engine fault detection based on grouped convolutional denoising autoencoders. Chin. J. Aeronaut. **32**(02), 86–97 (2019)
15. Dongliang, W., Xiaolei, F., Yiming, C., et al.: Research on evaluation method of plastic tensile yield strength testing laboratory. New Build. Mater. **46**(04), 128–131 (2019)

Online Monitoring Method of Big Data Load Anomaly Based on Deep Learning

Cao-Fang Long and Heng Xiao[(✉)]

Sanya University, Sanya 572022, China
hh356398632@163.com, xiaoheng564@163.com

Abstract. In the process of monitoring the abnormal load of big data in network behavior, more network traffic resources are consumed, which leads to the low efficiency of its operation. Therefore, an on-line monitoring method for the abnormal load of big data in network behavior based on deep learning is proposed. The online monitoring model of load anomaly is established, the network data distribution is analyzed, and the adaptive random link configuration is adopted to improve the channel balance and the positioning ability of the abnormal load. The load anomaly is identified through the load pattern and the online monitoring is completed. The experimental results show that the proposed method consumes about 50% of the traffic of the traditional method, which can effectively reduce the traffic consumption and improve the utilization rate of network resources. This method is more suitable for online monitoring of big data load anomalies in network behavior.

Keywords: Deep learning · Network behavior big data · Online monitoring

1 Introduction

Deep learning is a new field in machine learning. Its motivation is to build and simulate the neural network of human brain for analysis and learning. It imitates the mechanism of human brain to interpret data, such as images, sounds and texts. The concept of deep learning comes from the research of artificial neural network. Multilayer perceptron with multiple hidden layers is a kind of deep learning structure. In-depth learning, by combining low-level features to form a more abstract high-level representation of attribute categories or features, to discover the distributed feature representation of data [1]. Machine learning is a subject that studies how computer simulate or realize human learning behavior to acquire new knowledge or skills and reorganize the existing knowledge structure to improve its own performance. In 1959, Samuel of the United States designed a chess playing program that has the ability to learn and can improve his own chess skills in continuous playing. Four years later, the program beat Samuel. In 1966, the program defeated an unbeaten American player for eight years. This program shows people the ability of machine learning, and puts forward many thoughtful social and philosophical problems. The research of machine learning is based on the understanding of human learning mechanism in physiology, cognitive science, etc., to establish the computational model or cognitive model of human learning process, to develop various learning theories and learning methods, to study

S. Liu and L. Xia (Eds.): ADHIP 2020, LNICST 348, pp. 452–462, 2021.
https://doi.org/10.1007/978-3-030-67874-6_42

the general learning algorithm and carry out theoretical analysis, to build a task-oriented learning system with specific application [2]. All kinds of human sensory organs are receiving a large number of data at any time. Some of these data come from the human body itself, some come from the external environment, but the brain can always obtain or extract the most important information that deserves attention. How to represent and analyze information efficiently and accurately is the core goal of machine learning. Through anatomical knowledge, experts in neuroscience have found the way in which human brain expresses information: unlike previous inferences, the cerebral cortex does not directly extract the eigenvalues of data, but uses a hierarchical network model constructed by the brain to analyze and filter the stimulus signals received by neurons, and then can obtain the characteristics and rules of the perceived data [3]. The above conclusions are analyzed by measuring the transmission time of sensory signals in retina, prefrontal cortex and motor nerve. In short, for the visual system, the human brain does not directly process the "first-hand data" obtained by the eyes, but recognizes objects according to the results of aggregation and decomposition. It can be seen that a clear hierarchy greatly reduces the amount of data that human visual system needs to process, and can retain the required information to the maximum extent. Deep learning is to simulate the human visual system and extract the essential characteristics of a large number of data with potentially complex structure rules.

With the rapid development of the Internet, the load of the access network increases gradually, and the data information generated by these loads becomes the data source of various analysis in the network. Although these data have become an important value data source for decision-making analysis. Theoretically, the more load is, the more valuable it is to obtain data samples. However, a large number of load access to the network, bringing new challenges to the stability and functionality of the entire network system [4]. Under the big data platform, the online load of the access network is limited by the total load, data bandwidth, computing capacity, response time, data carrying capacity of the accessible network of the platform. If the online load needs to exchange data with the platform, it must establish an effective connection with the platform. In order to prevent the waste of resources caused by the load occupying the data channel without effective communication, an effective method must be adopted to monitor the online load effectively and improve the utilization rate of resources in a timely and effective manner. The traditional online load monitoring method mainly relies on the communication between the server and the load as the standard to check whether the load is online. This method is also the main method to check whether the other party is online between the network points. However, this method has certain limitations, needs to occupy more network resources, which is not conducive to the rational application of data bandwidth. In this paper, the neural network in the field of deep learning is used to monitor the abnormal load of big data in network behavior.

2 Online Monitoring of Network Behavior Big Data Load Anomaly

2.1 Establishment of Online Monitoring Model for Abnormal Load

In order to realize the on-line detection of big data abnormal load in network behavior, first of all, data structure analysis and abnormal load information flow model construction are needed. Statistical characteristic analysis method is used to calculate the characteristic quantity of abnormal load data. According to the distribution attribute of the characteristic quantity, the detection model design of abnormal load data is realized, the output link model of optical fiber network is established, and the network is designed The channel equalization control model adopts the irregular triangle network model to build the big data sampling node distribution model of cloud computing optical fiber network, as shown in the Fig. 1:

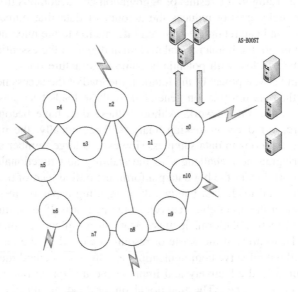

Fig. 1. Network data distribution

The network behavior big data load anomaly online detection model is composed of link layer, backbone node and sink node. The sampling data rule vector set of big data cluster head node in the network is distributed as follows:

$$SK_{i'} = \begin{cases} K_i & \text{if } i = 1 \\ New_{i'} & \text{otherwise} \end{cases} \tag{1}$$

Among them, $New_{i'} = (r_{i'1}, r_{i'2}, \ldots, r_{i'n})$ represents the Source node set, uses the deep learning method to mine the big data abnormal load $\left\{ M_h^{(0)}, h = 0, 1, \ldots, V - 1 \right\}$ in the network, sets the data flow of the time sequence sliding window, the sample clustering weight is $\left\{a_h^{(0)}, h = 0, 1, \ldots, V-1\right\}$, replaces the association rule items on each sub window to the clustering result set, and obtains the initial value $\left\{s_h^{(0)}, h = 0, 1, \ldots, V-1\right\}$ of the data clustering center. In order to improve the accuracy of load monitoring, it is necessary to carry out channel equalization design. The iterative equation for constructing network channel link offset correction is as follows:

$$f_{ih}(v+1) = f_{ih}(v) + \mu_{MCMA} \frac{\alpha H_{MCMA}(v)}{\alpha f_{ih}(v)} \tag{2}$$

Among them, μ_{MCMA} represents the initial routing location of the network, $f_{ih}(v)$ represents the initial sampling value of the abnormal load information flow, and constructs the big data abnormal load information flow model.

The nonlinear time series analysis method is used to model the information flow of big data abnormal load in the network, and the output of offset load in the channel is obtained as follows:

$$a_i(v) = \sum_{h=1}^{W} g_{ih}(v)^T s_h(v) + n_i(v) \tag{3}$$

The channel model of big data transmission in the network is:

$$y_h(v) = \sum_{h=1}^{Q} f_{ih}(v)^T a_i(v) \tag{4}$$

Where, f_{ih} indicates the DNS load frequency of big data. In the current snapshot window, the number of data categories that satisfy the decision of data classification is accurate DB_{ih}. The tuples in the classification space of big data abnormal load in the network $D[n+1]$ are deleted. $D[h] = D[h+1]$, In the phase space supported by the limited data set, the vector quantization decomposition formula of big data abnormal load in the network can be described as follows:

$$
\begin{aligned}
\min_{\phi} \|Y - X\phi\| &= \min_{\phi} \left\| O^T Y - \sum N^T \phi \right\| \\
&= \min_{\phi} \left\| \begin{bmatrix} O_1^T \\ O_2^T \end{bmatrix} Y - \begin{bmatrix} \Sigma_1 & 0 \\ 0 & 0 \end{bmatrix} \begin{bmatrix} N_1^T \\ N_2^T \end{bmatrix} \phi \right\| \\
&= \min_{\phi} \left\| \begin{bmatrix} O_1^T Y - \Sigma_1 N_1^T \phi \\ O_2^T Y \end{bmatrix} \right\| \\
&= \min_{\phi} \left\{ \left\| O_1^T Y - \sum_1 N_1^T \phi \right\| + X \right\}
\end{aligned} \tag{5}
$$

Among them, independent means the correlation coefficient X and ϕ aggregation coefficient of big data abnormal data in the network. The high-order statistics analysis method is used to reconstruct the characteristics of abnormal load information flow. When the size set of big data abnormal load in the network tends to infinity, it can be discarded, that is:

$$\min_{\phi} \|Y - X\phi\| = \min_{\phi} \left\| O_1^T Y - \sum {}_1 N_1^T \phi \right\| \tag{6}$$

In the link model of big data transmission in the network, the spatial distribution cluster of the t load sampling node h in the dimension space i is obtained by the second iteration d(t) calculation, then:

$$d_{ih}(t) = |a_{ih}(t) - j_{best}(t)| \tag{7}$$

Among them, the load time series is $a_{ih}(t)$ represented, and the fitness function is represented by j_{best}. The adaptive random link configuration method is used to detect the abnormal load and reorganize the data structure of the big data in the network, so as to improve the channel balance and the positioning ability of the abnormal load.

Then the exception detection and update strategy of the data i sampling node at the $(t+1)$ moment is:

$$\begin{cases} n_{id}^{(t+1)} = n_{id}^t + x_1 * u_1 \left(q_{id}^t - a_{id}^t \right) \\ \quad + x_2 * u_2 \left(q_{jd}^t - a_{id}^t \right) \\ a_{id}^{(t+1)} = a_{id}^t + n_{id}^{(t+1)} \end{cases} \tag{8}$$

Among them, $\{x_1, x_2\}$ the acceleration coefficient of single variable load detection is the random number between, which $\{u_1, u_2\}$ is $[0, 1]$ the lag detection coefficient of network big data abnormal load. m is the statistical characteristic quantity of abnormal load is extracted by time-frequency analysis method, and the correlation analysis is carried out according to the residual of detection model to build the network inspection regression analysis model. According to the residual of regression, the low inertia coefficient is adjusted adaptively

$$\begin{cases} d_{mean}(t) = \dfrac{\left| \sum\limits_{h=1}^{v} \sum\limits_{i=1}^{d} d_{ih}(t) \right|}{v*d} \\ d_{mean}(t) = |\max[d_{ih}(t)]| \\ k = \dfrac{|d_{\max}(t) - d_{mean}(t)|}{d_{\max}(t)} \end{cases} \tag{9}$$

Among them, $d_{mean}(t)$ is the average particle distance, $d_{\max}(t)$ is the maximum particle moment and k is the clustering degree of the network big data distribution are used to detect the abnormal load of big data, and the value range $[0, 1]$ is the statistical characteristic of the abnormal load extracted from them, and the abnormal data is diagnosed according to the feature extraction results.

2.2 Abnormal Data Diagnosis

In the phase of abnormal data diagnosis, the monitoring program will analyze the time sequence pattern of the obtained network data based on a small number of network characteristic parameters selected in advance, match the fault characteristics in the network database, and identify the suspicious network parameter characteristics based on the matching degree [5]. Compared with the traditional network fault monitoring stage, the method proposed in this chapter improves the time sequence of the original method, and more effectively predicts the precursor of the fault from the perspective of time. In the second stage, if the matching degree in the first stage is large, the model diagnosis stage procedure will be triggered [6]. The fault diagnosis model trained by the program using historical data will reapply the detailed parameter information of the network to the operation and maintenance management system. OAM will input the detailed parameter information of the network in the near time period into the model through data processing, and output the classification result (Fig. 2):

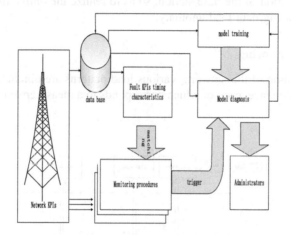

Fig. 2. Two stage fault diagnosis block diagram

In the traditional wireless cellular network, users at the boundary of base station are easily interfered by the neighbor base station. This effect can be based on frequency multiplexing technology or power control technology to reduce the degree of mutual interference [7]. But in the layered heterogeneous network, the interference generally includes the same layer network interference and the cross layer network interference. Cross layer network interference occurs between high-power macro base station and low-power base station. The low-power base station has the characteristics of intensive deployment. In a macro cell, there may be hundreds of home cell deployments. Moreover, due to the weak planning of the deployment, too many low-power base station deployments may make Acer station users included in the coverage of low-power base stations. On the one hand, the uplink signal of the user in the Acer Service will affect the performance of the low-power base station; on the other hand, the downlink signal of the low-power base station will also interfere with the user experience in the Acer station. The interference in the same layer is mainly reflected in the low-power base stations. The spatial distribution characteristics of low-power base stations are diverse, resulting in more complex interference environment [8]. Due to the weak planning of deployment, there will be overlapping coverage, which makes the interference everywhere. The complex interference environment is the main factor to reduce the performance of wireless network system. It not only reduces the network throughput, limits the network spectrum utilization, but also affects the stability of wireless link, causing frequent drop of users. Therefore, although the user access information can reflect the base station load from one point of view, it does not show the information of base station failover. Moreover, the load of low-power base stations is relatively small, and the user access of neighbor base stations may cause the load of neighbor base stations to be too heavy, again affecting the performance of neighbor base stations [9, 10]. To solve the problem of all users switching caused by the faulty base station, more comprehensive base station information is needed to represent, which can quickly find the fault of the original base station before the secondary performance pollution of the base station, so as to realize the online monitoring of the network behavior big data load anomaly.

2.3   Online Monitoring

During the operation of the network, various load patterns are obtained by monitoring the load vector sequence, each of which corresponds to a measurement space, as shown in the Fig. 3:

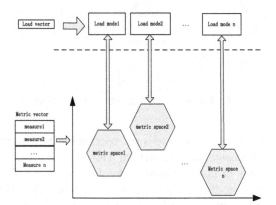

Fig. 3. Relationship between load pattern and metric space

As shown in the figure above, the corresponding relationship between the load pattern and the measurement space is represented by the stable operation state, which is defined as follows:

$$SS_i = \{KM_i, D_i, WS_i\} \tag{10}$$

Among them, KM_i is the center of the cluster i; D_i is the covariance matrix of all dimensions of the load vector in the cluster i; WS_i is the measurement space of the load pattern, which is composed of the set of measurement vectors $\{WN_1, WN_2, \ldots, WN_i, \ldots\}$ and the distance Euclidean.

In the process of running, we match the load vector of the running state with the known load pattern to obtain its corresponding measurement space, and use the local exception factor to calculate the exception score of the current measurement vector according to the measurement space [11, 12]; when determining the exception, we use the test to inspect each measurement separately to locate the exception measurement, and the exception monitoring method is shown in the Fig. 4:

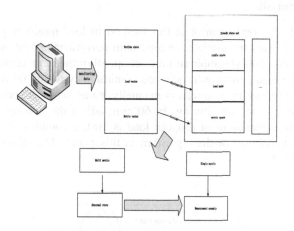

Fig. 4. Abnormal monitoring

By monitoring the changes of measurement, we can detect the abnormal state in the process of network behavior big data application running. At the same time, we can identify the abnormal load caused by the severe load fluctuation through the load pattern, and complete the online monitoring. In order to test the application effect of the proposed method, the following experiments are designed.

3   Case Simulation

3.1   Experiment Preparation

In view of the special problems that need to be solved in the process of online monitoring of big data load abnormality of network behavior, there are also special requirements for the construction of experimental environment. The construction of software environment for the experiment is shown in the table below (Table 1):

Table 1. List of software required for simulation experiment operation

	Operating system	Data base	Application software
Data server	Windows Server 2016	Oracle 11 g	ArcSDE 10.3
GIS server	Windows Server 2016		ArcGIS Server10.2
Web server	Windows Server 2016		JDK 1.6
Desktop client	Windows 10		IE browser 11.0
Mobile client	Android 10.0	SQLite	

At this point, the experimental preparation is completed, and the experimental platform can meet the needs of experimental operation. The specific results and analysis are as follows.

3.2 Result Analysis

In order to verify the performance of this method in load monitoring, cloudsim simulation platform is used to simulate the algorithm performance, and the data sample is 10000 loads. Firstly, 10000 sample sets are set up, all of them are connected to the data platform, then the load condition is adjusted continuously within 60 min, and some of the load is forced to lose the connection manually, so as to test the performance of this method in abnormal load monitoring. In 60 min, adjust the number of dropout load continuously. The upper limit of dropout load is 100, accounting for 1% of the total load. Test the adaptability of the algorithm in this paper. The simulation results are shown in the Fig. 5.

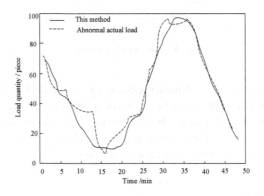

Fig. 5. On line monitoring of abnormal load in this method

As can be seen from the figure, this method can detect the abnormal load of the platform well. In the first 20 min, the two lines are coincident and fit well; but in the period of 20–30 min, some of the load is not detected; in the last 30 min, the method in this paper is better to detect the load.

In order to verify the advantages of the algorithm in data flow, compare the traditional method with the method in this paper, respectively simulate the monitoring flow consumed by the two monitoring algorithms when the number of loads is 5000, 10000, 15000, 20000 and 25000, and the results are shown in the figure below.

As you can see from Fig. 6, the monitoring flow will increase as the number of monitoring loads increases. When the load is 5000, the monitoring process of traditional method 1 is about 370 MB, and that of this method is 180 MB. The traditional method consumes about twice as much traffic as this method. Therefore, this method has an absolute advantage in the use of network affine resources, that is, the operation efficiency is high.

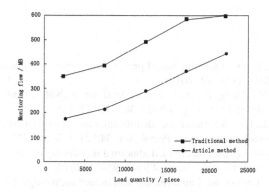

Fig. 6. Comparison of consumed flow

4 Summary and Prospect

When using the big data platform, workers should be clear: the online load of the access network is easily interfered by many factors, such as data bandwidth and computing power, etc., which requires workers to do data exchange processing between the online load and the platform to ensure that it can keep close contact with the platform. In order to avoid the problem of occupying data channels and wasting resources excessively, the staff should further monitor the online load in a practical way and make full use of resources.

Under the background of the rapid development of the Internet, the amount of network access load is increasing. The main analysis data of the network is generated from the resources of the load. And on this basis, referring to the specific changes of the load quantity in the adjacent time period, set up a load forecasting model that conforms to the status quo, and make a judgment on the load status according to the load status quo. For a period of time when the load changes greatly, the staff should make clear the load state through deep learning, and make forward and reverse operations on it. Although this kind of data is an essential resource for decision-making and analysis, and the more load, the higher data practicability, but a large number of loads connected with the network, the stability of the network system is undoubtedly a challenge.

5 Concluding Remarks

This paper proposes an online monitoring method of big data load anomaly based on deep learning. The experimental results show that the method in this paper can detect the load better within the limit of 30 min, and the load is 180 MB when the load is 5000, which shows that the method has absolute advantages in operation efficiency, which can better monitor the real connection of big data platform load and effectively reduce the occupation With the off-line load of network bandwidth, the efficiency of big data platform is improved, and less traffic is consumed in the use of network resources. Therefore, deep learning has strong applicability, online monitoring the application of big data load anomaly in network behavior.

References

1. Wang, Y., Tang, J.: Deep learning-based personalized paper recommender system. J. Chin. Inf. Process. **32**(04), 114–119 (2018)
2. Tang, C., Ling, Y., Zheng, K., et al.: Object detection method of multi-view SSD based on deep learning. Infrared Laser Eng. **47**(01), 302–310 (2018)
3. Zhang, Y., Li, M., Han, S.: Automatic identification and classification in lithology based on deep learning in rock images. Acta Petrol. Sin. **34**(02), 333–342 (2018)
4. Li, X.: Research on big data online load abnormal monitoring technology based on wavelet neural network. Mod. Electron. Tech. **42**(11), 95–97 (2019)
5. Liang, L., Li, J.: On-line load abnormality monitoring technology for large data based on wavelet neural network. Adhesion **40**(09), 94–96 + 116 (2019)
6. Chuili, H.U.: Distributed internet resources load balancing distribution simulation. Comput. Simul. **35**(07), 241–244 (2018)
7. Liu, S., Fu, W., He, L., et al.: Distribution of primary additional errors in fractal encoding method. Multimed. Tools Appl. **76**(4), 5787–5802 (2017). https://doi.org/10.1007/s11042-014-2408-1
8. Lu, Y., Zhang, T., He, E.: Probabilistic routing-based data fusion method in multi-source and multi-sink WSNs. Trans. Microsyst. Technol. **38**(07), 53–56 (2019)
9. Liu, S., Liu, G., Zhou, H.: A robust parallel object tracking method for illumination variations. Mob. Netw. Appl. **24**(1), 5–17 (2019). https://doi.org/10.1007/s11036-018-1134-8
10. Sun, L., Yu, K.: Research on big data analysis model of library user behavior based on Internet of Things. Comput. Eng. Softw. **40**(06), 113–118 (2019)
11. Liu, S., Liu, D., Srivastava, G., et al.: Overview and methods of correlation filter algorithms in object tracking. Complex Intell. Syst. (2020). https://doi.org/10.1007/s40747-020-00161-4
12. Lu, M., Liu, S.: Nucleosome positioning based on generalized relative entropy. Soft. Comput. **23**, 9175–9188 (2019). https://doi.org/10.1007/s00500-018-3602-2

Simulation Analysis of Building Energy Consumption Based on Big Data and BIM Technology

Ma Xiao[(✉)] and Qiu Xin

CCTEG Chongqing Engineering Co., Ltd., Chongqing 400042, China
mxiao2546@163.com

Abstract. In order to solve the problem of discrepancy between simulation results and measured results of building energy consumption simulation and analysis, a method based on big data and BIM technology is designed. The 3D building information model is constructed by BIM technology, and the factors affecting building energy consumption are obtained. The building energy consumption is simulated and predicted by heat balance method and Design Builder. Finally, data mining technology is used to modify the prediction results, and static energy analysis method is used to analyze the revised results. So far, the design of building energy consumption simulation and analysis method based on big data and BIM technology is completed. Compared with the original method, the simulation results of this method are close to the measured ones. In summary, the energy consumption simulation ability of this method is better than the original method.

Keywords: Big data · BIM technology · Building energy consumption · Numerical simulation

1 Introduction

With the rapid development of China's economy and the gradual improvement of living standards, the proportion of building energy consumption in total social energy consumption is becoming higher and higher. According to statistics, more than 50% of the material materials obtained by human beings from nature are used to construct buildings and their ancillary facilities, which consume about 50% of the global energy in the process of construction and use; at the same time, as a developing country, in the process of rapid economic development and modernization, it is necessary to coordinate the relationship between energy utilization and environmental protection, adhere to the people-oriented and sustainable development concept, and maximize the intensive use of resources [1, 2]. Since the global energy crisis broke out in the 1970s, more and more researchers have studied it in order to reduce the building energy consumption and have achieved some results. Since the emergence of BIM technology, there has been a BIM heat wave in the global construction industry. The function of BIM technology in the whole life cycle of buildings has been widely concerned.

S. Liu and L. Xia (Eds.): ADHIP 2020, LNICST 348, pp. 463–474, 2021.
https://doi.org/10.1007/978-3-030-67874-6_43

Energy-saving design of buildings needs to grasp the climate, including wind, light, rain, topography, landform and other local characteristics, but the complexity of today's buildings is much more than can be grasped by the subjective judgment or experience of architects, which requires the use of advanced computer technology to carry out complex data calculation and real-time dynamic simulation, to carry out preliminary energy consumption analysis of buildings, to achieve green energy-saving design of buildings [3]. Building energy consumption simulation is to use computer modeling and energy consumption simulation technology to analyze the physical performance and energy characteristics of buildings, which can be used in both new and existing buildings.

Based on the deep study of energy consumption data and the operation mode of buildings, an energy consumption simulation model based on big data and BIM technology is proposed.

2   Materials and Methods

Aiming at the problem that the simulation results of the original building energy consumption simulation analysis method are quite different from the measured results, the building energy consumption simulation analysis method based on big data and BIM Technology is set up. To ensure the effectiveness of this method design, the construction method design framework is shown in Fig. 1.

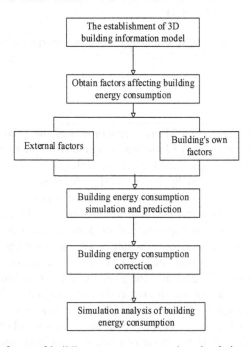

Fig. 1. Design frame of building energy consumption simulation analysis method

Through the design framework of Fig. 1, the method design in this study is completed. For the calculation part of the model, the calculation accuracy is set to two decimal places, so as to control the whole calculation process of the model.

2.1   Establishment of 3D Building Information Model

The first job of applying BIM technology is to establish 3D building information model by using related BIM information architecture design software according to physical and geometric information of construction project. The establishment of 3D model is the first condition to realize the function of BIM in the process of application, which includes all the digital information in the process of project construction. At the design stage, we can use different BIM design software to establish the relevant models, and then adjust the model to become the virtual model of the real project. The model can be used in the subsequent stage of the construction project. In this design, Revit Architecture, the core modeling software of BIM, is firstly used to build 3D building information model [4, 5], and then the professional software related to BIM technology is synthetically used to read the model information, check and analyze the model, and finally a perfect 3D information model is obtained. This is the basis of energy consumption simulation analysis of engineering project based on BIM technology. The building model of residential building with Revit architecture is shown in Fig. 2.

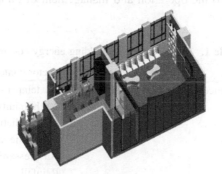

（a） Interior construction

（b）Overall appearance of the building

Fig. 2. Building information model

Using the software in Fig. 2 above, using the above software, the 3D model of the building to be measured is constructed, and the model is used as the model basis for building energy consumption simulation. The 3D building model based on BIM technology is different from the traditional 2D building model. All the components of the former have 3D information parameters, while the latter only have common 2D information. As mentioned above, BIM technology realizes the coordinated design of various specialties such as structure, HVAC and electromechanical in the whole life cycle of buildings. When the information of the component drawing element is modified or changed, the corresponding part of the building information model will be updated automatically, which reduces the tedious manual modification. This way of information sharing and the concept of coordinated design can make the designers master the overall situation, greatly reduce the errors and paper leaks in drawings, and ensure the completion of the project on schedule and with good quality and quantity (Fig. 3).

2.2 Access Factors Affecting Building Energy Consumption

According to the Study of Computer Evaluation System for Energy Conservation in Buildings, there are mainly three kinds of factors that affect the energy consumption in buildings. One is the external conditions of the surrounding environment and climatic factors of buildings; the other is the factors related to architectural design; and the third is the factors related to the operation and management of buildings [6]. This paper is

Table 1. Factors affecting building energy consumption

Index division	Types of factors	Factor content
External influence	Regional climate factors	Air temperature
		Air humidity
		Solar radiation
		Longitude and latitude
		Air pressure velocity
		Rainfall
Architectural design	Function use	The nature of the building
	Building survey data	Layer number
		Shape
		The measure of area
		Height
	Internal influence	Equipment fever
		Personnel Density
		Floor
		Home Furnishing
	Floor structure	Door and Window materials
		Vent size
		Exterior wall
		Interior wall
		Door and window area

based on the architectural design of the impact of factors on building energy consumption. Professional energy consumption simulation software is based on the impact of these three factors, and in the actual operation, the above factors also affect each other, and affect the building's energy consumption, the main factors affecting building energy consumption, as shown in Table 1.

If we want to reduce the energy consumption of buildings greatly, it is not enough to make qualitative analysis by light. It is very important to make quantitative analysis of energy consumption of buildings, to find out the relationship between energy consumption and various influencing factors, and to determine the influencing factors.

2.3 Building Energy Consumption Simulation Forecast

Building energy consumption simulation and analysis is a complex calculation process related to many aspects. It is necessary to record and calculate the design parameters of the building, such as the shape coefficient of the building, the ratio of windows and walls, the thermal performance and regional characteristics of the building, the heating and air conditioning system of the building and the various building equipment. Based on the above analysis, we can see that the energy consumption analysis method should consider many indexes related to building. There are three main analysis methods: building load calculation, digital analog calculation and actual measurement.

In recent years, the digital analog platform is used to calculate and test the building energy consumption, which provides a better platform for analyzing the building energy consumption control and energy saving. In this study, the synchronous simulation method is adopted, considering the synchronous effect of building heat and cold load and air conditioning system equipment on energy consumption, which increases the simulation accuracy.

The simulation begins with a Design Builder energy consumption analysis of the 3D model of the building that has been constructed [7, 8]. Using heat balance method and Design Builder software to simulate the building environment will produce a large number of data, such as outdoor meteorological conditions, indoor thermal environment parameters hourly, the whole building hourly cooling and heating load and the dynamic load of each room hourly. This paper mainly studies and analyzes the natural room temperature of different seasons, different layers of buildings, different rooms facing the same floor, and compares the building load under different window ratios, different heat transfer coefficients and different thicknesses of the same material. First, the building model can be pretreated, then the shadow lighting calculation, then the room temperature calculation and finally the load calculation. In the process of energy consumption simulation, mainly on the following aspects of the calculation, as shown in Table 2.

The energy consumption analysis software is used to simulate the above contents, and the data simulation results are stored in the form of a data table to carry out the building energy consumption analysis.

Table 2. Main contents of energy consumption simulation

Serial number	Type	Content
1	Natural room temperature	Room temperature in different horizons
2		Room temperature of different wall structures
3	Architectural coincidence	The influence of wall materials on building compliance
4		Influence of heat transfer coefficient of external wall on building load
5		Influence of same material and different thickness on building load
6	Air conditioning system	Energy consumption of air conditioning system
7		Personnel distribution in the building

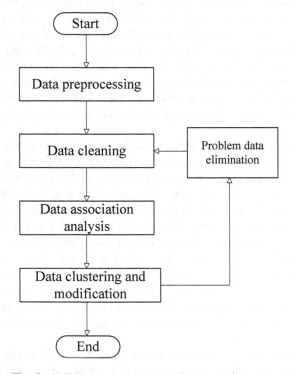

Fig. 3. Building energy consumption correction process

2.4   Building Energy Consumption Correction

In the process of data revision, the data mining technology will be used to revise the simulated building energy consumption data. The following process is used to complete the correction.

According to the process in Table 3, the problem data extraction and correction of energy consumption simulation results are completed. The data extraction part uses c-means clustering algorithm to complete the calculation process [9, 10]. Combine the modified data with the original data to form a complete energy consumption simulation data, and analyze it.

2.5   Simulation Analysis of Building Energy Consumption

In this part, the static energy consumption analysis method is used to analyze the modified energy consumption simulation results [11, 12]. First of all, the overall benefit of the building power and refrigeration power calculation, the specific formula as follows:

$$Q_R = \frac{o_i}{\partial_R} \sum_n (T - T_I) \tag{1}$$

$$Q_L = \frac{o}{\partial_L} \sum_{n_2} (T_I - T) \tag{2}$$

In the above formula, Q_R is set as the annual heating power; Q_L is the annual cooling power; T_L is the daily average outdoor temperature; T is the outdoor temperature corresponding to the balance of room heat gain and heat loss; o is the building heat loss coefficient; o_i is the heat transfer coefficient of the enclosure; ∂_R is the thermal efficiency of heating equipment; ∂_L is the efficiency of refrigeration equipment; n and n_2 are the days of using heating and refrigeration equipment respectively. Through the above formula, the obtained data simulation results are calculated, and the corresponding energy consumption characteristics of the building numerical simulation results are analyzed [13, 14]. In order to ensure the rationality and accuracy of the analysis, the above design results are combined with the effective heat transfer coefficient to calculate the energy consumption and heat of the building and improve the accuracy of the analysis [15, 16]. The specific calculation formula is as follows.

$$w_R = w_{RY} + w_{INF} - w_{U,H} \tag{3}$$

In the above formula, w_R is the heat consumption index of the building; w_{RY} is the heat transfer heat consumption of the enclosure structure of the unit building area; w_{INF} is the air penetration heat consumption of the unit building area; $w_{U,H}$ is the internal heat gain of the building of the unit building area.

3   Results

Aiming at the simulation analysis method of building energy consumption based on big data and BIM technology, the simulation experiment environment is designed in this research, and its application effect is studied.

3.1   Experimental Design

In this experiment, the method of building energy consumption simulation and analysis, which is designed in this paper, will be systematically studied by comparing with the original method. The experiment will be in the form of simulation experiments, comparing the differences between the two methods in use. A building in a city is chosen as the experimental object, and the energy consumption is simulated. The accuracy of energy consumption simulation of the two methods is compared and reflected by simulation curve. In order to improve the reliability of the experimental results, the simulated curves obtained by the two methods are compared with the measured curves, and the experimental process is completed.

Select the following experimental equipment and software to complete the experimental process, and the specific parameters are shown in Table 3.

Table 3. Experimental Equipment and Software

Serial number	Parameter	Model
1	CPU	Intel
2	Hard disk	16G
3	Memory	8G
4	Input device	Digital plate
5	Data base	SQL2016
6	Simulation software	Revit Architecture
7	Data analysis software	Design Builder

Use the above design to complete the design of the experimental environment, and use the above equipment to obtain the three-dimensional framework of the experimental target, as shown in Fig. 4.

The design method is used to analyze the energy consumption of the experimental object, and to compare the difference between the trend of energy consumption and the measured data. In this experiment, will be half a year as the experimental length, the completion of experimental comparison. In order to improve the precision of the experiment, the experiment is divided into two parts, and the differences between the original method and the design method are compared.

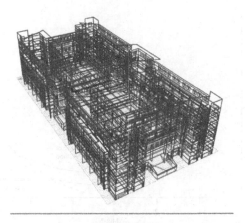

Fig. 4. Three-dimensional diagram of experimental object

3.2   Experimental Results Without Considering Environmental Factors

According to the experimental results in Fig. 5, on the premise of not taking into account the external environment, the trend of the design method and the measured curve is consistent, and the difference between the design method and the measured curve data is low, and the difference between the original method and the measured results is large (Fig. 6).

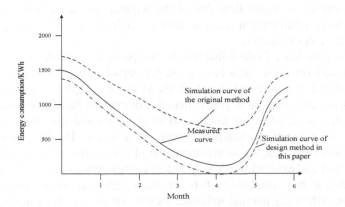

Fig. 5. Experimental results taking into account environmental factors.

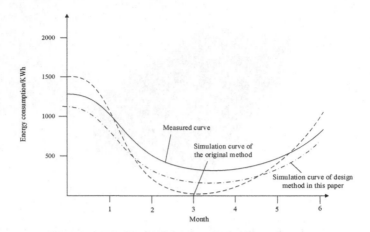

Fig. 6. Experimental results taking into account environmental factors

3.3   Experimental Results Taking into Account Environmental Factors

When environmental factors are considered, the design method is quite different from the original method. The trend of the original method is quite different from the measured results. The experimental results are combined with the previous experimental results. By comparison, the similarity between the design method and the measured results is high, while the similarity between the original method and the measured results is low. In conclusion, the energy consumption simulation accuracy of the design method is higher than that of the original design method, and the high-precision energy consumption analysis results can be obtained by using the design method in the culture medium.

It can be seen from Table 4 that for the comparison of various loads, the setting temperature and per capita area have a greater impact on the load, followed by the external wall and external window, which shows that the impact of internal factors on building energy consumption is greater than that of external factors on building energy consumption, that is, when the building is in operation and use stage, people's use mode and living habits on building energy consumption The influence of energy consumption is very great. As an important part of the external envelope, the influence of external wall and external window on indoor energy consumption is mainly realized by the real-time heat conduction between the external envelope and the outdoor environment. When the internal influence factors are certain, the external protection structure becomes the second major factor affecting the building energy consumption.

Table 4. Load simulation of combined scheme

Option	Cooling load (kwh/m2)	Thermal load (kwh/m2)	Total load (kwh/m2)
Option 1	21.40	4.01	25.63
Option 2	34.73	3.96	38.63
Option 3	39.54	6.42	46.21
Option 4	24.15	5.76	29.54
Option 5	37.59	6.25	55.35

4 Conclusion

Through the research, we can know that the traditional energy consumption analysis has corresponding shortcomings: because of the complexity of the project and the differences in form, function, content and operation of various analysis tools, it is not easy to select the analysis tools; because of the lack of deep understanding of professional issues and climate parameters, there are some differences between the building model and the actual project, which affect the accuracy of simulation results; and the combination with economy is not enough. Aided by BIM technology, it is very convenient to design, model and modify the scheme in BIM software after analyzing the energy consumption. The 3D model can be directly used to guide the field construction, and can directly output the architectural drawings in CAD format from the model. Using this research and design method can effectively reduce the cost and time of energy analysis, improve the speed and accuracy of analysis.

References

1. Zhang, S., Liu, Y., Hou, L., et al.: Comparative study on energy efficiency calculation methods of residential buildings in central heating areas of Northern China. J. Harbin Inst. Technol. **51**(10), 178–185 (2019)
2. Yang, X., Hu, H., Fang, F.: Simulation study on influence of ventilation period on energy consumption of malls in Hefei. J. Anhui Inst. Archit. Ind. **27**(03), 52–57 (2019)
3. Chen, J.: Simulation analysis of building energy consumption based on BIM technology take rural buildings in southern Shaanxi for an example. J. Shaanxi Shaanxi Univ. Technol. (Natl. Sci. Edution). **35**(02), 73–78 (2019)
4. Gao, H., Dang, T.: Energy consumption prediction model for CBD buildings based on energyplus. Build. Energ. Effi. **46**(12), 43–46 + 109 (2018)
5. Fu, L., Wang, G., Xia, J., et al.: Energy simulation and energy efficiency potential analysis of an office building in Qingdao. J. Qingdao Technol. Univ. **39**(05), 86–90 (2018)
6. Shuai, L., Gelan, Y.: Advanced Hybrid Information Processing. Springer International Publishing, USA, pp. 1–594
7. Wang, J., Li, J., Lin, J.: Comprehensive analysis and application of building energy conservation based on BIM technology. Guangdong Archit. Civil Eng. **25**(06), 55–58 (2018)
8. Xie, Y.: Energy consumption simulation of HVAC system and green building evaluation in an office building. J. Hunan City Univ. (Natl. Sci.) **27**(03), 32–36 (2018)

9. Liu, S., Li, Z., Zhang, Y., et al.: Introduction of key problems in long-distance learning and training. Mob. Netw. Appl. **24**(1), 1–4 (2019)
10. Zhao, Q., Lin, J., Huang, Z.: Simulation of overall energy consumption prediction of rural residential buildings. Comput. Simul. **35**(03), 436–439 (2018)
11. Amasyali, K., El-Gohary, N.M.: A review of data-driven building energy consumption prediction studies. Renew. Sustain. Energy Rev. **81**(1), 1192–1205 (2018)
12. Wei, Y., Zhang, X., Shi, Y., et al.: A review of data-driven approaches for prediction and classification of building energy consumption. Renew. Sustain. Energy Rev. **82**(1), 1027–1047 (2018)
13. Chengdong, L., Zixiang, D., Jianqiang, Y., et al.: Deep belief network based hybrid model for building energy consumption prediction. Energies **11**(1), 242–249 (2018)
14. Fu, W., Liu, S., Srivastava, G.: Optimization of big data scheduling in social networks. Entropy **21**(9), 902–903 (2019)
15. Kim, K., Yi, C., Lee, S.: Changes in urban temperature and building energy consumption due to urban park construction: the case of Gyeongui line forest in seoul. Energy Build. **12**(14), 259–263 (2019)
16. Lim, J., Lee, S.E.: Building energy consumption pattern analysis of detached housing for the policy decision simulator. IOP Conf. Ser. Mater. ence Eng. **317**(6), 12–17 (2018)

Author Index

Printed in the United States
By Bookmasters

Printed in the United States
By Bookmasters